Recent Progress in Energy Engineering

Recent Progress in Energy Engineering

Editor: Nora Ayling

www.callistoreference.com

Callisto Reference,
118-35 Queens Blvd., Suite 400,
Forest Hills, NY 11375, USA

Visit us on the World Wide Web at:
www.callistoreference.com

© Callisto Reference, 2017

ISBN: 978-1-63239-872-7 (Hardback)

The publisher's policy is to use permanent paper from mills that operate a sustainable forestry policy. Furthermore, the publisher ensures that the text paper and cover boards used have met acceptable environmental accreditation standards.

Trademark Notice: Registered trademark of products or corporate names are used only for explanation and identification without intent to infringe.

Printed in the United States of America.

Cataloging-in-Publication Data

Recent progress in energy engineering / edited by Nora Ayling.
   p. cm.
Includes bibliographical references and index.
ISBN 978-1-63239-872-7
1. Power (Mechanics). 2. Power resources. 3. Renewable energy sources. 4. Energy development.
I. Ayling, Nora.
TJ163.9 .R43 2017
621.042--dc23

Table of Contents

Permissions

List of Contributors

Index

Preface

This book aims to highlight the current researches and provides a platform to further the scope of innovations in this area. This book is a product of the combined efforts of many researchers and scientists from different parts of the world. The objective of this book is to provide the readers with the latest information in the field.

Energy engineering is defined as the design and development of electrical systems and circuits that enable transmission and distribution of electrical energy. This book on energy engineering discusses the various technological advances that have been made in the field of energy engineering. Renewable energy practices and efforts to increase work per unit of energy are methods that are followed in this field. It presents this complex subject in the most comprehensible and easy to understand language. This book aims to shed light on some of the unexplored aspects of energy engineering and the recent researches in this area. There has been rapid progress in this field and its applications are finding their way across multiple industries. Scientists and students actively engaged in this field will find this book full of crucial and unexplored concepts.

I would like to express my sincere thanks to the authors for their dedicated efforts in the completion of this book. I acknowledge the efforts of the publisher for providing constant support. Lastly, I would like to thank my family for their support in all academic endeavors.

Editor

Design, manufacturing and performance evaluation of house hold gasifier stove: A case study of Ethiopia

Dessie Tarekegn Bantelay[1], Nigus Gabbiye[2]

[1]School of Mechanical & Industrial Engineering, Bahir Dar Institute of Technology, Bahir Dar, 26, Ethiopia
[2]School of Food & Chemical Engineering, Bahir Dar Institute of Technology, Bahir Dar, 26, Ethiopia

Email address:
dessie2000ec@gmail.com (D. T. Bantelay), nigushabtu@gmail.com (N. Gabbiye)

Abstract: Biomass energy accounts for about 11% of the global primary energy supply, and it is estimated that about 2 billion people worldwide depend on biomass for their energy needs. Yet, most of the use of biomass is in an inefficient manner, primarily in developing countries, leading to a host of adverse implications on human health, environment, workplace conditions, and social wellbeing. Therefore, the utilization of biomass in a clean and efficient manner to deliver modern energy services to the world's poor remains an imperative for the development of community. One possible approach to do this is through the use of efficient biomass gasifiers. Although significant efforts have been directed towards developing and deploying biomass gasifiers in many countries, in Ethiopia especially the gasifiers have technical and economic barriers that hinder house hold gasifier dissemination and application. House hold gasifier stove design, manufacture and scaling up their dissemination remains an elusive goal. This article focuses on identification and evaluation of technical and economical limitations of house hold biomass gasifier stoves manufactured and distributed in Ethiopia. Then designing a family size standard gasifier stove and manufacture and evaluate the performance efficiency of the newly designed house hold gasifier stove. Finally device a mechanism that can improve the overall efficiency and applicability of house hold sized gasifier stoves.

Keywords: House Hold Gasifier Stove, Stove Design, Gasifier Stove Limitation, Improved Gasifier Stove

1. Introduction

Ethiopia is one of the developing countries with a high level of household energy consumption [1]. This consumption is primarily satisfied through excessive burning of biomass [1, 2]. The biomass consumption has vast implications both for deterioration of natural resources and the work load of rural women and girls charged with the responsibility of cooking in the kitchen [2].

The use of renewable energy sources in an efficient way are important indicator of a sustainable and environmental friendly energy supply. In Ethiopia still there is no access to adequate energy for their basic needs but the biomass, the primary energy source of the country, is used in a very inefficient way [3]. On one hand still Ethiopians do not have access to adequate energy for their basic needs [3]. On the other hand a large amount of energy is wasted by using enormous amount biomass fuel in a wasteful manner. The poor combustion technology of traditional stoves has serious negative impact on the health of rural women and small children as cooking traditionally takes place inside houses with very poor ventilation. This aggravated by a high population growth [4].

In Ethiopia still approximately 70-80% of the primary energy share is taken from biomass [4]. But the biomass, basically fuel wood which is directly used or processed to charcoal, neither cultivated in a sustainable way nor it is used efficiently contributing mainly for deforestation. In Ethiopia land degradation is the major cause of agricultural stagnation and poverty [5]. Deforestation, accelerated soil erosion, and land degradation are serious problems in Ethiopia [6]. Soil degradation is thus the most immediate environmental problem facing the country. The loss of soil and the deterioration in fertility, moisture storage capacity, and structure of the remaining soils all reduce the country's agricultural productivity [7]. Therefore, In Ethiopia improving energy efficiency can be seen as a further and a very important dimension to be addressed in the feature work.

2. Factors Affecting Gasification Process

Studies have shown that there are several factors influencing the gasification of wood [8,9,18].These include:

Energy content of Fuel: Fuel with high energy content provides easier combustion to sustain the endothermic gasification reactions. Beech wood chips have an energy content of approximately 20 MJ/kg. This is typical for most biomass sources and has been proved to be easy to gasify [8].

Fuel Moisture content of fuel: Since moisture is a non-burnable component in the biomass, it is important to keep as minimum as possible. All water in the feed stock must be vaporized in the drying phase before combustion. Otherwise there will be difficulty in sustaining combustion because the heat released will be used to evaporate moisture.

Wood with low moisture content can therefore be more readily gasified than that with high moisture. Wood with high moisture content should be dried first before it can be used as fuel for the gasifier. The beech wood chips used in the experiments have been factory dried to a moisture content of 10% prior to packaging. This makes it suitable as a fuel for the gasifier .Updraft gasifiers are also capable of operating with fuels that have moisture contents of up to 50% [8,9].

Size of the Fuel: Fuel should be of a form that will not lead to bridging within the reactor. Bridging occurs when unscreened fuels do not flow freely axially downwards in the gasifier. Therefore particle size is an important parameter in biomass gasification because it determines the bed porosity and thus the fluid-dynamic characteristics of the bed. On the other hand, fine grained fuels lead to substantial pressure drops in fixed bed reactors. The experimental wood chips are approximately 7.7 to 38mm^3 and regular in shape [8,9]. This size is not fine grained when compared to the micron scale and thus no substantial pressure drops occur in the reactor.

Temperature of the Reactor: There is a need to properly insulate the reactor so that heat losses are reduced. If heat losses are higher than the heat requirement of the endothermic reactions, the gasification reactions will not occur [8].The reactor should be insulated to keep heat losses minimal [10]. As the temperature increases the formation of combustible gases will increase and the yield of char and liquids will increase and leads to more complete conversion of the fuel. Hydrocarbon gases (especially methane and ethylene) increase with temperature while the yields of higher hydrocarbons (C_3-C_8: organic chemicals having 3 to 8 carbons) decrease at temperatures above 1202° F. The energy content of the syngas increases steadily up to 1292° F then decreases at higher temperatures. The rate of char gasification and yields of methane increase with increasing pressure, and the impacts are most significant at high temperatures (1652°-1742° F) [9].

Residence Time: For a given reactor temperature, higher fuel bed heights increase the time fuels are available for reactions to occur (residence time), which increases total syngas yields and increases the concentrations of hydrogen, carbon monoxide, carbon dioxide, methane and ethylene in the syngas.

Fluidization velocity: fluidization is the processing technique employing a suspension of a small solid particle in a vertically rising stream of fluid – usually gas – so that fluid and solid come into intimate contact affects the mixing of particles within the reactor. Higher velocities increase the temperature of the fuel bed and lead to the production of lower energy syngas.

The equivalence ratio: actual fuel-to-air ratio divided by the stoichiometric fuel-to-air ratio affects the temperature of the fuel bed. High ratios increase the rate of syngas production, and low ratios result in the production of lower syngas yields and energy content and more tar. As the steam-to-air ratio increases the energy content of the syngas also Increases.

3. Methods

After different gasifier stoves collected from end user and manufacturers their technical and economical limitations were assessed thoroughly. The new standard gasifier stove were designed by determining each design parameter as following And each design parameters were computed using the following equations.

Energy demand: This refers to the amount of heat that needs to be supplied by the stove. It was determined based on the amount of food to be cooked and /or water to be boiled and their corresponding specific heat energy. The amount of energy needed to cook food was calculated using equation 3:1[9].

$$Q_n = \frac{M_f \times E_s}{T} \qquad (3:1)$$

Where:
Q_n - energy needed, kJ/hr
M_f - mass of food, kg
E_s - specific energy, kJ/kg
T - Cooking time, hr

Energy input: This refers to the amount of energy needed in terms of fuel to be fed into the stove and computed using the formula [9].

$$FCR = \frac{Q_n}{HV_f \times \xi_s} \qquad (3:2)$$

Where:
FCR - fuel consumption rate, kg/hr
Q_n - heat energy needed, kJ/hr
HV_f - heating value of fuel, kJ/kg
ξ_g - gasifier stove efficiency, %

Reactor diameter: This refers to the size of the reactor and a function of fuel consumed per unit time (FCR) to the specific gasification rate (SGR) of biomass material, which is in the range of 110 to 210 kg/m^2-hr as revealed by the results of several test on biomass material gas stoves [9]. As shown below, the reactor diameter computed using equation 3:3 [9].

$$D = \left(\frac{1.27 \times FCR}{SGR} \right)^{0.5}$$ (3:3)

Where:
D - Diameter of reactor, m
FCR - fuel consumption rate, kg/hr
SGR - specific gasification rate of biomass material, 50-210 kg/m^2 -hr

Height of the reactor: This refers to the total distance from the top to bottom end of the reactor. This determines how long would the stove be operated in on loading of fuel. Basically, it is a function of a number of variables such as the required time to operate the gasifier (T), the specific gasification rate (SGR) and the density of biomass material (ρ). It empirical calculated using equation 3:4 [9].

$$H = \left(\frac{SGR \times T}{\rho} \right)$$ (3:4)

Where:
H - Length of the reactor, m
SGR - specific gasification rate of biomass, kg/m2 -hr
T - Time required to consume biomass, hr
ρ - Biomass material density, kg/m^3

Time to consume biomass: This refers to the total time required to completely gasify the biomass material inside the reactor. This includes the time to ignite the fuel and the time to generate gas, plus the time to completely burn all the fuel in the reactor. The density of the biomass material (ρ), the volume of the reactor (Vr), and the fuel consumption rate (FCR) are the factors used in determining the total time to consume the biomass material fuel in the reactor. As shown below it was computed using equation 3:5 [9].

$$T = \frac{\rho \times V_r}{FCR}$$ (3:5)

Where:
T - Time required for consuming the biomass, hr
V_r - volume of the reactor, m^3
ρ - Biomass material density, kg/m^3
FCR - rate of consumption of biomass material, kg/hr

Economical insulation thickness: it is the minimum insulation thickness of the stove and computed using equation 3:6 [9].

$$2\pi L (T_o - T_i)(1/kr_o - 1/kr_o^2) = [(\ln(r_o/r_i))/k + \ln(r_o h_o)]^2$$ (3:6)

Where:
L – Reactor height, m
K – Thermal conductivity of insulation material, J/m
h – Convective heat transfer, J/m^2
T_i – insulation's inside temperature, k
To – insulation's outside temperature, k
r_i – insulation's inside radius, m

r_o – insulation's outside radius, m

4. Results and Discussion

4.1. Limitations

Like other developing countries, In Ethiopia most of the people cook using Biomass fuel by direct combustion and inefficient wood stoves. This causes high waste of wood, health problems and destruction of forest. In the last few decades, many improved wood stoves have been developed but the stoves are;

- Often more difficult to manufacture,
- Often more heat goes to the stove than to the food,
- Do not offer good control of cooking rate, Too short chimney
- Too long chimney
- Poorly insulated chimney and combustion chamber
- Narrow Primary air opening
- Unbalanced Secondary air opening
- Using mild steel for combustion chamber
- Expensive and they are not always accepted by the cooks for whom they are developed.
- Does not consider the traditional cooking method.

Because of these limitations, people often preferred to cook by direct combustion of wood on three stone based combustion process and over charcoal. However, it is very wasteful and very polluting that needs urgent solution. Designing standard house hold gasifier stove that can solve the above limitation and improve manufacturing difficulty, reduce heat lose, having controlled cooking rate, acceptable by user and Operated indoors at economical price was the focus of this work.

4.2. Design Principles

In order to improve the overall efficiency and applicability of house hold biomass gasifier stoves to alleviate the prescribed limitations listed above is a must. These can be achieved through

1. Use of locally available raw material
2. Simplify the manufacturing process and reduce the cost of production
3. Offers good control of cooking rate
4. Reduces heat loss to the surrounding during operation
5. Improve indoor air quality
6. Lead to the development of local skills and jobs
7. Improve performance efficiency through

Place an insulated short chimney right above the fire: The combustion chamber chimney will have proportional height. Placing a short chimney above the fire increases draft and helps the fire burn hot and violent. Smoke will contact flame in the chimney and combust, reducing emissions. A taller combustion chamber chimney will clean up more smoke, but a shorter chimney will bring hotter gases to the pot. The very tall combustion chamber chimney can develop too much draft bringing in too much cold air that will decrease heat transfer. So a careful analysis using appropriate software packages like Fluent is necessary[11].

The opening into the fire: the size of the spaces within the stove through which hot air flows, and the chimney should all be about the same size. This is called maintaining constant cross sectional area, and helps to keep good draft throughout the stove. Good draft not only keeps the fire hot; it is also essential so that the hot air created by the fire can effectively transfer its heat into the pot. Air does not carry very much energy, so a lot of it needs to go through the stove in order to accomplish the task of heating food or water. The size of the openings is larger in more powerful stoves that burn more wood and make more heat. So to design appropriate cross sectional area throughout the stove it is better to use Catia software.

Insulate the heat flow path: Using insulative materials in the stove design and manufacturing keeps the flue gases hot so that they can more effectively heat the pot. Insulation is full of air holes and is very light. Dense materials are not insulation they soak up heat and divert it from cooking food. The economical thickness of insulation (ETI) is defined as the thickness of insulation for which the cost of insulation is balanced by increased energy savings over the life of the project.

Maximize heat transfer to the pot with properly sized gaps: Getting heat into pots is best done with small channels. The hot flue gases from the fire are forced through these narrow channels, or gaps, where it is forced to scrape against the pot or griddle. If the gap is too large the hot flue gases mostly stay in the middle of the channel and do not pass their heat to the desired cooking surface. If the gaps are too small, the draft diminishes, causing the fire to be cooler, the emissions to go up, and less heat to enter the pot. The two most important factors for getting large amounts of heat into a pot are:

- Keep the flue gases that touch the pot as hot as possible; and,
- Force the hot gases to scrape against the surface quickly, not slowly.

Air does not hold much heat. Faster hot flue gases scraping against the pot or griddle will transfer much more heat than slow moving cooler air. The size of the channel could be estimated by keeping the cross sectional area constant throughout the stove. This will achieve through systematic design and modeling.

Rate of combustion: the rate of gas production depends on the amount of primary air admitted to the bottom combustion chamber. For this reason it is very important to design a tight sealing valve on the bottom which permits a wide range of air adjustments and a means to control rate of combustion.

5. Standard House Hold Gasifier Stove Design

Energy demand Determination: Considering seven family members and "the traditional Ethiopian Derowote" as it is the most energy intensive food type in Ethiopia, to cook within one hour. A maximum of 10302.1Kcal/hr required at house hold level needed.

Energy input determination: using the maximum energy required to cook "Derowote" (10302.1Kcal/hr) and assuming the new gasifier stove will have an overall efficiency of 39% (maximum theoretical efficiency) and heating value of eucalyptus Specific fuel consumption of the stove is 6kg/hr or 0.0167kg/min of eucalyptus. Since all types of eucalyptus species grow uniformly throughout Ethiopia, an average calorific value considered in the analysis.

Reactor diameter: The reactor diameter computed using equation 3:3.For Gasifier stove of 6kg/hr fuel consumption, the computed internal diameter of the reaction chamber using an average specific gasification rate of 130 kg/m -hris 0.24m.

Height of the reactor: The height of the reactor computed using equation 3:4. Considering the desired operating time of the gasifier stove to be one hour and a biomass density of 250 kg/m^3 the height of the reactor should be 0.52m.

Insulation material: Even though there are numerous types of well-known stove insulations materials, most of them are not available in the local market, needs special manufacturing technique and knowledge and too expensive to use in house hold gasifier stove manufacturing. As a result of this assessing locally available insulation material is one of the main targets of the project. Based on the investigation the following insulation materials are identified as a potential candidate and presented in table 1. The insulation materials are selected and prioritized based on their low thermal property, relative availability throughout the country, inexpensive, it doesn't require special knowledge, skill and equipment to manufacture, light weight and acceptable by the community.

Table 1. *Relative properties of locally available stove insulation materials* [30]

No	Insulation material	Thermal Conductivity K – J/m	Relative Availability	Relative cost	Workability
1	Wood ash	0.092 –0.17	High	Low	High
2	Clay	0.15 – 1.8	Medium	Low	High
3	Saw dust	0.08	Low	Medium	High
4	Gypsum	0.09	Low	High	Medium
6	Dry River sand	0.15 – 0.25	Medium	Low	High
7	Cement mortar	1.73	Low	High	Medium
8	Light weight concrete	0.1 – 0.3	Low	High	Medium
9	Brick work common	0.6 - 1	Low	High	Medium
10	Fire clay brick	1.4	Low	High	Medium
11	Dry soil/earth	1.5	High	Low	High

Insulation Thickness Determination: The wall of the reactor and chimney consists of three different materials. The first inner layer is composed of 1cm thick clay (the thickness is determined from the local manufacturing experience) that has a thermal conductivity of 1.8W/m^{o}K. The second layer is made from wood ash with a thermal conductivity of 0.017 W/m^{o}K. The exterior layer is made from 2mm thick mild steel sheet. The maximum inside temperate of the stove is expected to be 670^{0}C and the outside air temperate near to the stove surface is 46°c with a convection heat transfer coefficient of 40 W/m^{2o}K and also it is better to neglect the heat lost at the outer surface.

Using economical insulation thickness equation 3:6the thickness of the insulation for all insulation materials listed in table 4.4 including wood ash are less than 3cm. But for manufacturing simplicity and operator safety it is better to use 3cm thick insulation around the combustion chamber. As a result of this 3cm thick wood ash used as insulation material around the combustion chamber of the new gasifier stove.

Time to Consume Biomass: For 24cm diameter and 52cm height reactor biomass gasifire stove operated at a fuel consumption rate of 6kg/hr, the time required for a single cycle operation would be 60 minutes. Within this period all the eucalyptus wood fuel will burnt out and the wood gas driven out. Only the charcoal will left and by removing the upper part of the stove safely the charcoal will utilized for same or other cooking purpose.

The Size of Primary Air Inlet: The feedstock enters the reactor undergo combustion with the limited oxygen present in the chamber. To gasify 25.42gm of wood 24.48gm of oxygen needed.to create the appropriate thermo-dynamic environment at 2m/s wind speed 13.76cm^2 primary air inlet required

6. Manufacturing of the Stove

For the manufacturing of the stove any crafts man shop is sufficient enough. It is manufactured by participating end users and traditional stove makers and has the following basic futures illustrated in the figure 1 below.

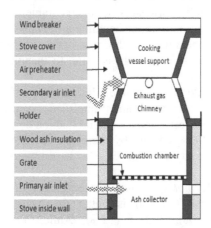

Figure 1. Schematic diagram of the new stove

The stove made of two parts. The upper part (figure 2) and lower part (figure 3) of the stove. This made the stove easy for operation. After the gasification process completed by removing the upper part safely the bottom part used as charcoal stove.

The upper part of the stove serves to oxidize the gas generated as a result of gasification process and serves to transfer the heat generated as a result of secondary oxidation to the cooking vessel (food) efficiently. The upper part (figure 2) includes:

- *1*-2mm thick sheet metal wind breaker to prevent the diffusion of heat to surrounding prior to transferring the heat to the cooking vessel.
- 1-2mm thick sheet metal Stove cover.
- Air preheater to warm up the secondary air prior to combustion. The preheater utilizes the waste heat energy and helps to improve combustion efficiency and maintaining stove body at low temperature.
- V-shaped cooking vessel support to accommodate different size and shape cooking appliance
- Exhaust gas chimney
- Secondary air inlet

Figure 2. Upper part of the new stove

On the other hand, the lower part of the stove serves as gasifier and charcoal stove. The lower part (figure 3) includes:

- Holder to fix the upper part with the lower one and to prevent side fire.
- Wood ash to pervert heat loss around the combustion chamber
- 2cm thick clay Grate having enough primary air inlets used to support wood fuel. The air holes total area enough to primary air inlet and ash fail down but small in size to prevent small sized charcoals.
- Combustion chamber made of clay in which gasification process takes place.

Figure 3. Lower part of the stove

7. Experimental Results

Efficient and complete combustion is a prerequisite of utilizing wood as an environmentally desirable fuel. In addition to a high rate of energy utilization, the combustion process should therefore ensure the complete conversation of wood in to heat and avoid the formation of environmentally undesirable compounds. As shown in the figure 4 below water boiling test carried out to measure and analyze the performance efficiency of the newly designed stove.

Figure 4. Water boiling test of the new stove

The result obtained from water boiling test summarized in figure 5 below. as shown in the figure 5 below the stove takes an average of 60min to boil, have an average specific fuel consumption of 47.04g/min, with an average thermal efficiency of 17.2% and the stove average external body temperature recorded as 46^0c.

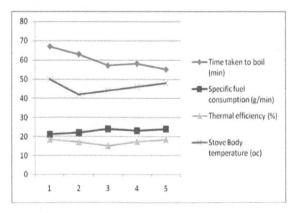

Figure 5. Summary of water boiling test results of the new stove

8. Improvements

Compared to the latest and widely used stoves the new design has the following improvements.

1 *Use of cheap and locally available materials:* the existing house hold stoves made from stain less steel and to some extent from mild steel. But it is expensive and unavailable in rural parts of Ethiopia. The new stove design uses clay as refractory material, ash as insulation material and any hard materials including mild steel or wood as external cover.

2 *Construction simplicity:* since most gasifier stoves are complicated and made of stainless steel, they require better manufacturing technology. The new design excludes the use of stainless steel and considers the local manufacturing capability of the society. In most crafts man use of Mild steel, wood, ash, saw dust and clay is common practice. As a result it doesn't require new skill and technology. These simplify the manufacturing process of the stove.

3 *Use of different size pots:* the size of cooking apparatus used depends on the type of food and volume of food to be cooked. As a result the house hold gasifier stoves have to support different size cooking vessels. But most of the stoves available today are designed for single size vessels. If they serve out of the designed size they will exposed for additional heat loss.

4 *Control of cooking rate:* cooking rate varies from location to location and based on the type of service. But most gasifier stoves of today have no primary air control mechanism to regulate rate of gasification. The new design considers the importance of controlling rate of gasification by controlling primary air inlet. It incorporates primary air controlling mechanism.

5 *Elimination of pot support:* the current house hold gasifier stoves use additional support structure for the cooking vessels. But the new design eliminates this additional feature. As a result it makes compact and manageable.

6 *Elimination of charcoal stove:* the champion-2008 stove needs additional charcoal stove to combust the charcoal produced after gasification. But in this stove after gasification process completed the upper part of the stove removed and the lower part of the gasifier serves as charcoal stove.

7 *Safety:* The existing Updraft gasifier stove outside body temperature as presented by Dr Simon was 93^0C, which can accidentally burn users, so it is recommended to use a handle when moving or handling a hot stove. But the outside temperature of the new gasifier stove was recorded as 46^oc, which is safe to handle during operation.

8 *Low production cost:* the design Use of locally available raw material and Simplified manufacturing process. As a result the cost of production compared to champion-2008 reduced by more than 8 times. Based on the current material cost it is estimated to be 1106.67birr.

9 *Improve thermal efficiency:* the efficiency of the new gasifier stove (17.2%) is better thanAnila1 (14%), Sampad (12%), TLUD (12%), Anila2 (15%) and compatible with Everything (19%)[11].

9. Conclusions

Energy is one of the most important basic commodities that determine the progress and status as well as the wellbeing of the community. A country's socio-economy cannot show progressive development unless energy is explored, developed, distributed and utilized in an efficient and appropriate way. Based on this facts developing, designing, manufacturing and performance evaluation have been employed in this study.

This study provides the development of enclosed biomass gasifier stove specifically designed for house hold cooking. It provides a mechanism in which household gasifier stoves efficiency and applicability could be maximized. The gasifire reactor has an internal diameter of 24cm and an overall dimension of 30.4cm diameter by 70cm height and a specific fuel consumption of 1.3kg/hr. The performance efficiency of the stove was evaluated using water boiling test and a thermal efficiency of 17.2% was obtained.

The results obtained from this study shows that the gasifier performance and operating conditions are good. So the stove can provides modern energy services for basic needs and productive applications in the areas.

References

[1] Ethiopian Electric Agency, 2002, Symposium Proceedings of Rural Electrification, Addis Ababa, Ethiopia.

[2] Richard, Wright.T Bernard and Nebel.J, 2002, Environmental Science, to Ward a Sustainable Future, New Delhi, India.

[3] Ethiopian Forestry Action Program (1996)

[4] DestaMaberatu and MulugetaTamire, 2002, Energy in Ethiopia; Status, Challenges and Prospects, Proceedings of Energy Conference 2002, Addis Ababa Ethiopia.

[5] Improved rural woodstoves. October 2003. Ethiopian Rural Energy Development and Promotion Center. BTG biomass technology group BV

[6] Ethiopia Rural Biomass Supply-side Strategy. March 2003. Ethiopian Rural Energy Development and Promotion Center. Biomass Technology Group BTG.BV

[7] W. Gashie, 2005. Factors controlling Households Energy Use: Implication for the Conservation of the Environment. Addis Ababa University, Ethiopia:

[8] Reddy, A.K.N, R.H Williams, 1997, New Opportunity in Energy Demand , Supply and Systems in energy After Rio-Prospective and challenges UNDP

[9] YohannesShiferawSherka, June, 2011; Design and Performance Evaluation of Biomass Gasifier Stove, Addis Ababa University, Addis Ababa

[10] Alexis T. Belonio, 2005, RICE HUSK GAS STOVE Hand BOOK, APPROPRIATE TECHNOLOGY CENTER, Iloilo City, Philippines,

[11] Dr Simon Shackley: Biochar stoves, an innovative studies perspective; UK Biochar Research (UKBRC), school of Geosciences, Univrsity of Edinburgh. 8th April 2011.

Impact of Atmospheric Parameters on Power Generation of Wind Turbine

Ravindra B. Sholapurkar[1], Yogesh S. Mahajan[2]

[1]Department of Chemical Engineering, DR. Babasaheb Ambedkar Technological University, Lonere, Tal Mangaon, Dist. Raigad, Maharashtra, India

[2]DR. Babasaheb Ambedkar Technological University, Lonere, Tal Mangaon, Dist. Raigad, Maharashtra, India

Email address:

rbsat68@gmail.com (R. B. Sholapurkar), yogesh_mahajan66@yahoo.com (Y. S. Mahajan), ysmahajan@dbatu.ac.in (Y. S. Mahajan)
[*]Corresponding author

Abstract: Economic growth of any country depends on access to reliable energy. Wind energy is fast gaining importance among non-conventional sources, which is a function of parameters like topography of the terrain, weather conditions etc. The present work explores the potential of a 225 MW turbine located in a mountainous site in Maharashtra, India. Values of wind velocity, air temperature, density and power generation were recorded for one complete year. Analysis was done using power curves. The results show that the energy output of wind turbine is based on power curves of a specific site. The conceptual features such as energy per rated power, efficiency of wind turbine and average energy per hour are calculated. It is useful for the investor to access the wind turbine pay-back period and adopt of new optimizing technique.

Keywords: Wind Energy, Atmospheric Parameters, Wind Power Generation

1. Introduction

Owing to the exponential escalation of world population, the requirement for energy is increasing at a huge rate. The exhaustion of fossil fuels as well as the upcoming realization of the environmental degradation has given precedence to the use of conventional and renewable alternative energy sources like solar, wind and solar-hydrogen energy [1]. In recent years, the fast development in wind energy technology has made it a promising alternative to the conventional energy system. Wind energy systems have made a noteworthy contribution to everyday life in developing countries [2, 3]. As the developing countries are preparing to meet their needs during the 21st century, they are confronting many challenges, in the use and deployment of renewable energy sources.

To suffice the world's electricity needs, the wind energy potential is vast [4]. Almost every country has sites with regular wind speeds of more than 5 m/s measured at a height of 10 m, which are adequate for development [5, 6]. The production of wind energy depends on the geographic and environmental conditions of the wind turbine location.

Furthermore, wind data having real statistical significance, i.e. data obtained over long periods is a requisite for the assessment of generation of specific plants and the relative technical-economic considerations [7]. Such data is only scarcely available in the literature [8, 9]. Through systematic recordings of wind speed and direction is a potential site for the development of wind energy. [10]. Wind surveys and installations have up to now been concerned with on-land sites [11]. Better wind velocity will lead to more amount of wind power generation. The offshore sites and shallow water sites are better for windmill installation but maintenance and installation [12].

The low energy concentration of wind, its variations and unsystematic availability over time, collide with the needs of electricity production [13]. Above all, the low energy concentration in wind signifies that, a big number of wind turbines need to be used where a wind plant of major power is to be built, which should be huge in size taking into account of the installed generating capacity [14]. However, as an additional source of electricity under suitable conditions, wind can still be put to a good use [15], chiefly since it is easily accessible, with no emissions and since it allows additional diversification of energy sources. World watch Institute (USA) has reported that wind power plants are liable to provide more employment opportunities per Terawatt-hour produced per year than any other energy technology used by the electricity utilities [16]. Also noteworthy is the fact that, in growth with other kinds of generating plants, the so-called "energy payback time" is less for wind energy, as it differs from a few months to one year at most, as it relies on the size of the wind turbine. While a machine has functioned for this period, it has produced enough energy to provide for another wind turbine similar to it [17]. Energy is the chief measure of all kinds of work by human beings, nature and all that happens in the world is the flow of energy in one to the other form.

2. Importance of Wind Turbine as Alternate Source of Energy

Wind gets its movement as a result of its energy. A device competent of slowing down the mass of moving air can haul out part of the energy and thus this energy can be switched into useful work [18]. Following aspects direct the output of wind energy converter: the wind speed, cross-section of the wind swept by rotor, conversion efficiency of rotor and generator transmission system. Hypothetically, it is possible to get 100% efficiency by arresting and averting the channel of air through the rotor. But in reality, a rotor is proficient to slow down the air column merely to one third of its free velocity or so [19]. A competent wind generator is capable to switch maximum up to 60% of the accessible energy in wind into mechanical energy. Additionally, the whole efficiency of the power generation can go down to, say, 35% owing to the losses incurred in the generator or pump [20, 21].

3. Wind as Secondary Source Derived from Sun

The sun's non-uniform heating of the earth's surface results in the circulation of air in the atmosphere. As the temperature of the earth's surface heats up, the air present over the earth's surface will become light in weight and expand; it is compelled upwards by cool, denser air which blows in from surrounding areas causing the wind to flow. The strength of wind and the nature of wind depend upon the nature of the land; the quantity of cloud over and the angle of the sun in the sky. Generally, the air above the earth's surface tends to heat up more during the day time than air over the water surface. In coastal areas, during the day time, earth surface will become hot swiftly than water surface. In the nighttime however, this course of action is reversed, and cool air flows off shore [22].

As the solar input keeps on differing, the change and course of these terrestrial winds alter with the seasons. The terrestrial winds are caused when the breezy surface air sweeps downward from the poles powering the warm air

over the tropics to mount [23]. However, the course of these huge air movements affected by the revolving of the earth and the net outcome is a large counter-clockwise movement of air around low-pressure regions in the northern hemisphere and clockwise movement in the southern hemisphere [24]. It is known that the regular wind speeds are superior in hilly and coastal areas than they are in the inlands. The winds also have a tendency to blow more consistently and with superior potency above the surface of the water where there is a less surface drag. However the wind patterns are remarkably constant despite the alternating nature of the wind [25]. Wind velocity amplifies in magnitude with elevation. It has been shown that at a height of ten meters, velocity is found to be 20-25% greater than at the surface. At an elevation of 60 m, it may be 30-60% higher on account of the reduction in the drag effect of the earth's surface [26].

4. Wind Power

Wind energy is possessed by the movement of air. The moving air is cut by a device like propeller and the kinetic energy is converted into useful work, the process being known as harnessing of wind energy [27].

There are three factors on which the wind energy depends:
- The wind speed
- The cross-section of wind swept by rotor
- The overall conversion efficiency of the rotor, transmission system and generator or pump

German physicist Albert Betz concluded in1919 that no wind turbine can convert more than16/27 (59.3%) of the kinetic energy of the wind into mechanical energy turning a rotor. To this day, this is known as the Betz Limit or Betz' Law. The theoretical maximum power efficiency of *any* design of wind turbine is 0.59 (i.e. no more than 59% of the energy carried by the wind can be extracted by a wind turbine). This is called the "power coefficient" and is defined as: *Cp max* = 0.59 [15].

$$p = \frac{1}{2}\rho A v^3\, C_p, \text{-----------------------1}$$

Where p = power generation, ρ = air density, A = swept area, v = wind velocity and
C_p = Power co-efficient

5. Case Study

In this work we have selected a site with wide variation of wind speed (from 5 – 13 m/s) and the site itself is windy. The location is at Dhebewadi, Taluka Karad, District Satara, in the state of Maharashtra, India, which is at a mean sea level (MSL) of 1012 m. The average rainfall during monsoon (June – September) is 36 mm. The average percentage of humidity is 60 %. There are several wind turbines at this place by private players viz. VESTAS, SUZULON, TATA POWER etc. The data obtained for one particular turbine run by VESTAS turbine is used in this work. The turbine details are: Rotor diameter: 82 m, Effective diameter of blade: 78 m,

Gearbox rated power: 1800 KW, Gearbox Ratio: 1:84.3, Rated Power: 1650 KW, Speed: 1200 rpm, Apparent Power: 1808 KVA, Rated current: 1740 A.

6. Methodology

The data pertaining to various parameters and power generated for the particular wind turbine was noted for one complete year (January – December 2014). Parameters like average temperature, average velocity and average density were calculated along with number of hours, the turbine worked and the average power generation in the given month was calculated from year – long data. From this the average energy, capacity of the plant, energy per swept area and energy per rated power were also calculated and are shown in (Table 1). Sample calculations for the month of January are shown below in the equation 2 to equation7. Also since power generation depends on factors like wind speed, air temperature and air density, so effect of these on power generation are plotted Fig. (1 - 6). Also the overall effect of various parameters on power generation is as shown in Fig. (7). the turbines rated or nominal power is 1650 KW. The complete analysis of the VESTAS wind turbine for the year 2014 is made with the help of equation from 8 to equation 13. By this we can come know that, weather the present turbine is working efficiently and many more important calculations like energy per rated power and energy per swept area.

6.1. Monthly Calculation for Example January 2014

1 $\quad$ Swept Area $= \frac{\pi d^2}{4} = \frac{\pi}{4}(82)^2 = 5281$ m $\qquad$ ----2

2 $\quad$ Average hour turbine worked $= \frac{No.of\ hours\ turbine\ worked\ in\ month}{No.of\ days\ in\ month} = \frac{456.5}{31} = 14.72$ hours /day $\qquad$ ----3

3 $\quad$ Average Energy Kw/hr. $= \frac{Average\ Power\ generation}{Average\ hour\ turbine\ worked} = \frac{4697.02}{14.72} = 319.09$ Kw / hours $\qquad$ ----4

4 $\quad$ Energy per swept area $= \frac{Average\ Power\ generation}{Total\ Swept\ Area} = \frac{4697.02}{5281} = 0.89$ Kw / m^2 $\qquad$ ----5

5 $\quad$ Energy per rated (nominal) Power $= \frac{Average\ Power\ generation}{Rated\ Power} = \frac{4697.02}{1650} = 2.84$ Kwh / Kw $\qquad$ ----6

6 $\quad$ Capacity factor or efficiency $= \frac{Average\ Power\ generation}{Rated\ Power} x\ 100 = \frac{4697.02}{1650\ x\ 24\ x\ 31} x\ 100 = 0.38$ $\qquad$ ----7

6.2. Detailed Analysis of the Turbine for the Year 2014

7 $\quad$ Average turbine worked $= \frac{Total\ No.of\ hours\ turbine\ worked\ in\ year}{Total\ No.of\ days\ in\ year} \frac{7118.7}{365} = 19.5$ hours/day $\qquad$ ---8

8 $\quad$ Average Energy of turbine Kw/hr $= \frac{Average\ Power\ generation\ in\ year}{Average\ hour\ turbine\ worked}$

$\quad = \frac{1,39,989}{365} = 383.53$ Kw/day (generated per day) $\qquad$ ---9

9 $\quad$ Energy per swept Area $= \frac{Power\ Production\ Per\ Year}{Swept\ Area} = \frac{1,39,989}{5281} = 26.5$ K / m^2 $\qquad$ ---10

10 $\quad$ Energy per rated Power $= \frac{Production\ Per\ year}{Rated\ Power} = \frac{1,39,989}{1650} = 84.84$ Kwh / Kw $\qquad$ ---11

11 $\quad$ Capacity Factor $= \frac{Production\ Per\ year}{Rated\ Power} = \frac{1,39,989}{1650\ x\ 24\ x\ 365} x\ 100 = 0.97\%$ $\qquad$ ---12

12 $\quad$ Tip Speed $= \frac{2\ \pi\ n\ r}{60} = \frac{2\pi\ x\ 14\ x\ 41}{60} = 60.10$ m/s, When n = 14 rpm, Radius r = 41m $\qquad$ ---13

Table 1. Monthly average temperature, average velocity and average density for year 2014.

Sr No	Month	Average Temperature C	Average Velocity m/s	Average Air density Kg/m³	No of hours turbine worked in month	Average Power generation Kw
1	January	23.7	5.4	1.20	456.5	4697.02
2	February	30.8	5.2	1.16	653.9	7131.2
3	March	34.5	5.2	1.14	727	10189.9
4	April	36.7	4.83	1.14	701	10906.7
5	May	38.4	8.6	1.14	718.3	12093.8
6	June	29.0	11.7	1.16	693.3	22610.2
7	July	26.4	11.4	1.18	703	32191.1
8	August	25.1	11.48	1.18	713.2	28537.3
9	September	24.6	4.34	1.18	357.1	4026.53
10	October	25.9	6.62	1.18	427.7	2017
11	November	26.1	6.8	1.18	482	2822.25
12	December	22.3	5.66	1.20	485.7	2766
					7118.7	1,39,989

Table 1. Contuine.

Sr No	Month	Average hour turbine worked per day (hours)	Average energy kw/hr	Energy per swept area	Energy per rated power	Capacity of plant
1	January	14.72	319.09	0.89	2.84	0.38
2	February	22.54	316.38	1.35	4.32	0.64
3	March	23.45	434.54	1.93	6.17	0.83
4	April	23.36	466.89	2.07	6.61	0.91
5	May	23.17	521.96	2.29	7.33	0.98
6	June	23.11	978.37	4.28	13.70	0.90
7	July	22.67	1419.99	6.09	19.50	2.62
8	August	23.00	1240.75	5.41	17.29	2.32
9	September	11.90	338.36	0.76	0.20	0.33
10	October	13.79	146.26	0.38	0.9	0.16
11	November	16.06	175.73	0.53	1.71	0.24
12	December	15.66	176.63	0.52	1.68	0.23

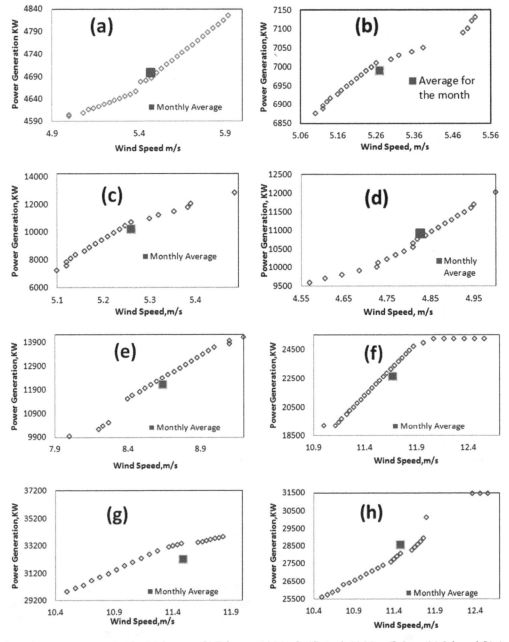

Fig. 1. Effect of wind speed on power generation for (a) January, (b) February, (c) March, (d) April, (e) May, (f) June, (g) July and (h) August 2014.

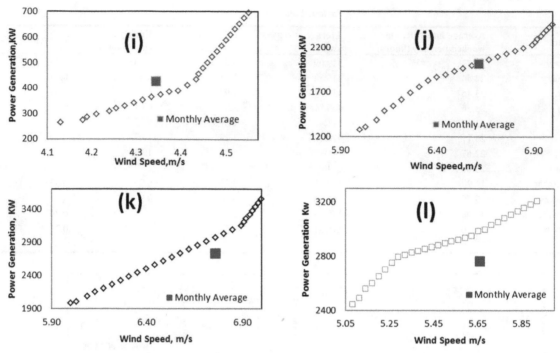

Fig. 2. *Effect of wind speed on power generation for (i) September, (j) October, (k) November and (l) December 2014.*

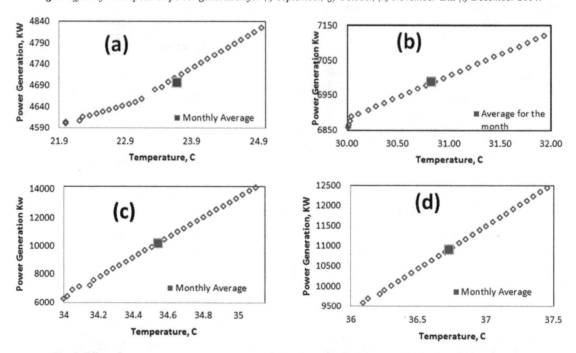

Fig. 3. *Effect of temperature on power generation for (a) January, (b) February, (c) March and (d) April 2014.*

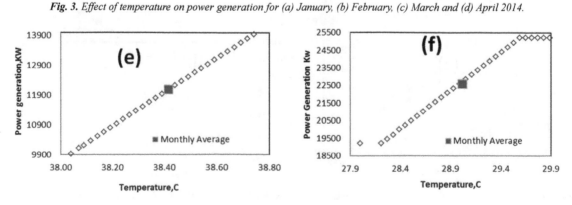

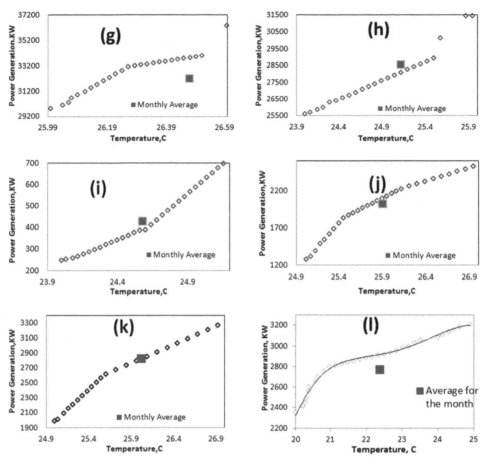

Fig. 4. Effect of temperature on power generation for (e) May, (f) June, (g) July, (h) August, (i) September, (j) October, (k) November and (l) December 2014.

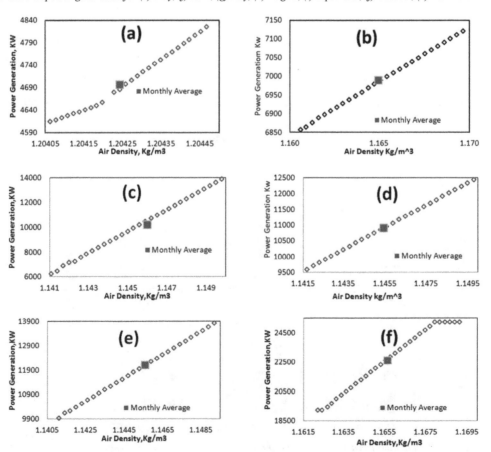

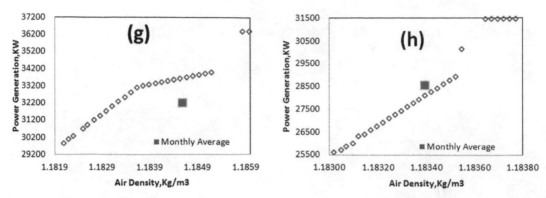

Fig. 5. *Effect of air density on power generation for (a) January, (b) February, (c) March, (d) April, (e) May, (f) June, (g) July and (h) August 2014.*

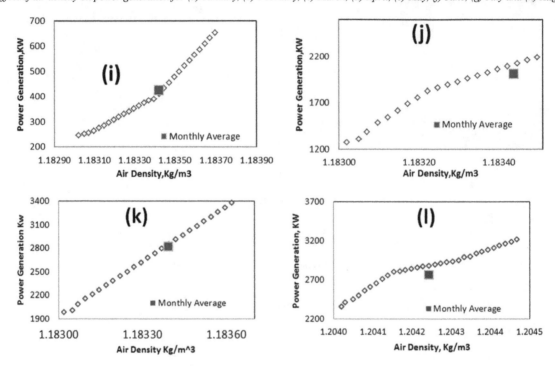

Fig. 6. *Effect of air density on power generation for (i) September, (j) October, (k) November and (l) December 2014.*

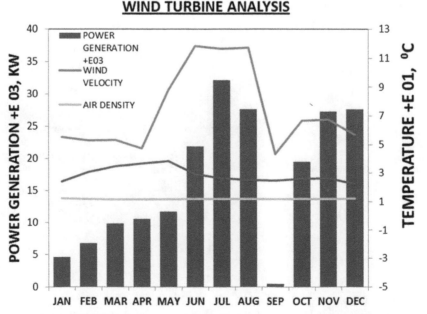

Fig. 7. *Effect of wind speed, temperature and air density on power generation, combined effect.*

7. Results and Discussion

As shown by the figures above (fig. 1 to 7) the power generation for the entire year 2014 was monitored and special emphasis was given to understand and interpret the effect of various parameters on power generation. Three such factors have been identified it, namely wind speed (m/s), temperature of air (°C) and density of air (Kg/m^3). The natural variation of parameters has been recorded, plotted and analyzed for the entire year 2014. The following can be observed:

1. When the wind speeds are larger, large power generation is possible. This is evident especially in the months of June, July and August. Very low power generations have been observed in September and November where wind speeds are quite low. According to Betz's law, the power generation is proportional to 3rd power of wind speed. Hence in comparison with any other parameter effect of wind speed will predominate the power generation.

2. It is observed further that air density is comparatively large in the month of June-August with the corresponding increase in power generation. Air density is also large in September-December. but the wind speed is low and hence the power generation values are low.

3. The effect of temperature on power generation was observed and it was seen that on dry days (non-monsoon period) larger temperature causes more power generation. However since the variation of temperature throughout the year is only marginal (20°C to 39°C) no profound effect is noted. In countries where winters are sever, temperature might be a very important parameter. It is also seen that in tropical countries like India, in part of western mountainous region of Maharashtra, which is hotter than cold throughout the year, the effect of temperature is not seen much.

4. The Betz's equation $P = 1/2 \, \rho A V^3 Cp$ watts indicates that wind velocity will have most profound effect on power generation, since the velocity has power of three (V^3). This has also reflected in this work to be true

8. Conclusion

The result of a variety of parameters on the power generation in the mountainous regions in Maharashtra, India was scrutinized and calculated for the entire year 2014. Analysis of this result was done with the aid of power curves in which the most profound effect of wind speed on power generation for the duration of this entire period was examined and noted. On the other hand, at the end of the findings and this study and we come to a conclusion that we have to also consider the fact that these findings are and will be difficult to generalize for wind power generation; because as we all are aware a lot of a times it may demonstrate a discrepancy with local parameters.

References

[1] Demirbas A. Biomass and the other renewable and sustainable energy options for Turkey in the twenty-first century. *Energy Sources,* V. 23 pp. 177–87, (2001).

[2] Cavallo A. J, Grubb M. J. Renewable energy sources for fuels and electricity. London: *Earthscan Publications Ltd,* (1993).

[3] Celik A. N. A simplified model for estimating the monthly performance of autonomous wind energy systems with battery storage. *Renewable Energy,* V. 28, pp. 561–72, (2003).

[4] Ravindra B Sholapurkar and Yogesh S Mahajan, Review of wind energy development and policy in India, *Energy Technology and Policy,* V. 2, pp. 122-132, (2015).

[5] Renewable Energy Resources: Opportunities and Constraints 1990—2020, study performed and published by the *World Energy Council,* London, UK, (1993).

[6] Lazar Lazic, Goran Pejanovic, Momcilo Zivkovic and Luca Llic, Improved wind forecasts for wind power generation using the Eta model and MOS (Model Output Statistics) method, *Energy,* pp. 567-574, (2014).

[7] Akpinar E. K, Akpinar S, An assessment on seasonal analysis of wind energy characteristics and wind turbine characteristics, *Energy Conservation and Management,* V. 46, pp. 1848-67, (2004).

[8] Lun I. Y. F., J. C. Lam, A study of Weibull parameters using long-term wind observations, *Renewable Energy* V.20, pp. 145–153, (2000).

[9] Seguro J. V., T. W. Lambert, Modern estimation of the parameters of the Weibull wind speed distribution for wind energy analysis, *Journal of Wind Engineering and Industrial Aerodynamics,* V.85 pp.75–84, (2000).

[10] Pechak Olena, Mavrotas George, Diakoulaki Danae., Role and contribution of the clean development mechanisms to the development of wind energy, *Energy,* V.7, pp.75-85, (2011).

[11] Ali Naci Celik, Energy output estimation for small-scale wind power generators using weibull-representative data, *Journal of Wind Engineering and Industrial Aerodynamics,* V. 91, pp. 693-707, (2003).

[12] Gipe P., Wind Energy Comes of Age, Wiley, New York, NY, (1995).

[13] Olayinka S. Ohunakin, Olaolu O. Akinnawonu., Assessment of wind energy potential and the economics of power generation in Jos Plateau state, Nigeria, *Sustainable for Energy,* V.16, pp.78-83. (2012).

[14] Tow Leong Tiang, Dahaman Ishak., Technical review of wind energy potential as small-scale power generation sources in Penang Island Malasia, *Renewable and sustainable energy review.* V. 16, pp. 3034-3042, (2012).

[15] Casale C., E. Sesto, Wind Power Systems for Power Utility Grid Connection, *Advances in Solar Energy,* V. 9, USA, (1994).

[16] Sesto E., D. F. Ancona, Present and prospective role of wind energy in electricity supply, in: Proc. Of New Electricity 21, Paris, OECD Publications, Paris, France, (1995).

[17] Marco Rauge, Ilvia Bargigli, Sergio Ulgiati, Life cycle assessment and energy payback time of advanced photovoltaic modules: Cd Te and CIS compared to poly-Si, *Energy*, V. 32, pp. 1310-1318, (2007).

[18] Keith M Sundeland, Gerald Mills, Michael F Conlon, Estimating the wind resource in an urban area: A case study of micro-wind generation potential in Dubin, Ireland, *Journal of Wind Engineering and Industrial Aerodynamics.*, V. 118, pp. 44-53, (2013).

[19] Landberg L., Short-term prediction of the power production from wind farms, *Journal of Wind Engineering and Industrial Aerodynamics.* V. 80, pp. 207-220, (1999).

[20] https://nccr.iitm.ac.in/ebook%20final.pdf B. Viswanathan., An introduction to energy sources, national center for catalysis research department of chemistry Indian institute of technology, Madras, (2006).

[21] Sfetsos A., A comparison of various forecasting techniques applied to mean hourly wind speed time- series, *Renewable Energy.* V. 21, pp. 23-35, (2000).

[22] http://onlinelibrary.wiley.com/doi/10.1002/9783527646289.ch 1/summary, Edward L. Wolf, A Survey of Long-Term Energy Resources, *Nanophysics of Solar and Renewable Energy*, Wiley publication, (2012).

[23] Thor S, Taylor P W., Long-term research and development needs for wind energy for the time frame 2000–2020. *Wind Energy.* V. 5, pp. 73–75, (2002).

[24] Trivedi M P, 1999. Environmental factors affecting wind energy generation in western Coastal region of India. *Renewable Energy*, V. 16, pp. 894-898, (1999).

[25] Zhigang Lu, Shoulong He, Tao Feng, Xueping Li, Xiaoqiang Guo and Xiaofeng Sun, Robust economic/emission dispatch considering wind power uncertainties and flexible operation of carbon capture and storage, *International Journal of Electrical power and Energy Systems*, V. 63, pp. 285-292, (2014).

[26] Rai G. D. Non-Conventional Energy sources, *Khanna Publishers*, (2011).

[27] Masseran, A M, Razali K Ibrahim, An analysis of wind power density derived from several wind speed density functions: the regional assessment on wind power in Malaysia, *Renewable and Sustainable Energy Reviews*, V. 16, pp. 6476-6487, (2012).

3

Oil-Independent Life Style: Solar Oasis

Huseyin Murat Cekirge

Department of Mechanical Engineering, Prince Mohamed Bin Fahd University, Al Khobar, KSA

E-mail address:

hmcekirge@usa.net (H. M. Cekirge)

Abstract: The objective of this study is to introduce solar energy as an electric and water source to distant communities in desert areas. Fossil fuel dominates world as the major source of energy that can be easily transformed into electricity and used for water reclamation in many societies. However, ambiguities and questions for supplying necessary fuels, such as fossil fuels, leads to more reliable energy sources. The sun is the major source of an alternative energy source and water, and establishing of solar energy dependent community will be explained in the paper, solar oasis. CSP, solar energy generation systems and their comparisons are introduced.

Keywords: CSP, Tower Systems, Parabolic Trough Systems, Solar Desalination, Heliostat

1. Introduction

With continuously rising energy costs, the solar power has become an increasingly attractive option for electricity production and water distillation. Solar power has the direct ability to drastically lower energy and operating costs. The significant amount of solar energy is available in many regions of the world, see Figure 1. Solar energy generation systems can be classified into two categories: thermal (Concentrated Solar Power) and electric (photovoltaic) systems see Table 1.

Recent advances in solar technology in power systems and water distillation need to be utilized to offset the continuously increasing electricity and water demand in Saudi Arabia. These technologies can be applied to save conventional energy consumed by inhabitants as well as energy used for water distillation. This will assist in satisfying the consumption demands of electricity in Saudi Arabia. Present energy sources are based on oils, which leads in overpopulated cities to pollution that is emissions of greenhouse gases, economy and depletion of resources of fossil fuels can be considered as a problems; and also all other difficulties of urban life.

In the present study, an appropriate Concentrated Solar Power (CSP) Steam/Electricity Generation Systems for production electricity and water by using solar energy are considered [1]. The proposed research will design solar power plant and water distillation system; establish appropriate principles and improvement of design concept of solar power plants. It should be noted that areas of at least 2000 kWh/m²/y are usable for CSP plants due to profitability of investments [2]. This solar system can be established in any area with low quality of water resources. In many desert areas, the brackish water exists as groundwater which could be used after desalination, which can be performed as a solar desalination process; see, Figures 2, 3 and 4.

Table 1. Comparison of solar energy systems.

	Solar Tower	Parabolic Trough	PV
Cost of Electricity Production	Optical heat transfer, high operating temperatures, better steam quality	Oil-based heat transfer medium, restricted operating temperatures	Comparable
Operating and Maintenance Costs	Commercial components and no heat transfer fluid circulation in the solar held	Specialized components and circulation of oil-based heat transfer fluid throughout the solar field	Comparable
Efficiency	Direct steam generation (DSG) with high temperature and high pressure steam	Steam production through heat exchanger	Lower
Thermal Storage	Heat tracing is as long as the tower height, therefore very efficient and simple design	Heat tracing is extremely longer and expensive with many complicated pumps,	Non-existent

	Solar Tower	Parabolic Trough	PV
Capital Cost	Commercialized components, simpler and widespread supply chain	extra energy demand to keep the system running Specialized components complex and limited supply chain	Higher
Water Consumption	More efficient thermodynamic cycle with reduced water requirements for cooling per electricity produced, closed loop water circulation	Less efficient thermodynamic cycle with more waste heat produced generated per electricity produced	Not relevant
Terrain Requirements	More flexible in terrain requirements as each heliostat is independent from one another	More stringent terrain requirements as placement of troughs poses many challenges on uneven surfaces	More land required

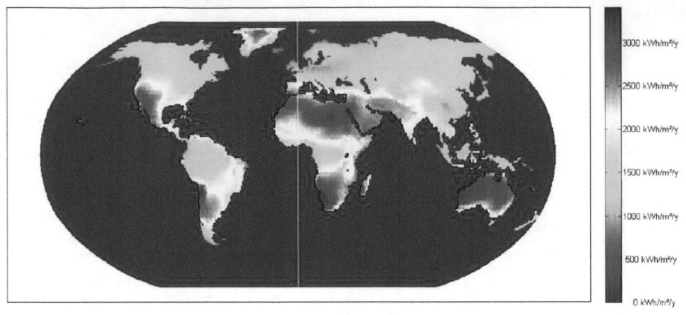

Figure 1. Direct normal irradiance in the world, [2] and [3].

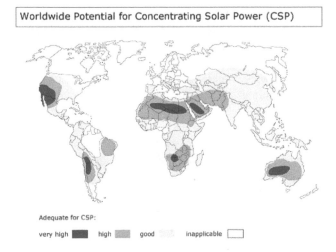

Figure 2. Worldwide Potential for Solar Power, CSP, [5].

Figure 3. CSP Tower plant, [5].

Renewable energy such as solar thermal, particularly CSP Tower, can be used in remote area, due to

- Increasing overall energy efficiency;
- Utilizing an abundant renewable source of energy such as sun.

Large cities and energy intensive life style, restricted agricultural production and increasing pollution, the existing successful CSP plants in world are considered as quite encouraging due to the fact that they use the solar energy [4, 5], see Figure 5.

Electricity

Water desalination

**Super Steam
(up to 550°C, 200 bar)**

Figure 4. Schematic CSP Tower plant and products, [5].

MAJOR CSP PROJECTS IN THE WORLD

Solana Generating
Station.

Ivanpah Solar
Power facilities.

377 MW Ivanpah Solar power Facility, USA
354 MW Solar Energy Generating Systems, USA
280 MW Solana Energy Generating Systems (SEGS), US
250 MW Genesis Solar Energy, USA
200 MW Casablanca, Extresol, Manchasol, Andasol, Sp
160 MW Noor 1, Morocco
110 MW Crescent Dunes Dolar Energy project, USA
110 MW Planta Solar Cero Dominadaor, Chile
100 MW KaXu Solar One, South Africa
100 MW Shams 1, United Arab Emirates
50 MW Khi Solar One, South Africa
44 MW Kogan Creek Solar Boost, Australia*
25 MW Hassi-R'mel, Algeria
25 MW Gujarat Solar One, India*
20 MW Gemasolar, Spain
 4 MW Greenway CSP, Turkey

*Under construction

Crescent Dunes
Solar Energy
Facility.

SEGS.

Figure 5. The existing CSP plants in world, [4, 5].

Figure 6. A sample solar oasis, [5].

Small CSP plants 5-20 MW is quite appropriate in remote areas of Saudi Arabia, the brackish water is quite abundant in desert environment. Electricity and water brings life to the location. The agricultural and live stocks can be grown easily; and a sample solar oasis is presented in Figure 6. Concentrated Solar Power (CSP) Tower System efficiently generates energy with zero pollution or zero emission, and 100% clean energy in a wide variety of markets and applications world-wide. The application of small power plants will lead small energy loses in the system, which leads to more efficient energy uses for consumption instead of large plants.

The information presented is for clear and concise explanations of basic concepts, associated with

Concentrated Solar Power (CSP) Tower System which enables the people to use inexpensive and hundred percent clean energy world-wide.

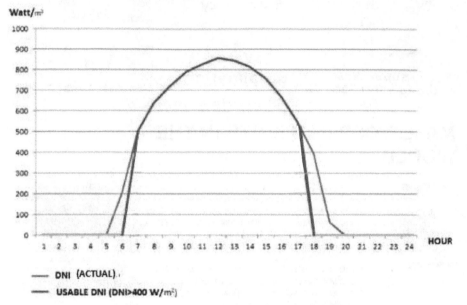

Figure 7. Direct Normal Irradiation per hour, {5}.

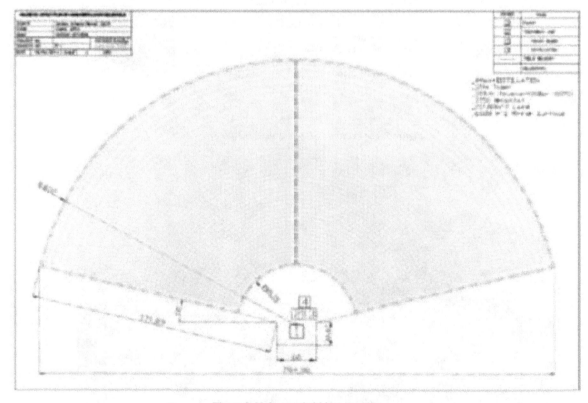

Figure 8. Heliostat field layout, [5].

2. Design Features

In CSP projects, solar beam is concentrated on the thermal receiver surface by reflective mirrors, called heliostats. Mirrors can only focus the solar beams which

come perpendicular to the mirror's surface; therefore Direct Normal Irradiance (DNI) values are used for output calculations in CSP systems, see Figure 7.

The CSP Tower systems generate high volume and high quality steam that is ready to use in steam turbines for various industrial applications, including drinking water production. CSP tower's operational internal electric consumption is significantly less to compare to parabola troughs. The system is highly efficient flexible, reliable, consistent, and customer oriented. The environmental problems are minimal and can be solved within its company limits and simplistic. The technology is reputable and has several research and supply chain partners. The system is self sustaining, low cost, light weight, and high efficient, it offers heliostats which are easy to install and maintain. Heliostat field and control system have an efficient communication which reduces field erection cost and complexity as well as maintenance costs. The other important features are the modular design and infrastructure (both software and hardware) for quick customer support and feature modifications. The plan footprint is relatively small due to a compact design.

ELECTRICITY

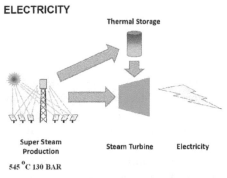

Figure 9. The CSP solar power plant, [5].

The heliostats are computer controlled for reflector positioning and wireless system for maximum optical efficiency. The heliostats and the CSP plant are presented by Figures 8 and 9, respectively.

Available solar energy is calculated as follows:

Available Solar Energy = Number of Heliostats × Surface Area (per Heliostat) × DNI

Tower is made as strong and durable construction, and the receiver is a heat recovery boiler which proven design of 150 year old and 50 years life time. Tower system consumes less area respect to other CSP systems of the same energy output. The Tower system has very short piping; there is very small probability for pipeline failures, which is a common problem in parabolic troughs. As a result of thermal stresses on pipes. There is no toxic thermo fluids in Tower systems as in parabolic troughs and these fluids are quite expensive fluids and increase production cost of electricity and water and they must be replaced every two years. Tower systems are designed for these high thermal stresses. It is much easy and less costly cleaning of mirrors since heliostats are flat. The heliostats are self tracking and self calibrating for collecting solar energy at maximum level. Even more, initial installation of Tower system is extremely simple with respect to other systems and the lifetime operation is straightforward, see the comparison in Table 2, [6, 7].

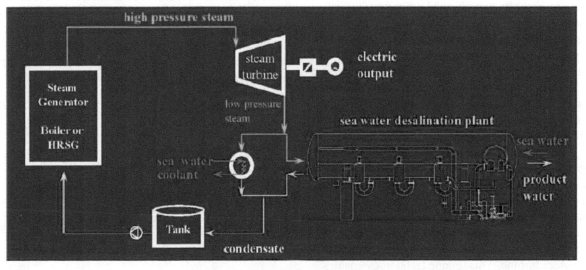

Figure 10. Water Desalination System, [5].

Table 2. Comparison of CSP Tower and CSP Parabola Trough systems.

	CSP TOWER SYSTEM	CSP PARABOLA TROUGH SYSTEM
Cost (Capital)	Lower	Higher
Cost (Operational)	Lower	Higher
Efficiency	Higher	Lower
Molten Salt Storage	Easy and low cost	Very complex and expensive
Maintenance	Easy and low cost	Complex and expensive
Cleaning	Easy	Complex
Land Requirement	Less	More
Operational Robustness	Strong	Fragile

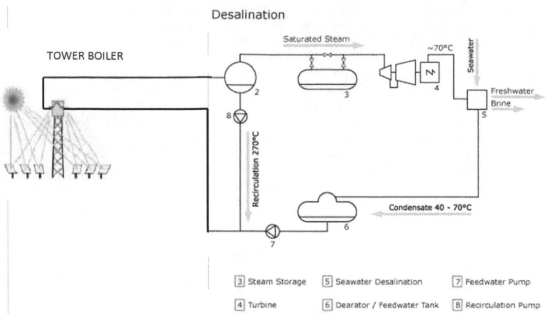

Figure 11. *Schematic Water Desalination System, [5]*

CSP Parabola Trough system is considered proven technology; however it has no further improvement left to move forward. This system's molten salt operation is very risky and expensive, and heat tracing is expensive and energy consuming. This system is very difficult to erect and line up for the first time; and needs expensive calibrations.

CSP Tower system is a new and highly efficient method; and plenty room to excel its capacity and decreasing in cost. Piping and heat tracing requirements are very short; only between tower and storage. Molten salt moves only within tower height with a very low operational cost. Erections of independent heliostats are easy; less costly; and calibrated automatically and require minimal maintenance.

It becomes clear that seawater desalination has already been understood as potential source to resolve the fresh water problems in numerous countries. However, it should be noted that in spite of the good reliability and favorable economic aspects of desalination processes, the problem of high energy consumption needs to be resolved. The solar Tower energy system can be considered as an answer to this problem. A most efficient steam turbine (back pressure turbine) must be used for maximum efficiency. A "large amount compact water distillation system" can be adapted back to power block, back pressure turbine. This unique technology enables fresh water distillation available while electricity production continues, see Figure 10 and 11. The Tower system needs small area and standalone configuration, and the capacity can be increased at reasonable small cost and leads to lower energy cost.

During day-time, necessary heat for the boiler to generate steam is provided by focusing solar energy onto the receiver using solar concentrating mirrors (heliostats). CSP's innovative technology must include an optimization algorithm to most efficiently track the sun throughout the day for maximum energy output.

Molten salt is selected in solar power tower systems for heat or energy storage, since it is liquid at atmospheric pressure at high temperatures; it provides an efficient, less expensive medium in which to store sun's thermal energy, its operating temperatures are compatible with high-pressure and high-temperature steam turbines, and it is non-flammable and nontoxic, [8]. The molten salt is a mixture of 60 percent sodium nitrate and 40 percent potassium nitrate, and calcium nitrate in some cases, [9], [10] and [11]. Electricity can be generated in periods of less solar radiation days and at night using the stored thermal energy in the hot salt tank. Normally tanks are well insulated and can efficiently store energy for up to seven days, see Figures 12 and 13.

TANK MOLTEN SALT

LOW COST AND MAINTENANCE FREE THERMAL STORAGE ENABLES 24 HOUR OPERATION

Figure 12. *Molten salt tank, [5].*

3. Investments

Production of electricity, drinking and agricultural water from brackish water are final product of the investment. This water can be used also for animals. The investment cost of tower systems is thirty or thirthtfive percent lower CSP parabolic trough systems [12, 13 and 14]. These plants are free of large energy losses, and the feasibility studies are extremely encouraging. CSP tower plants of 10-40 MW systems are excellent power and water source for desert dwellings. The CSP tower systems are not fragile and their maintenance is not costly and efficient respect [1, 4 and 5], see Figures 14 and 15. It should be mentioned that construction of linear Fresnel systems were stopped due various technical and economical reasons, [1]. Due to long piping arrangements with thermal shocks and thermal insulations, CSP parabolic trough systems are not suitable for the moving tracking systems. Moreover, construction and operating expenses of these systems are incomparable to CSP

tower systems. These plants are free of large energy losses, and the feasibility studies are extremely encouraging. CSP tower plants of 10-40 MW are excellent power and water source for desert dwellings. These systems are not fragile. Their maintenance is very minimal and efficient respect CSP parabolic trough systems, [1], see Figures 14 and 15; [1], [4] and [5].

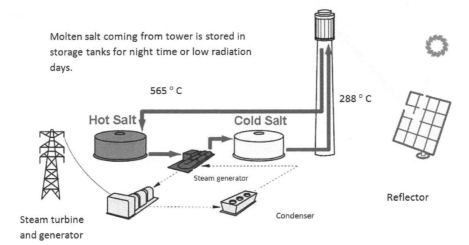

Figure 13. *Separate molten salt tower. [5].*

Parabolictrough

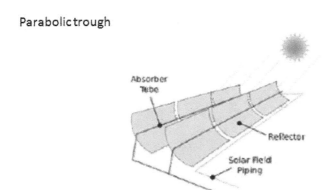

Figure 14. *Parabolic troughs, [5].*

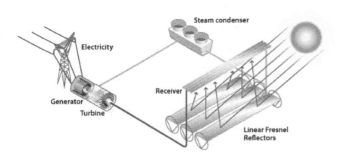

Figure 15. *Linear Fresnel System, [5].*

4. Conclusion

It is obvious that the CSP tower power plans are pollution free and excludes problems due to thermal fluids as in CSP parabolic tough systems. The produced water can be used for

agriculture and possible industrial activities. The initial investment is very reasonable and risk free. The end product of CSP tower systems is non-oil dependent continuous energy and water from sun. The system can be used in remote areas and islands. These areas will gain economical value which does not have potable water and energy sources in place. It is known that these ideas may be known, however the paper is putting out the oil free society is a survival many areas without energy and water.

References

[1]   S. Erturan, private communication, 2015.

[2]   S. Lohmann et al, Validation of DLR-ISIS data, German Aerospace Center, Oberpfaffenhofen, 2006, (www.pa.op.dlr.de/ISIS/).

[3]   F. Trieb et al, Concentrating Solar Power for the Mediterranean Region, German Aerospace Center by order of Federal Ministry for the Environment, Berlin, 2005, (www.dlr.de/tt/med-csp).

[4]   Christian Breyer and Gerhard Knies, Global Energy Supply Potential Of Concentrating Solar Power, Proceedings Solarpaces 2009, Berlin, September, 15 – 18, 2009.

[5]   S. Erturan, Presentation of Greenway Solar, Istanbul, 2014, (http://greenwaycsp.com).

[6]   Robert Foster, Majid Ghassemi and Alma Cota, Solar Energy Renewable Energy and the Environment, CRC Press, Boca Raton, 2010.

[7]   Soteris Kalogiorou, Solar Energy Engineering Processes and Systems, Second Edition Academic Press, Boston, 2013.

[8] Ozie Waite Advances in Solar Power, The English Press, Delhi, 2011.

[9] Robert Ehrlich, 2013, Renewable Energy: A First Course, CRC Press, Boca Raton, 2013.

[10] David Biello, "How to Use Solar Energy at Night," Scientific American, a Division of Nature America, Inc., 19 June 2011.

[11] Tom Mancini, "Advantages of Using Molten Salt," Sandia National Laboratories, archived from the original on 2011-07-14, 10 January, 2006.

[12] G. J. Kolb, C. K. Ho, T. R. Mancini and J. A. Gary, Power Tower Technology Roadmap and Cost Reduction Plan. Sandia National Laboratories, Draft, Version 18, December 2010.

[13] C. Turchi, M. Mehos, C. K. Ho and G. J. Kolb, Current and future costs for parabolic trough and power tower systems in the US market. *SolarPACES 2010*. Perpignan, France, 2010.

[14] Jim Hinkley, Bryan Curtin, Jenny Hayward, Alex Wonhas, Rod Boyd, Charles Grima, Amir Tadros, Ross Hall, Kevin Naicker and Adeeb Mikhail, Concentrating solar power – drivers and opportunities for cost-competitive electricity, CSIRO Energy, 2011.

4

Qualitative Risk of Gas Pipelines

Huseyin Murat Cekirge

Department of Mechanical Engineering, Prince Mohamed Bin Fahd University, Al Khobar, KSA

Email address:

hmcekirge@usa.net

Abstract: A qualitative analysis for determining axis of a gas pipeline in view of general risk is outlined in the paper. The increasing building of gas pipelines for energy transportation vehicle, the importance of risk associated to this transport is gained strong importance. Risk assessment is an extremely useful tool in providing a framework in which to identify the possible hazards and determine the risks associated with gas pipelines. The determination of route, thickness, diameter and segment lengths is the major factor of constructing gas pipelines. The paper is presenting this risk in general and in a qualitative manner.

Keywords: Gas Pipelines, Quantitative Risk Analysis, Quantitative Risk Assessment, Fire Gas Pipelines, Explosion Gas Pipelines, Toxication Gas Pipelines, Gas Pipeline Segments

1. Introduction

For securing energy sources in view of changing political situations, new gas pipelines will be constructed from production areas to the consumption zones.

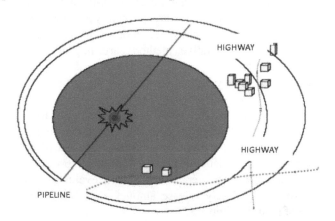

Figure 1. The danger zone, [1].

One of the major difficulties of building these pipelines is overcoming of risk problems on environment where the pipelines are built, [2-6]. Emergency response plan and risk analysis must be prepared for the pipeline's building and operation. These plans must cover environmental compliance of the gas pipeline, investigation of gas pipeline's socio-economic aspects and prevention of all problems during operation and minimization of their effects calculation of risk factors.

Calculation of risk or risk assessment consists of technical and environmental specifications and must include the risk factors originated from environmental events and operational activities. The risk factors must be in suitable details and risk probability must be in calculation basis, reasonable, regular and documentable.

The operational risks are pipe age, terrestrial location of the pipeline, precautions against corrosion, periodic maintenance, supervision of the line and other factors. Operational risks that might be caused by leakage are fire (pool, jet and flare up), gas distribution (concentration), explosion and toxicity. It is necessary to define the danger zone for determination of payable losses and compensations, see Figure 1. Figure 2 and 3 present the movement of the plume from the gas pipeline. In Figures 4, 5 and 6, the influence zones are presented, [7 - 11].

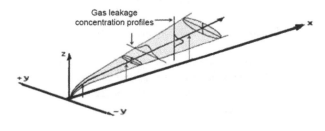

Figure 2. Gas plume from the pipeline, [11].

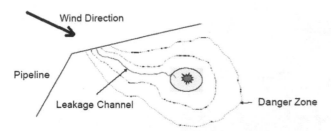

Figure 3. Effects of meteorology and topography on gas plume, [1].

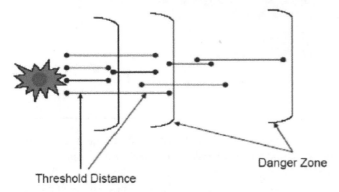

Figure 4. Influence zone and various distances of gas leakage, [1].

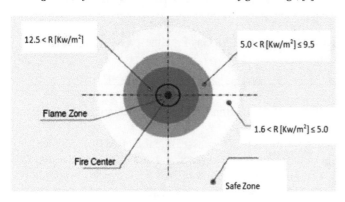

Figure 5. Impact areas for fire, [1].

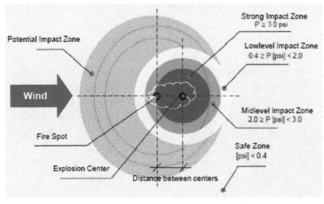

Figure 6. Impact areas for explosion, [1].

The receptors must include count and type; and these receptors must be quantified in units. A unit can be assigned a fixed money value and that simplifies remediation or compensation to injured values.

2. Risk Analysis of Hazardous Events

The environmental risks are affected be terrestrial location of the pipeline, settlements, flora and animals, agriculture and livestock, terrorism and other factors, [11-18],

Basic principlesof calculations are

- Calculate on every point
- Comparison of segment lengths
- Hazards
- Calculation of all collapse factors for pipeline
- Calculation of thickness in terms of time factor
- addition of all risks

The risk assessment factors are pipe age, the land on which the pipeline is placed, precautions against corrosion, position of settlements near the pipeline and others factors. Indexing methods is used to quantify the risk.

Risk assessment results determination of pipeline route, calculation of pipe segments' length, calculation of pipe diameter and thickness and other details.The foresights for risk analysis are conervative calculations, uncertain details, past events, expert assessments, relative assessments, scenarioassessments, probabilistic assessments

Determination risk term by using absolute results or relative result's model.The models should contain in, probabilistic methods, scenarios, trees, results and verification, and determination of formation frequency

Basic risk scenarios are leakage jet (pool) fire, tear jet (pool) fire, leakage flare fire, tear flare fire, leakage explosion, tear explosion, underground leakages in embedded pipes If sea passages are considered the toxicity of beam risingin the sea and its effects, the effects of sea water's density change, especially on surface.

The pipeline risk analysis factors are size, age, type and operating pressure of the pipeline, product transported and location of the line, relative to natural and man-made threats. Consequences of of gas release are proximity of the line, meteorological and topographical conditions, local terrain and land use. The qualitative methods are improved to quantitative methods.

The general steps in a pipeline risk analysis are as follows:

Data Compilation – The first step is to compile all pertinent data for the risk analysis. This includes the location and characteristics of the pipeline and the area site of interest.

Hazard identification – The pipeline system must be characterized in sufficient detail to formulate potential accident scenarios and to pen-lit subsequent evaluation of accident probability, likely release amount and nature and magnitude of resulting impacts:

Probability analysis – Probability analysis determines the likelihood of an event expressed in relative (typically referred to as likelihood) or quantitative teens (typically referred to as probability).

Consequence analysis – Consequence analysis examines the potential physical impacts and derivative consequences (e.g. harm to people. or the environment) of a pipeline failure and accidental release of product.

Risk evaluation – Risk evaluation creates a numerical

combination of both the probability of all and its consequences.

Risk control – Risk control consists of prevention and mitigation measures respectively to reduce the probability that a release of pipeline product will occur and to minimize the impacts of any release that might occur.

Reporting – When the risk analysis is completed and results are reported to management. The report contains information on the area pipeline, method, data and assumptions and results.

The causes of pipeline failure are corrosion (internal and external), excavation damage, natural forces ground movement, flooding, and displacement, etc.), other outside forces (e.g. fire or explosion near the pipeline); material and weld defects, equipment and operations (e.g. such as over pressuring and an inadequately protected system through inappropriate operating settings); and other (i.e. not included above or unknown).

Risk Control can be done through prevention and mitigation. Prevention measures are used to control risk by reducing the likelihood of a risk event occurring. Traditionally; codes, standards, regulations and an operator's own good practices comprise prevention activities. Specific prevention activities generally focus on specific causes of pipeline failures For example, prevention measures associated with excavation damage including pipeline makers, patrols, and one-call notifications. Mitigation measures are

pre-engineered systems. Procedures and practices that reduce the consequences of a pipeline product release, should a release occur. Emergency preparedness and emergency response plans are one of the most basic elements of mitigation. Some mitigation measures are common to all pipelines; some depend on whether the line is a gas or liquid pipeline and whether the issue is product flammability, toxicity or both. Risk analysis has three stages; these are risk screening, qualitative and detailed qualitative analysis.

For a given length of pipeline within X of a site property line, each of these hazards has a unique length of pipe from which the impacts could reach a receptor. Outside of this length the impacts could not reach the receptor. The segment length for which a hazard can have an impact is the length XSEG; the hazard impact distance and XSEG, see Figure 7.

The steps of an analysis, in sequence, to determine the hazard impact distance XSEG length for each of the, three hazard types based on the distance between the receptor and the pipeline hazard source, and the hazard impact distance which is maximum mortality impact from the closest approach of the pipeline to the receptor.

- Average mortality at the receptor for each XSEG.
- Base adjusted failure probability for the pipeline.
- Base probability for each XSEG.
- Conditional probability factor for each event scenario
- Conditional probability of individual exposure.

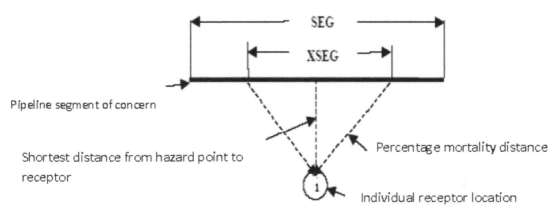

Figure 7. Segment and safety XSEG, [1].

3. Discussion

The complications and detail of risk analysis of a gas pipeline is introduced. The fire, explosion and toxicity are main issues for a building and operating a gas pipeline. Importance of selection of segment length is the important part of the designing a gas pipeline. The paper is the guidance will give basic preparation rules of risk analysis of gas pipelines.

4. Conclusion

This study of risk of a pipeline transport system activity is presented. Total risks of infrastructures are based on the infrastructure and operation. These quantitative risks are based on threatening consequences of events identified. Quantitative risk assessment has its foundation in the application of procedures used in other transport infrastructure of hydrocarbon sector and industry. By that way, the risk of an infrastructure can be determined in the area of tolerable range. In this procedure, spill events must be included that have occurred to date.

In considering risk characteristics of a pipeline transport system and operation, the risk of each infrastructure must be considered. The risks located in acceptable levels are considered tolerable. Acceptable levels of risk are tolerable to the human settlements which fall within the acceptable zones. In the area of human settlements, divulgation of activities and preparation of communities are considered as preventive measures, [19 – 22].

The ground motions in the integrity of the infrastructure should be monitored by annually scheduled activities and new assessments should be scheduled with idle paths, in which to identify new high sensitivity points for modifying maintenance programs.

References

[1]   H. M. Cekirge, Safety Analysis, Maltepe Uni., Internal Rep. 1/1, Istanbul, 2010.

[2]   G. Pluvinage and M. H. Elwany (Editors) Safety, Reliability and Risks Associated with Water, Oil and Gas Pipelines, NATO Science for Peace and Security Series C: Environmental Security, Springer, 2007.

[3]   United States Congress Senate, To reduce risk to public safety and the environment associated with pipeline transportation of natural gas and hazardous liquids, and for other purposes, US Congress and Senate, Paperbackshop-US, Secaucus, NJ, 2010.

[4]   J.L. Kennedy, Oil and Gas Pipeline Fundamentals, Pinacle Books, Richford, VT, 1984.

[5]   Rizkalla, M. (editor), Pipeline Geo-Environmental Design and Geohazard Management, ASME, 2008.

[6]   W. K. Muhlbauer, Pipeline Risk Management Manual: Ideas, Techniques, and Resources, Gulf Publishing Services, Burlington, Mass., 2004

[7]   GIUSP, The Gas Industry Unsafe Situations Procedures– Edition 6, Including 1 & 2, 2 April 2012.

[8]   Young-Do Jo and Daniel A. Crowl, Individual risk analysis of high-pressure natural gas pipelines, Journal of Loss Prevention in the Process Industries, Volume 21, Issue 6, Pages 589–595, November 2008

[9]   IGEM, Institution of Gas Engineers & Managers, Assessing the risks from high pressure Natural Gas pipelines, IGEM/TD/2 Edition 2, Communication 1764, 2013.

[10] E. W. McAllister, Pipeline Rules of Thumb Handbook, Elsevier Inc., 2005.

[11] U.S. Department of Transportation, Natural Gas Pipeline Systems,(https://primis.phmsa.dot.gov/comm/NaturalGasPipel ineSystems.htm), 2011.

[12] R. Michael Reynolds, ALOHA (Areal Locations of Hazardous Atmospheres) 5.0 Theoretical Description, NOAA Technical Memorandum NOS ORCA-65, Seattle, Washington 98115, August 1992.

[13] Pipeline and Hazardous Materials Safety Adminstration, Gathering Pipelines: Frequently Asked Questions, (http://phmsa.dot.gov/portal/site/PHMSA/menuitem.ebdc7a8a 7e39f2e55cf2031050248a0c/?vgnextoid=0a69c4d9aebb6310 VgnVCM1000001ecb7898RCRD&vgnextchannel=623b1433 89d8c010VgnVCM1000008049a8c0RCRD&vgnextfmt=print), 2015.

[14] Government Accountability Office, Pipeline Safety, (http://gao.gov/assets/590/589514.pdf), 2012.

[15] W. E. Baker, P. A. Cox, P. S. Westine, J. J. Kulesz and R. A. Strehlow. Explosion Hazards and Evaluation. Elsevier Scientific Publishing Company, Amsterdam - Oxford - New York, 1983.

[16] M. R. Hankinson, G. Ashworth and B. P. Mohsen Sanai and James D. Colton, "A Full Scale Experimental Study of Fires Following the Rupture of Natural Gas Transmission Lines," ASME Conference Proceedings, 2000.

[17] Design Institute for Physical Property Data (DIPPR), American Institute of Chemical Engineers (AIChE),given in DIPPR computer software, Version March 1995, Technical Database Services, Inc., 1995.

[18] B31.8S 2001, Supplement to B31.8 on Managing System Integrity of Gas Pipelines. ASME International, New York, New York, January 31, 2002, Center for Chemical Process Safety (CCPS), Guidelines for Chemical Transportation, 2002.

[19] Y. D. Jo and D. A. Crowl, Individual Risk Analysis of High-Pressure Natural Gaspipelines. Journal of LossPrevention in the Process Industries, 21, 589-595, (http://dx.doi.org/10.1016/j.jlp.2008.04.006), 2008.

[20] D. P. Nolan, Applications of HAZOP and What-If Safety Reviews to the Petroleum, Petrochemical and ChemicalIndustries, Noyes Publications, Saddle River, 7, 1994.

[21] Y. Y. Haimes, Risk Modelling, Assessment, and Management. 3rd Edition, John Wiley & Sons Inc., Hoboken, 2008.

[22] 7th Report of the European Gas Pipeline Incident Data Group, Gas Pipeline Incidents, December 2008, Document No. EGIG 08 TV-B 0502, 2008.

5

Analysis of Energy Cost Savings by Substituting Heavy Fuel Oil with Alternative Fuel for a Pozzolana Dryer: Case Study of Bamburi Cement

Veronica Kavila Ngunzi

Department of Engineering and Innovative Technology, Kisii University, Kisii, Kenya

Email address:

vngunzi@yahoo.com

Abstract: The research study was carried out with the aim of analyzing the energy cost saving achieved by substituting heavy fuel oil with alternative fuel for a pozzolana dryer. This was carried out on an existing dryer where data from reports for previous years on energy requirements, that is, heavy fuel oil cost and usage was collected. An auxiliary system to handle biomass was designed and fabricated. Further a projected substitution scenario was determined through the use of excel worksheet which was set as the benchmark of evaluation on the expectations of the actual substitution. Comparison of fuel composition and cost of both actual and projected substitution scenarios was carried out. Further an economic analysis was carried out to establish the viability of the project. From the study findings of both the projected and actual substitution, the cost of energy was reducing with an increase in alternative fuel substitution with coefficients of correlation (R^2) of 1 and 0.5422 respectively. Again the projected and actual savings were increasing with an increase in alternative fuel substitution with coefficients of correlation (R^2) of 1 and 0.6288 respectively. From the economic analysis, the cost benefit analysis gave a positive net present value of 67,409,041. IRR was 4.10 %, simple payback period was 12 days and return on investment was 29.72%. Using these four techniques of capital budgeting, the investment was worthwhile to undertake. Further on economic analysis substitution effect was carried out. On the substitution effect, there was gradual cost drop of the energy used to dry pozzolana from 357491491 Kenya shillings with increasing percentage alternative fuel substituted to 106,269975 Kenya shillings when heavy fuel oil is completely substituted by alternative fuel. From the study, the high and fluctuating cost of heavy fuel oil used in pozzolana drying can be achieved through substitution with alternative fuel.

Keywords: Heavy Fuel Oil, Alternative Fuel, Projected Substitution, Actual Substitution, Existing Dryer, Auxiliary System

1. Introduction

There has been overreliance on fossil fuels in many manufacturing industries over the years. This has led to evident increase in cost of fuel and increasing production cost. The cost increase is caused by hidden costs which are not paid for by the companies that produce and sell energy but are passed on to the consumers of the energy. These costs include climate change adaptation costs, climate change damage costs, and fossil fuel dependence costs. These costs are indirect and difficult to determine, therefore they have traditionally remained external to the energy pricing system, and are thus often referred to as externalities. Hence the overreliance on fossil fuels results in damage to human health, the environment, and the economy. (www.ucsusa.org,

19.09.2013). Again the fossil fuels being relied on for industrial energy supply will most probably be depleted within a few hundred years.

With the growing realization of the impact of fossil fuels on global warming, there is a renewed interest in the utilization of biomass as a renewable and carbon-neutral energy source. The use of biomass and waste fuels is a growing area based on sound economic and environmental benefits. Biomass fuel-switching is possible, achievable and beneficial to the environment and companies that are willing to embrace it. Once implemented, companies can also benefit from the generation of carbon credits through the Clean Development Mechanism (United Nations Development Programme, 2009).

The production of cement is also an energy-intensive process. The typical energy consumption of a modern cement plant is about 110-120 kWh per ton of produced cement (Alsop, 2001). The energy consumption in the cement mills contributes roughly 50 kg CO_2 emissions per tonne to the overall greenhouse gas emissions of the industry (MIT – Research, 2011). The most energy-consuming cement manufacturing process is finish grinding drawing on average 40% of the total energy required to produce a ton of cement (Alsop, 2001).

The cement manufacturing industry is therefore under increasing pressure to reduce emissions. Cement manufacturing releases a lot of emissions such as carbon dioxide (CO_2) and nitrogen oxide (NOx). It is estimated that 5 percent of global carbon dioxide emissions originate from cement production (Hendriks, et al, 1998). The use of alternative fuels in cement manufacturing, therefore do not only afford considerable energy cost reduction, but they also have significant ecological benefits of conserving non-renewable resources, the reduction of waste disposal requirements and reduction of emissions. Use of low-grade alternative fuels in some kiln systems reduces NOx emissions due to re-burn reactions. There is an increased net global reduction in CO_2 emissions when waste is combusted in the cement kiln systems as opposed to dedicated incinerators.

Pozzolana is one of the main components of pozzolanic cement accounting for 35% of the mass of cement. This pozzolana has to be dried before inter-grinding with clinker in order to maintain cement to clinker ratio and to maintain higher grinding efficiency. The drying process uses a couple of dryers which are traditionally equipped with hot gas generators (HGG) fired by either diesel oil or heavy fuel oil (HFO). This increases the energy per tonne of cement produced. This is due to the energy required to reduce the moisture content to about two to three percent. However heavy fuel oil is facing high and fluctuating cost and the price gap between the fossil fuels in use today to dry pozzolana and the possible price of the biomass is in the range 8 - 10€/GJ (Bamburi Cement Annual Report, 2012). Therefore, there is a clear interest to study the possibility of converting the existing HGGs to use biomass in order to reduce cost of fuel for drying pozzolana and dependence on and the use of fossil fuels. Currently the use of biomass instead of fossil fuel is gaining acceptance as a cost effective form of renewable energy. Beside the lower costs, biomass fuel results in lower emissions and residues.

According to Kurchania et al.(2006), biomass energy or "bio-energy" includes any solid, liquid or gaseous fuel, or any electric power or useful chemical product derived from organic matter, whether directly from plants or indirectly from plant-derived industrial, commercial or urban wastes, or agricultural and forestry residues. Thus bio-energy can be derived from a wide range of raw materials and produced in a variety of ways. Because of the wide range of potential feed stocks and the variety of technologies to produce them and process them, bio-energy is usually considered as a series of many different feedstock/technology combinations.

Previous studies carried out to address this concern have aimed at reducing CO^2 emission by substitution and focused on price elasticity of the inter-fuel substitution using mathematical models. The previous studies have used data obtained from entire production process involved in cement manufacturing industries. This however faces the challenge of generalization given that the different operational areas of the manufacturing system for cement are likely to have different energy consumption patterns and requirements. There is however a need to apply the lessons learned from the studies using the mathematical models to study the inter-fuel substitution in specific operational areas of the cement manufacturing sectors that consume large quantities of fossil fuels and observe the behavior of the different processes. Such an observation can be done when an experiment is designed to assess the variation in energy cost behavior at different levels when the fossil fuels are substituted with alternative fuels. At the cement grinding stage of the process, it is possible to carry out this substitution since pozzolana drying falls in this category of sectors that consumes large quantities of fossil fuels. The stage is also recognized as an important source of CO_2 emissions.

Substantial potential for energy efficiency improvement exists in the pozzolana drying. a portion of this potential can be achieved as part of modification and expansion of existing facilities. At Bamburi Cement Limited Nairobi grinding plant, an opportunity exists where pozzolana dryer can be modified to accommodate biomass for substitution. This is because biomass is the most cost-effective and practical and therefore offers the most realistic and sustainable energy strategy. This study analyses the energy cost savings by substituting heavy fuel oil with biomass for a pozzolana dryer in order to achieve sustainable energy strategy by improving the existing dryer to accommodate the use of alternative fuels.

2. Materials and Methods

2.1. Description of the Experiment Site

The dryer studied is at the Nairobi Grinding Plant (NGP) in Athi-river about 26km from Nairobi along the old Mombasa road and next to the Namanga junction. This plant is part of the Bamburi Cement Company which belongs to the Lafarge Group (the world largest manufacturer of building materials). On average the plant produces 100, 000 tonnes of cement consuming about 150, 000 litres of HFO per month. The HFO is used in drying pozzolana before inter-grinding with the clinker.

2.2. Description of Cement Drying Process

Figure 1 below shows the cement drying process. The existing pozzolana dryer installation basically consists of HGG fired with HFO and waste oil drum dryer, filter and exhaust fan. HFO is transferred to the air-fuel mixing chamber of the burner. LPG is also introduced in the mixing chamber to improve the ignition of the fuel. Atomizing compressed air at 31°C is introduced to the atomizing unit where it meets primary and secondary air. Atomized air and fuel then mix and ignition and combustion take place while flue gases are generated. The dryer slopes slightly so that the discharge end is lower than the material feed end in order to convey the material through the dryer under gravity. Material to be dried enters the dryer, and as the dryer rotates, the material is lifted up by a series of internal fins lining the inner wall of the dryer. When the material gets high enough to roll back off the fins, it falls back down to the bottom of

the dryer, passing through the hot gas stream as it falls. This gas stream is moving towards the discharge end from the feed end (known as co-current flow) by help of a suction fan. The gas stream is made up of a mixture of air and combustion gases from a burner, in which case the dryer is called a direct heated dryer. Wet gypsum and pozzolana are dried then conveyed through conveyor and elevator system to their storage silos.

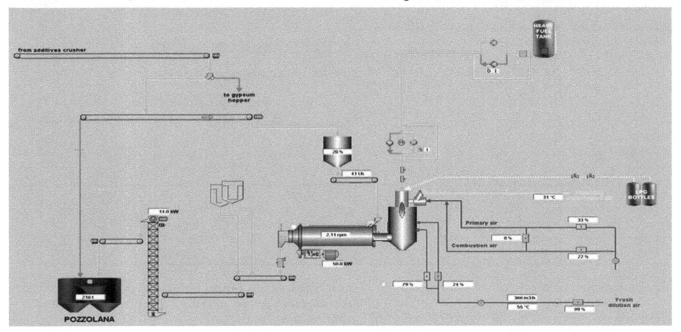

Figure 1. *Cement Drying Process.*

2.3. Description of the Pilot Auxiliary System to Handle Biomass

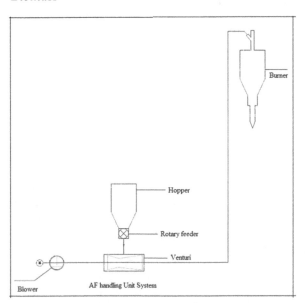

Figure 2. *Auxiliary System.*

An auxiliary system was designed and fabricated to handle and deliver the AF fuel. It consisted of a blower run by a 30kW motor, venturi, rotary feeder run by a 20kW motor, a hopper of 2 tonne capacity and piping system with a diameter

of 150mm to the burner. The blower through centrifugal force propels air forward giving it some velocity. When the air reaches the venturi there is a pressure drop and increase of velocity of the air. At the same time rice husks flow down the hopper and discharged through the rotary feeder. They are then blown though the piping system into the burner where they are mixed with HFO. The rice husks are introduced at various percentages of substitution and data. The line presentation of the auxiliary system is as shown in figure 2.

2.4. Data Acquisition

Data was collected for a period of 20 days where GJ of HFO and AF used for a number of hours of running the dryer for different percentages of substitution were obtained. This data was analyzed to get the total cost of HFO, AF and energy per year which was presented inform of graphs. Again a projected substitution scenario was carried out for the purposes of comparison and drawing of conclusion on the viability of this project.

The procedure below was carried out for the year 2014.
Given;

   i. HFO price Kes/kl= 76599.79 = A

HFO density ton/kl =0.92 = B

HFO LHV GJ/ton = 39.77 =C

Therefore;

HFO Kes/GJ $=\frac{A \div B}{C} = D$

$$\frac{76{,}599.79 \div 0.92}{39.79} = 2{,}093.47$$

ii. Assuming 1 € =116 Kes

Therefore HFO €/GJ $=\frac{D}{116} = \frac{2093.47}{116} = 18.05 = E$

iii. Budget MJ/t Cement =125 =F

Budget ton of cement in 2014 =1366120.6 =G

Budget GJ/Yr $=\frac{F \times G}{1000} = \frac{125 \times 1366120.0}{1000} = 170765.08 = I$

iv. Assuming there was additional cost of labour to handle alternative fuel at 12%

Alternative fuel LHV GJ/t = 12.70

Alternative fuel €/GJ = (1+12%) x (4.39+0.4) = 5.36

Where 4.39= cost of rice husks per Giga joule

0.4 = cost of bags per giga joule

Alternative fuel kes/GJ =5.36 x 116 = 622.32

Where

1 € = 116 kes.

v. Therefore

$$\text{HFO Cost} = I \times (1 - \%AF) \times E$$

Where;

I = budget GJ/yr

E= HFO kes/GJ

AF fuel cost = (Budget GJ/yr x AF substitution %) x AF cost in Kes/ GJ

(Source of Costs - NGP annual Report, 2012)

3. Results and Discussion

3.1. Projected Substitution Scenario

The projected substitution scenarios were calculated using excel program and tabulated as shown shown in table 1 below.

Table 1. Projected Substitution Scenarios.

DESCRIPTION	0%	5%	10%	15%	20%	25%
HFO Cost (Kes)	357,491,491.33	339,616,916.77	321,742,342.20	303,867,767.63	285,993,193.07	268,118,618.50
AF Cost (Kes)	-	5,313,498.75	10,626,997.50	15,940,496.25	21,253,995.01	26,567,493.76
Total Cost (Kes)	357,491,491.33	344,930,415.52	332,369,339.70	319,808,263.89	307,247,188.07	294,686,112.26
Savings (Kes)	-	12,561,075.82	25,122,151.63	37,683,227.45	50,244,303.26	62,805,379.08

3.2. Actual Substitution Scenarios

An actual test of the substitution was carried out at various percentages for twenty days to establish GJ of HFO and AF used. The data was further analyzed to establish the amount of energy used per day per hour and per year and tabulated in table 2 below.

Table 2. Actual Substitution Data.

Day	1	2	3	4	5	6	7	8	9	10
%substitution	0.00	1.26	3.74	4.15	5.18	5.95	6.76	7.70	8.02	9.55
GJ of HFO Used	425.87	515.80	628.28	468.05	286.26	648.70	573.73	587.74	504.75	543.27
GJ of AF used	0.00	7.01	18.80	19.30	22.91	41.02	47.54	57.49	54.31	64.25
Total GJ	425.87	522.81	647.08	487.35	309.17	689.72	621.27	645.23	559.06	607.52
Hours of running dryer	11.85	14.78	19.12	15.98	10.22	22.92	20.52	21.72	18.75	21.00
GJ/hr of HFO	35.94	34.90	32.86	29.29	28.01	28.30	27.96	27.06	26.92	25.87
GJ/hr of AF	0.00	0.47	0.98	1.21	2.24	1.79	2.32	2.65	2.90	3.06
Total GJ/hr	35.94	35.37	33.84	30.50	30.25	30.09	30.28	29.71	29.82	28.93
GJ/day of HFO	862.52	837.56	788.64	702.95	672.23	679.27	671.03	649.44	646.08	620.88
GJ/day of AF	0.00	11.38	23.60	28.99	53.80	42.95	55.60	63.52	69.52	73.43
Total GJ/day	862.52	848.95	812.23	731.94	726.04	722.22	726.63	712.96	715.60	694.31
GJ/year of HFO	314820.35	305710.96	287852.13	256578.10	245365.71	247932.46	244925.67	237044.31	235819.20	226621.20
GJ/year of AF	0.00	4154.78	8613.39	10579.97	19637.14	15677.80	20294.85	23186.57	25373.63	26801.43
Total GJ/Yr	314820.35	309865.74	296465.52	267158.07	265002.86	263610.26	265220.53	260230.88	261192.83	253422.63
Cost of HFO/Year	658919001.82	639853040.87	602474516.23	537017958.32	513550440.00	518922640.31	512629432.57	496133739.56	493569585.60	474318171.60
Cost of AF/Year	0.00	2584271.12	5357528.03	6580744.43	12214302.86	9751592.25	12623399.06	14422049.39	15782399.10	16670488.57
Total of energy Cost /year	658919001.82	642437311.99	607832044.27	543598702.75	525764742.86	528674232.57	525252831.64	510555788.95	509351984.70	490988660.17
Cost savings (kes) / year	0.00	16481689.83	51086957.55	115320299.07	133154258.96	130244769.25	133666170.18	148363212.87	149567017.12	167930341.65
Cost savings (euro) / year	0.00	138501.60	429302.16	969078.14	1118943.35	1094493.86	1123245.13	1246749.69	1256865.69	1411179.34

Day	11	12	13	14	15	16	17	18	19	20
%substitution	10.81	11.69	12.86	13.49	14.51	15.87	16.12	20.17	20.60	21.28
GJ of HFO Used	612.12	540.87	480.04	468.13	623.09	348.48	576.22	450.79	210.61	501.35
GJ of AF used	72.15	71.56	71.56	73.03	105.74	133.78	110.71	113.93	68.61	135.53
Total GJ	684.27	612.43	551.60	541.16	728.83	482.26	686.93	564.72	279.22	636.88
Hours of running dryer	24.00	22.12	18.88	17.93	22.32	14.88	20.82	18.75	8.93	20.25
GJ/hr of HFO	25.51	24.45	25.43	26.11	27.92	23.42	27.68	24.04	23.58	24.76
GJ/hr of AF	3.01	3.24	3.79	4.07	4.74	8.99	5.32	6.08	7.68	6.69
Total GJ/hr	28.51	27.69	29.22	30.18	32.65	32.41	32.99	30.12	31.27	31.45
GJ/day of HFO	612.12	586.84	610.22	626.61	669.99	562.06	664.23	577.01	566.03	594.19
GJ/day of AF	72.15	77.64	90.97	97.75	113.70	215.77	127.62	145.83	184.39	160.63
Total GJ/day	684.27	664.48	701.19	724.36	783.69	777.84	791.85	722.84	750.42	754.82
GJ/year of HFO	223423.80	214196.26	222730.42	228712.70	244546.08	205153.55	242444.15	210609.09	206600.63	216880.30
GJ/year of AF	26334.75	28339.31	33202.63	35680.02	41500.11	78757.58	46581.15	53228.10	67303.87	58629.27
Total GJ/Yr	249758.55	242535.57	255933.05	264392.73	286046.18	283911.13	289025.30	263837.18	273904.50	275509.57
Cost of HFO/Year	467626013.40	448312765.44	461174776.86	478695691.49	511834935.54	429386376.7	507435605.65	440804821.18	432415112.52	45393046 0.15
Cost of AF/Year	16380214.50	17627052.59	20652034.07	22192973.88	25813066.88	48987215.16	28973477.00	33107875.71	41863009.99	36467408.47
Total of energy Cost /year	484006227.90	465939818.03	486826810.93	500888665.37	537648002.42	478373591.94	536409082.65	473912696.90	474278122.51	49039786 8.62
Cost savings (kes) / year	174912773.92	192979183.79	172092190.89	158030336.45	121270999.40	180545409.88	122509919.17	185006304.92	184640879.31	16852113 3.20
Cost savings (euro) / year	1469855.24	1621673.81	1446152.86	1327986.02	1019084.03	1517188.32	1029495.12	1554674.83	1551604.03	1416143.98

3.3. Projected and Actual Total Energy Cost per Year

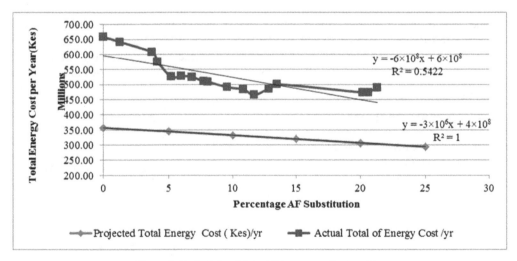

Figure 3. *Projected and Actual Total Energy Cost per Year.*

From the projected substitution scenario in table 1 and figure 3 the total energy cost was decreasing with an increase in AF substitution. This is because AF costs are lower than HFO and therefore energy mix cost cheaper than when only HFO is used. The relationship of total energy cost against percentage substitution is linear given by:

$$y = -3 \times 10^6 x + 4 \times 10^8 \qquad (1)$$

Where:
y = total cost of energy/year in Kenya shillings
x = percentage of AF substitution.
The above equation can be rewritten as:

$$\text{Total energy cost/yr} = -3 \times 10^6 \%AF + 4 \times 10^8 \qquad (2)$$

The degree of correlation of the total energy cost of energy and percentage AF substitution indicated by R^2 was 1 because this was an ideal scenario giving a perfect relation. On the other hand of actual substitution scenario the total energy cost per year was also decreasing with an increase in percentage AF substitution as shown in table 2 and figure 3. This was because the energy mix used was cheaper as opposed to using only HFO for drying. There was a fairly strong correlation of the total energy cost of energy and percentage AF substitution indicated by R^2 of 0.5422. The curve was also not smooth because of technical errors during the operation of the dryer. The equation of the trend line of the total cost of energy against percentage AF substitution was linear given by:

$$y = -6 \times 10^8 x + 6 \times 10^8 \qquad (3)$$

This implied that:

$$\text{Total energy cost/yr} = -6 \times 10^8 \% AF + 6 \times 10^8 \qquad (4)$$

From the experimental results the actual total energy costs were higher than the projected total energy cost.

3.4. Projected and Actual HFO Cost per Year

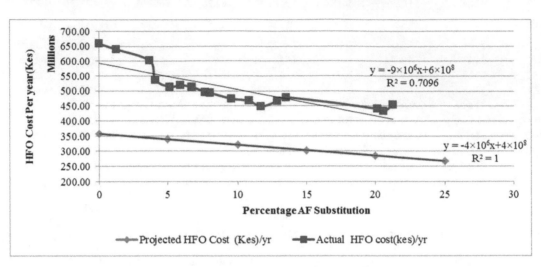

Figure 4. Projected and Actual HFO Cost per Year.

A comparison of projected HFO cost per year was done against actual HFO cost per year. From table 1 and figure 4 the cost of HFO was decreasing with an increase in percentage AF substitution in both projected and actual substitution scenarios. The relationship for the projected substitution scenario was expressed as:

$$y = -4 \times 10^6 x + 4 \times 10^8 \qquad (5)$$

Equation 5 can be rewritten as:

$$\text{HFO cost/yr} = -4 \times 10^6 \% \, AF + 4 \times 10^8 \qquad (6)$$

The correlation coefficient of $R^2 = 1$ because this situation was a perfect scenario. The cost of HFO cost was decreasing because the cost of the energy mix was lower than the cost of using HFO only in the dryer.

From table 2 and figure 4, the actual cost of HFO was decreasing with an increase in percentage substitution. There was a fairly strong linear corellation between the cost of HFO and percentage AF substitution with R^2 of 0.7096. The curve of cost of HFO per year against percentage substitution was however not smooth because the scenario was real and therefore affected by the operating conditions. The equation for the trendline of the relationship between actual cost of HFO and percentage substitution was given by:

$$y = -9 \times 10^6 x + 6 \times 10^8 \qquad (7)$$

Equation 7 was rewritten as;

$$\text{HFO cost/yr} = -9 \times 10^6 \% AF + 6 \times 10^8 \qquad (8)$$

3.5. Projected and Actual AF Cost

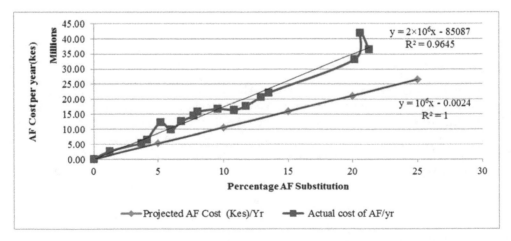

Figure 5. Projected and Actual AF Cost per Year.

For the projected substitution scenario from table 1 and figure 5, the cost of AF was increasing with an increase in percentage AF substitution. The relationship was expressed as:

$$y = 1 \times 10^6 x - 0.0024. \qquad (9)$$

This would further be expressed as:

$$\text{AF cost/yr} = 1 \times 10^6 \%\text{AF} - 0.0024. \qquad (10)$$

The coefficient of correlation R^2 was 1 because the scenario was ideal. The slope graph was increasing because more AF fuel was used as the percentage AF substitution increased.

From Table 2 and figure 5, the actual cost of AF per year was increasing with an increase in percentage AF substitution. This is because more AF was used with increasing percentage substitution. From figure 4.5, there was

a strong linear correlation between actual cost of AF and percentage AF substitution with $R^2 = 0.9645$. However the curve was not smooth because substitution was real and therefore affected by the operating conditions of the system. The relationship of the actual substitution was expressed by a linear trend line of:

$$y = 2 \times 10^6 x - 85087 \qquad (11)$$

This equation can further be expressed as:

$$\text{AF cost/yr} = 2 \times 10^6 \%\text{AF} - 85087 \qquad (12)$$

3.6. Projected and Actual Savings per Year

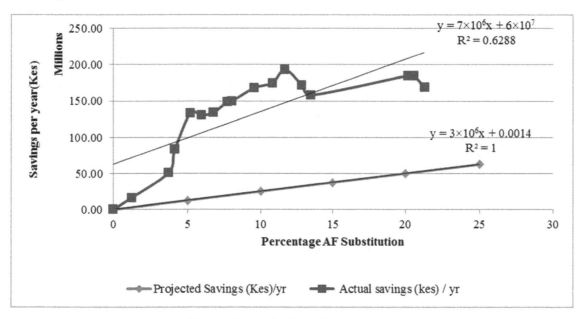

Figure 6. Projected and Actual Savings per Year.

In the projected substitution from table 1 and figure 6, the percentage savings were increasing with an increase in percentage AF substitution. There was a linear relationship between percentage savings and percentage AF substitution expressed as:

$$y = 3 \times 10^6 x + 0.0014 \qquad (13)$$

This implied that:

$$\text{Savings} = 3 \times 10^6 \%\text{AF} + 0.0014 \qquad (14)$$

The coefficient of correlation of $R^2 = 1$ because the scenario was ideal. More savings were made with an increase in percentage AF substitution because the energy mix was cheaper than using HFO only for drying. For the actual substitution scenario from table 2 and figure 6 the costs savings per year increased with an increase in percentage substitution. This was because of the lower cost of the energy mix from HFO and AF. There was also fairly strong correlation of percentage savings and percentage AF substitution with $R^2 = 0.6288$. The curve was not smooth because the substitution experiment was a trial and we experienced technical problems such as clogging of hopper

with rice husks. There was a linear trade line relationship between the percentage savings and percentage AF substitution given by:

$$y = 7 \times 10^6 x + 6 \times 10^7 \qquad (15)$$

This implied that:

$$\text{Savings} = 7 \times 10^6 \%\text{ AF} + 6 \times 10^7 \qquad (16)$$

3.7. Economic Analysis

3.7.1. Cost Benefit Analysis

The cost of installing the pilot project was as indicated in table 3.

Table 3. Installation Cost Breakdown.

Cost Breakdown	Amount(KES)
Steel structures material cost	372,000.00
Mechanical/Electrical installation	469,918.00
Materials cost(blower/electrical motor/rotary feeder/electrical cables/ panels & automation)	2,469,200.00
Trials(Labour & rice husks)	494,970.00
TOTAL	3,806,088.00

Table 4. *Fuel Handling Cost.*

Rice husks	Monthly tonnage	Cost per ton (KES)	Transport cost/t to collection center,(KES)	Bagging, Handling cost/t, (KES)	Total cost per ton,(KES)
	200	6000	800	1500	8300
Total fuel handling cost					1660000

The maintenance cost was assumed to be at 5% in the first and second year, doubling in the third year and three times in the fourth and fifth year of the initial maintenance cost. The discounting rate was at 10%.

Table 5. *Cost Benefit Analysis.*

COSTS / YEAR	1	2	3	4	5
Installation costs	3,806,088.00	0.00	0.00	0.00	0.00
maintenance cost	190,304.00	190,304.00	380,608.00	570,912.00	570,912.00
fuel handling cost	1,660,000.00	1,660,000.00	1,660,000.00	1,660,000.00	1,660,000.00
Total cost per year	5,656,392.00	1,850,304.00	2,040,608.00	2,230,912.00	2,230,912.00
Benefits					
Fuel cost Reduction	0.00	25,122,151.63	25,122,151.63	25,122,151.63	25,122,151.63
Net Cash flow	-5,656,392.00	23,271,847.63	23,081,543.63	22,891,239.63	22,891,239.63
Discount rate	10%				
Discount factors	1.00	0.91	0.83	0.75	0.68
Discounted cash flows					
Total cost per year	5,656,392.00	1,683,776.64	1,693,704.64	1,673,184.00	1,517,020.16
Benefits per year	0.00	22,861,157.98	20,851,385.85	18,841,613.72	17,083,063.11
Net cash flow	-5,656,392.00	21,177,381.34	19,157,681.21	17,168,429.72	15,566,042.95
Cumulative	-5,656,392.00	15,520,989.34	34,678,670.56	51,847,100.28	67,413,143.23
NPV	KES 67,409,040.84				
IRR	4.10				

This analysis was done to come up with the total costs incurred in the projects and the benefits to be gained from the implementation of the project to establish if the substitution was worthwhile. The analysis was done at a 10% alternative fuel substitution. Net present value and internal rate of return were calculated in order to take into account the time value of money. This was done using the excel program. NPV is normally calculated as:

$$NPV = I_1 + \frac{I_2}{1+r} + \frac{I_3}{(1+r)^2} + \cdots + \frac{I_n}{(1+r)^n} \quad (17)$$

Where I's= cash flow for each year
The subscript = year number
r = the discount rate.
The internal rate of return is the interest rate that makes the Net Present Value zero.

$$0 = P_0 + \frac{P_1}{1 + IRR} + \frac{P_2}{(1 + IRR)^2}$$
$$+P_3/(1 + IRR)^3 + \ldots +P_n/(1 + IRR)^n \quad (18)$$

Where;
$P_0, P_1, P_2, P_3 \ldots P_n$ is the cash flows in periods 1, 2, 3. . . n,

respectively; and IRR is the project's internal rate of return.
But from the excel function NPV was calculated as;

$$NPV = NPV(rate, value1, value2, \ldots) \quad (19)$$

And

$IRR = IRR$ (Net cash flow at year 1: Net cash flow at year 5, 0.1) (20)

The cash flows were discounted at 10 percent in order to cater for the risks associated with the project. From the analysis a positive net present value of 67,409,040.84 was realised which was an indicator that the substitution was worthwhile. IRR was calculated to be 4.10 %. This was the discount rate often that made the net present value of all cash flows from the substitution project equal to zero. The internal rate of return was a rate quantity which was an indicator of the efficiency, quality and yield of an investment.

3.7.2. Effect of Substitution
Projected substitution data was used to establish the effect of substitution. A graph of total energy cost and cost of using HFO only were plotted against % AF substitution to establish the effect of substitution.

Table 6. Substitution Effect.

DESCRIPTION	0%	5%	10%	15%	20%	25%	30%	35%	40%	45%	50%
HFO Cost (Kes)	357491491.33	339616916.77	321742342.20	303867767.63	285993193.07	268118618.50	250244043.93	232369469.37	214494894.80	196620320.23	178745745.67
AF Cost (Kes)	0.00	5313498.75	10626997.50	15940496.25	21253995.01	26567493.76	31880992.51	37194491.26	42507990.01	47821488.76	53134987.51
Total Cost (Kes)	357491491.33	344930415.52	332369339.70	319808263.89	307247188.07	294686112.26	282125036.44	269563960.63	257002884.81	244441809.00	231880733.18
Savings (Kes)	0.00	12561075.82	25122151.63	37683227.45	50244303.26	62805379.08	75366454.89	87927530.71	100488606.52	113049682.34	125610758.15

DESCRIPTION	55%	60%	65%	70%	75%	80%	85%	90%	95%	100%
HFO Cost (Kes)	160871171.10	142996596.53	125122021.97	107247447.40	89372872.83	71498298.27	53623723.70	35749149.17	17874574.57	0.00
AF Cost (Kes)	58448486.26	63761985.02	69075483.77	74388982.52	79702481.27	85015980.02	90329478.77	95642977.52	100956476.27	106269975.03
Total Cost (Kes)	219319657.36	206758581.55	194197505.73	181636429.92	169075354.10	156514278.29	143953202.47	131392126.66	118831050.84	106269975.03
Savings (Kes)	138171834	150732910	163293986	175855061	188416137	200977213	213538289	226099365	238660440	251221516

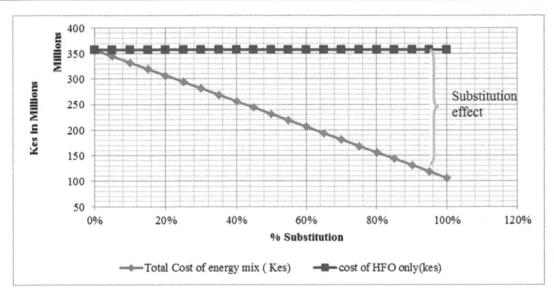

Figure 7. Substitution Effect.

The substitution effect measures how much higher price encourages consumers to use other goods, assuming the same level of income. Table 6 Figure 7 show a gradual cost drop of the energy used to dry pozzolana from 357,491,491.33 Kenya shillings with increasing percentage AF substituted to 106,269,975.03 Kenya shillings when HFO is completely substituted by AF. This effect is caused by the relatively high cost of HFO that induces the use of more of a relatively lower priced energy, that is, AF and less on high priced HFO. This is due the rise the cost of fossil fuels. This is a positive scenario in economics, but the degree of substitution can only be justified by the availability of AF to completely substitute HFO and the efficiency of the dryer to run on AF alone. This is an area for further research to determine the efficiency of the dryer in relation to the percentage substitution with HFO.

3.7.3. Operational Expenditure Analysis

Both simple payback period and return on investment were carried out to determine the viability of the investment. The analysis was carried out using the pilot substitution scenario with annual savings at 9.55% AF fuel substitution.

Table 7. Operational Expenditure.

Pilot substitution project	
Capital invested	
Installation costs	3,806,088.00
fuel handling cost	1,660,000.00
Total cost per year	5,466,088.00
Annual savings at 10 % = 168000000 (From table 2)	

$$\text{Simple payback period} = \frac{\text{capital invested}}{\text{annual savings}}$$

$$= \frac{5466088}{168000000} = 0.0325 \text{ years} = 0.39 \text{ months} = 12 \text{ days}$$

$$ROI = \frac{\text{Gain from investment} - \text{cost of investment}}{\text{cost of investment}}$$

$$\frac{168000000-5466088}{5466088} = 29.72\%$$

From the operational expenditure analysis simple payback period was 12 days and return on investment was 29.72%. The short payback period and high return on investment indicate that this project is of high yielding benefit to the investor. From the four capital budgeting techniques i.e. NPV, IRR, Simple payback period and ROI the investment was worthwhile to undertake.

4. Conclusion

From the findings reported in this study regarding the substitution of HFO with biomass in a pozzolana dryer, it can be concluded that Substitution led to a reduction of the cost of energy used and therefore savings increased with the increase of percentage substitution. Secondly, using the four techniques of capital budgeting, i.e. NPV, IRR, Simple payback period and ROI the investment was worthwhile to undertake. Researchers need to investigate further and determine the efficiency of the dryer in relation to the percentage substitution with HFO to determine the maximum efficiency. Future research can expand on substitution in relation on capital and labour employed and establish the percent savings per unit of cement produced.

Nomenclature

€	Euro
AF	Alternative Fuel
CO	Carbon Monoxide
CO_2	Carbon Dioxide
GHG	Green House Gas
GJ	Giga Joule
HFO	Heavy Fuel Oil
HGG	Hot Gas Generator
kWH	Kilowatt Hour
MJ	Mega Joule
NGP	Nairobi Grinding Plant
NO_x	Nitrogen Oxides

References

[1] Alsop, P. (2001). *Cement plant operations handbook for dry process plants*. 3rd Edition, Trade ship Publications Ltd, Portsmouth, United Kingdom.

[2] Bamburi cement (2012), *Annual Report and Financial Statements,* Bamburi Cement Corporate Office Nairobi.

[3] Boden T.A, Marland.G, and Andres R.J. (2010), *Global, Regional, and National Fossil-Fuel CO_2 Emissions,* Carbon Dioxide Information Analysis Centre, Oak Ridge National Laboratory, U.S. Department of Energy, Oak Ridge, Tenn., U.S.A. doi 10.3334/CDIAC/00001_V2010.

[4] Hendriks (1998), *Reduction of Greenhouse Gases from the Cement Industry.* Conference Proceedings, Switzerland.

[5] http://www.ucsusa.org/clean_energy/our-energy-choices/coal-and-other-fossil-fuels/the-hidden-cost-of-fossil.html (accessed 21.09.13)

[6] Kurchania A.K. Rathore N.S., Panwar N.L (2006), *Renewable Energy Theory & Practice,* Himanshu Publications.

[7] MIT Research Profile Letter, (2011), *Clinker Grinding at Breaking Point,* Concrete Sustainability Hub, 2011.

[8] NGP (2013), Bamburi Annual Report and Financial Statements.

[9] NGP (20112), Bamburi Annual Report and Financial Statements.

[10] United Nations Development Programme, (2009).*Biomass Energy for Cement Production: Opportunities in Ethiopia.*

6

Design and implementation of operation and maintenance of information systems for Yalong River Hydropower Development Company

Hualong Cai[1,2], Yuanshou Liu[2], Yunfeng Liu[2]

[1]State Key Laboratory of Water Resource and Hydropower Engineering Science, Wuhan University, Wuhan, China
[2]Yalong River Hydropower Development Co., Ltd., Chengdu, China

Email address:

whucad@whu.edu.cn (Hualong Cai), Liuyuanshou@ylhdc.com.cn (Yuanshou Liu), Liuyunfeng@ylhdc.com.cn (Yunfeng Liu)

Abstract: Based on the information system of Yalong Company, some analysis are carried out that operation and maintenance of information systems are described covering hardware systems, software systems and the content of individual client behavior management, and analyzes the business logic structure and software architecture system of operation and maintenance, and achieve performance monitoring, security monitoring, supervision and operation and services Desktop system as a whole functions for the overall operation and maintenance information system.

Keywords: Yalong River Hydropower Development Co., Ltd., Operation and Maintenance Information System, Design and Implementation

1. Introduction

With the putting into production of Ertan, Guandi and the first-second cascade Jinping Hydropower Station and with the construction of Tongzilin, Lianghekou and other power plant, Yalong River Hydropower Development Co., Ltd. (hereinafter referred to as: Yalong Company) gradually comes into a period of rapid development. The information systems of engineering construction, electricity production, operating management as well as other types of business is in use gradually with depth integration of information technology, then the company's information technology operation system has become a dynamic, open, continuous developing and evolving complex system. The operation and maintenance of information technology has very distinctive features with a wide geographic distribution, real-time requirements, diverse content of operation and maintenance. Therefore, the urgent need for information operation and maintenance systems provides an efficient, standardized means of support, and meets some new demands for the mission of systems.

2. Management Category of Operation and Maintenance Information Systems

Information system of Yalongcompany includes: financial management systems, human resources management system, OA system, file systems, electrical production management system deployed in a centralized manner; project management system deployed in each construction administration deployed in a distributed manner. Centralized deployment system is released in Chengdu headquarters room, while distributed systems are deployed in each room of construction administration camp. The information operation and maintenance systems covers condition monitoring of hardware equipment such as servers, network, etc.; application monitoring of system software running, monitoring and management on the PC client and the user's personal behavior that is divided into two layers with the centralized deployment and distribution deployment management.

2.1. Operation and Maintenance Management of Hardware Monitoring

Operation and maintenance management of hardware monitoring involves: running status monitoring of services, status and performance monitoring of network devices and condition monitoring of data flow as shown in Fig.1.

The CPU temperature and usage, memory usage, network cards occupancy rate and other technology parameters are obtained through the program installed on the server, and passed to the background operation and maintenance system that monitors real-timely the value of the switch backplane bandwidth, switching capacity, packet forwarding rate and other performance indicators ; monitors such real-time performance metrics as storage speed storage capacity, CL, SPD chip, parity, memory bandwidth. Set the threshold by the background operation and maintenance management software to provide an early warning mechanism.

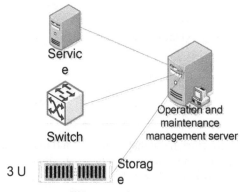

Fig. 1. Real-time hardware monitoring.

2.2. Operation and Maintenance Management of software Monitoring

Operation and maintenance management of software monitoring involves: running status monitoring of database, function monitoring of implication layer and condition monitoring of email services, middleware condition monitoring, web service status monitoring, etc. as shown in Fig.2.

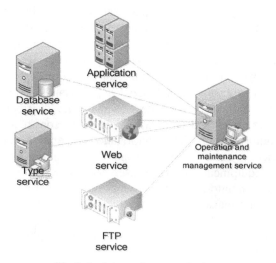

Fig. 2. Real-time software monitoring.

The index such as database response time, throughput performance, Oracle EBS, Maximo, Lotus Domino can be monitored real-timely through software and transmitted to the

operation and maintenance systems. By setting the threshold, operation and maintenance systems report to operation and maintenance personnel according to an early warning mechanism.

2.3. Operation and Maintenance Management of Monitoring User'S Behavior

Operation and maintenance management of monitoring user's behavior involves: software installation and version control monitoring of PC client, individual user behavior monitoring system and security monitoring of PC client.

For personal computers, laptops, tablet PCs, they are installed with the security software company purchasing, by which some information can be collected example for the software version of terminal device installed, the behavior of individual users of business systems and safety information. The operation and maintenance of information systems analyze and access the indicators of health degree of individual terminals by information collected and send unhealthy terminal information to the mailboxes of the employee using the device and alert maintenance of equipment terminals.

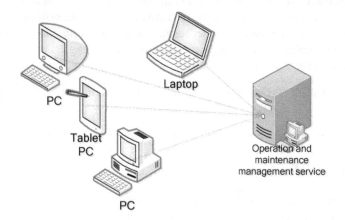

Fig. 3. User's behavior monitoring.

3. Logical Structure of the System

The customer service of operation and maintenance information system can be divided into the customer services of ware maintenance, the customer service of equipment maintenance and the desktop customer service. The software customer service is responsible for the operation and maintenance of software applications of trouble ticket resolution; the equipment operation and maintenance customer is responsible for resolving trouble tickets of servers, storage, network and other hardware devices; the desktop customer is responsible for such trouble tickets as system operation and maintenance of the PC client and software configurations. In the presentation layer, in accordance with the contents of the different sub-portals, the customer service can be divided into the operation and maintenance sub-portal, equipment sub-portal, security sub-portal and desktop sub-portal to provide information services for different operation and maintenance teams. In the business layer, the configuration, re-optimized work flow can be come true; real-time monitoring of application software, hardware, data acquisition equipment can be applied, desktop information gathering and the company's overall security information gathering can be also achieved along with external systems and data exchange. Business layer achieves operation of the database through the data access layer with storage / operation and maintenance of information, and analyze the results to put them into the database again.

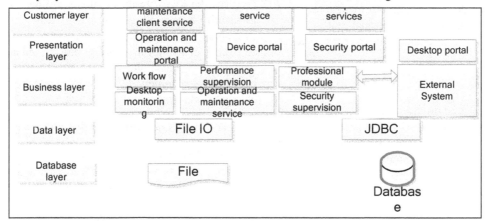

Fig. 4. Logical framework of system.

4. Software Structure of the System

Operation and maintenance of information systems is divided into the presentation layer, business layer, database access layer and the database layer. The presentation layer shows the operation and maintenance information through the static and dynamic pages; users interact with the operation and maintenance system via the presentation layer. Business layer achieves communication with presentation layer by Struts MVC Action; reconstruction and implement of work flow comes true through Workflow, business functions can achieve the realization by the Spring framework. Business layer carries out operation of the database by FileIO, JDBC, IBATIS.

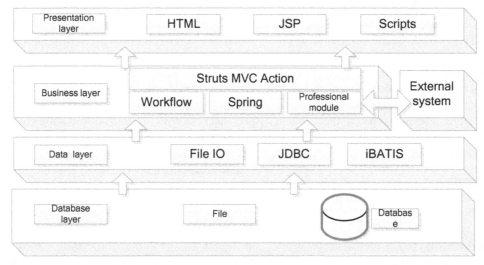

Fig. 5. Software framework of system.

5. Realization of System Functions

The functions of information operation and maintenance system of Watershed Company includes performance monitoring, security supervision, Desktop supervision, operation and maintenance supervision,etc.

5.1. Performance Monitoring

Performance monitoring refers to the real-time monitoring and management for business systems, middleware, database and network, server rooms and other hardware equipment, facilities. Business regulation includes financial management systems based on Oracle EBS, human resources management system, Maximo-based electricity production management system, file management system based on Unis, Oracle Portal-based portal management system, Lotus Domino based OA system, in which the system real-timely monitors the operation indicators of applications.

Network monitoring can carry out supervision on the network topology, running condition of network switches and other equipment, the port operation of the regulatory company-wide link. The running status of device can be monitored and obtained with cross-device type and device manufacturers and the threshold for the health degree of is set as an early warning. Room supervision can get real-time such key indicators as room temperature, humidity, power supply and regulate the host running parameters of the engine room, storage and network devices. Three-dimensional and multi-media technology shows the operational status of the engine room and the ability to achieve an interactive room management.

5.2. Security Supervision

Security supervision system provide a security information and status of the entire company network, server, PC client for the operation and maintenance personnel monitor , including the risk warning, failure analysis, strategic release, alarm events, reports, data analysis and reporting capabilities show. By the security software presetting in the server and the client as well as the hardware modules of network equipment, gathering security state information on a company-wide devices, can provide real-time security data and generate early warning reports, so that operation and maintenance personnel can take measures to prevent and handle security incidents.

5.3. Desktop Supervision

Desktop supervision includes desktop security management, desktop patch management, software management and desktop assets management. Desktop security management involves the lifecycle terminal-supporting management such as security access, illegal outreach, virus protection, security assessments, user behavior. Software and patch refers to software distribution, configuration management, patch management, alarm and updates. Desktop asset management involves management of software and hardware assets and changes in the status quo.

5.4. Services of Operation and Maintenance

5.4.1. Work Order Management

Work order is the carrier of operation and maintenance services for the event handling, can be divided into four categories: be accepting work orders, work order processing, archiving pending work orders, work orders . Work order process is as follows:

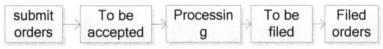

Fig. 6. *Process flow of work order.*

If users identify problems during running of the system, fill out and submit the issue work orders. The problems in work orders are listed into a single work queue according to the classification, waiting for the operation and maintenance personnel accepted who handle issues raised in each work order. After processing, work orders are waiting for archiving. Confirmed by the records management staff, the work order is for archiving and for future reference. When they find a similar problem, operation and maintenance personnel can call the archived work orders and processing programs reuse.

5.4.2. Information Distribution

Operation and maintenance personnel choose the range (the whole company, the Authority, power plants, departments) to publish important notes, spirit of the meeting, personnel and other information of the Authority, power plants, departments in the form of static pages and accessories.

5.4.3. Operation and Maintenance Assessment

Classification for the type of work orders and making a statistical analysis for work order processing time, user evaluation and the user visits can make an assessment to operation and maintenance personnel. By the examination, their service quality has been greatly improved. Statistical analysis of data can find the weak link, improve and enhance the overall controlling efficiency of the operation and maintenance system.

5.4.4. Dissemination of Maintenance Information

The statistical analysis for operation and maintenance information, the report of operation and maintenance of the quality of service submitted monthly, quarterly, semi-annual, annual are published in the prescribed information bar so that all operation and maintenance personnel can clearly get dynamic operation and maintenance management. Colleagues can also publish information bar operation and maintenance related procedures, norms, easy operation and maintenance personnel can easily find the information you need.Meanwhile, the bar can also publish information related to the operation and maintenance procedures and norms,so operation and maintenance personnel can easily find the information needed.

5.4.5. Duty Management

Duty management includes: management on duty arrangements, log management on duty, duty patrol management, room inspection management, shift management, access management, content management, inspection duty. Duty management is able to make a regular basis, qualitative and quantitative operation and maintenance management for the centralized deployment of software and hardware facilities and services of headquarters rooms, data centers, business systems. It is also able to carry out an efficient supervision for related facilities and services of distributed project management system.

5.4.6. Scheduling Management

Scheduling management can be divided into the plan preparation, plan generation, plan distributions, implementation of the plan, implementation assessment of the plan. By the plan management module of operation and maintenance system, operation and maintenance resources can be configured in the overall company , and gradually formed applications, room management, network management, staffing and other professional jobs plan outline through the accumulation of time, providing a reference for the management of daily operation programming.

5.4.7. System Management

The administrators of operation and maintenance system carry out information management of operation and maintenance personnel through the backstage management system, depending on the professional work of each person with different tasks, define different roles within the system, given the appropriate administrative privileges, so that each operation and maintenance resources are able to maximize resource efficiency through operation and maintenance information system.

6. Conclusion

Based on running status information systems of the Yalong Company, combined with ITIL framework system, operation and maintenance information systems can achieve business management, service management, infrastructure management, service planning and implementation, application management and security management and so on. The implementation of the operation and maintenance system improves the quality of information systems services to provide an efficient platform support for business entities.

References

[1]   Chen Yuhui. Create a distinctive grid integrated management information system of operation and maintenance [J], power information,2011(2):165-169

[2]   You Zhenwei. Design and implementation of operation and maintenance of electric information systems [J],Jiangxi Communication technology,2011(9): 5-10

[3]   Zhao Zhendong. Design and implementation of operation and maintenance of electric information systems [D],Beijing: Beijing University ofposts and telecommunications,2012

[4]   Wei Ming. Scheme of ITIL-based information management system of operation and maintenance [J], power information, 2011(4):83-86.

[5]   Li Zhen. Modeling of operation and maintenance of national indicatorsbased on high-dimensional multi-objective optimization of network [J], Computer and Modernization, 2013(6):23-26.

[6]   Zhu Ruoming. Design and implementation of monitoring system for internet behavior [D]. Changsha: Central South University,2011

7

Optimization of Multiple Effect Evaporators Designed for Fruit Juice Concentrate

Manal A. Sorour

Food Eng. and Packaging Dept., Food Tech. Res. Institute, Agric. Res. Center, Giza, Egypt

Email address:

manal.sorour@yahoo.com

Abstract: The most important running cost item in evaporation is energy. However, energy consumption per unit production can be reduced considerably through the use of multi effect evaporation. A multiple effect evaporation scheme was studied and economically analyzed. Different economic items, including cost of steam and cost of evaporator in addition to annual total cost for single, double, three, four, five and six effects of evaporators were estimated and analyzed. The results indicated that minimum annual total cost is obtained by using a double effect evaporator.

Keywords: Multiple Effect Evaporators, Optimum Number of Effects, Annual Cost, Evaporator Cost, Steam Cost

1. Introduction

Multiple effect evaporator control is a problem that has been widely reported in the pulp and sugar industries. Therefore evaporation is a very important unit operation and must be controlled smoothly, [1].

Process optimization has always been a noble objective of engineers entrusted with the responsibility for process development and improvement throughout the food industry, [2]. On the other hand, the chemical industry has used cost analysis in several cases in relation to design and process optimization. A classic example in the chemical industry is the determination of the optimal number of effects in evaporation system, were the optimum is found when there is an economic balance between energy saving and added investment, this is, a minimization of the total cost, [3].

Evaporation is the removal of solvent as vapor from a solution. It is the operation which is used for concentration of solution. There could be single effect evaporator or multiple effect evaporators. With addition of each effect steam economy of the system also increases. Evaporators are integral part of a number of process industries like Pulp and Paper, Sugar, Caustic Soda, Pharmaceuticals, Desalination, Dairy and Food Processing etc., [4] described a steady state model of multiple effect evaporators for simulation purpose. The model includes overall as well as component mass balance equations, energy balance equations and heat transfer rate equations for area calculation for all the effects. Each effect in the process is represented by a number of variables which are related by the energy and material balance equations for the feed, product and vapor flow for forward feed. The code has been developed using SCILAB. Results of the present approach are validated with industrial data.

Evaporators are kind of heat transfer equipment where the transfer mechanism is controlled by natural convection or forced convection. A solution containing a desired product is fed into the evaporator and it is heated by a heat source like steam. Because of the applied heat, the water in the solution is converted into vapor and is condensed while the concentrated solution is either removed or fed into a second evaporator for further concentration, [5].

The increase in awareness to cost curtailment of energy consumption coupled with the increase in the consumption of the concentrated and dehydrated food products, energy management has become indispensable in food industries. One such process of obtaining these food products is by employing evaporation in food industries. The process of evaporation is employed in the food industry primarily as a means of bulk and weight reduction for fluids. Therefore, evaporation is one of the most large scale operations in food industry. Energy is the most important running cost item in evaporation. The operating cost of an evaporator system is mostly determined by the energy that is required to achieve certain evaporation rate. Under steady state condition there must be a balance between the energy entering and leaving the system. Energy can be saved by reusing vapor formed from the boiling product. One of the few ways of accomplishing such is by multiple-effect evaporation, [6].

Boiling point elevation at a certain external pressure can be determined from a thermodynamic equation using the latent heat of vaporization and molar fraction of the food. However, the use of these equations requires knowledge of the proportions of specific components of the foods that cause changes in the boiling points, [7].

The aim of the work is to explain how total cost of evaporation can be reduced by using multi-effect evaporator and achieve an economically optimal number of effects. This can be proved through application of mass and heat-balance analysis for tomato juice concentration from 5% to tomato paste 36%.

2. Experimental Set Up

2.1. Determination of Boiling Point Rise

An experimental procedure was conducted to determine the BPR (boiling point rise). A tomato juice solution of 5% (wt%) concentration was boiled. The concentration was measured at equal intervals of time (15 minutes). The measured concentration was plotted versus temperature difference between boiling point of solution and boiling point of water.

2.2. Single Effect Evaporator

Single effect evaporator use more than 1kg of steam to evaporate 1kg of water. The general configuration for evaporation in this work is schematically represented in (Fig. 1). According to this scheme, the vapor leaving the evaporator is waste, so a single effect evaporator may be considered wasteful of energy, since the latent heat of the leaving vapor is not used but discarded.

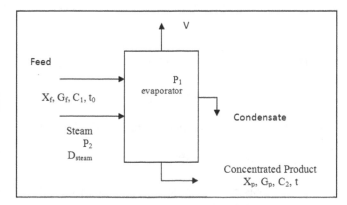

Fig. 1. *Simplified Block Flow Diagram for Single Effect Evaporator*

2.3. Simulation Model

The mathematical model representing the process may be formulated as follows, [8]:-

The amount of concentrated product and the amount of the vapor leaving the evaporator could be calculated by using an overall material balance as follows:

$$G_f = G_p + W \qquad (1)$$

Where, G_f is the flow rate of the feed solution, kg/hr, G_p is the flow rate of the concentrate product, kg/hr, W is flow rate of the vapor leaving the evaporator, kg/h

To determine G_p a component mass balance on solids is needed:-

$$G_f X_f = G_p X_p \qquad (2)$$

Where, X_f is the concentration of tomato juice concentrate entering the evaporator (by weight), X_p is the concentration of tomato concentrate product stream (by weight).

The heat load of the evaporator (Q) may be represented by making heat balance on the evaporator:

$$Q + G_f C_1 t_o = W H_v + G_p C_2 t + Q_{loss} \qquad (3)$$

Where C_1, C_2 is the specific heat of feed and product solution, kJ/kg deg .

Q is the heat load of the evaporator, kJ/kg, H_v is the enthalpy of secondary vapor, kJ/kg, t is the boiling point of the product, °C, and t_0 is the temperature of the feed solution, °C

Considering the feed solution as a mixture of concentrated solution (product) and evaporated water, we can write:

$$G_f C_1 t = G_p C_2 t + W C_{H2O} t \qquad (4)$$

$$G_p C_2 = G_f C_1 - W C_{H2O} \qquad (5)$$

Where, C_{H2O} is the specific heat of the liquid water, kJ/kg deg. substituting the value of $G_p C_2$ in equation (3) we obtain:

$$Q = G_f C_1 (t-to) + W (H_v - C_{H2O} t) + Q_{loss} \qquad (6)$$

If the heat losses are neglected ($Q_{loss} = 0$), we can write:

$$Q = G_f C_1 (t - to) + W (H_v - C_{H2O}t) \qquad (7)$$

2.4. Calculation Methods for Multiple-Effect Evaporators

Multiple-effect evaporation is evaporation in multiple stages, whereby the vapors generated in one stage serve as heating ' steam ' to the next stage. Thus, the first stage acts as a 'steam generator' for the second which acts as a condenser to the first and so on. The number of 'effects' is the number of stages thus arranged. The first effect is heated with boiler steam. The vapors from the last effect are sent to the condenser Fig. 2.

The economic and environmental consequences of this result are obvious. In a multi effect evaporator for each kilogram of water evaporated:

1. The quantity of steam consumed is inversely proportional to the number of effects.

2. The quantity of cooling water utilized in the condenser is inversely proportional to the number of effects.

For heat transfer to occur, a temperature drop must exist at each effect. In other words, the temperature of the vapors generated in a given effect must be superior to the boiling temperature in the following effect:

$$T_1 > T_2 > \ldots \ldots > T_n, \text{ hence: } P_1 > P_2 > \ldots \ldots \gg\gg\gg P_n$$

In doing calculations for a multiple-effect evaporator system, the values to be obtained are usually the area of the heating surface in each effect, the mass (kg) of steam per hour to be supplied and the amount of vapor leaving each effect. The given or known values are usually as follows:

(1) Steam pressure into first effect,

(2) Final pressure in vapor space of the last effect,

(3) Feed conditions and flow to first effect, ... (n)

(4) the final concentration in the liquid leaving the last effect,

(5) Physical properties such as enthalpies and/or heat capacities of the liquid and vapors, and

(6) the overall heat transfer coefficients in each effect. Usually the areas of each effect are assumed equal. The calculations are done using material balances, heat balances, and the capacity

$q = UA\Delta T$ for each effect. A convenient way to solve these equations is by trial and error.

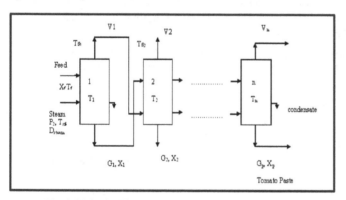

Fig. 2. *Multiple Effect Evaporation Equipment (Co-Current)*

The Equations representing the process, of multiple effect evaporators may be formulated as follows:

1 - From the known outlet concentration and pressure in the last effect, the boiling point in the last effect can be determined by overall mass balance and component mass balance we can calculate $L_1, L_2, \ldots \ldots L_n$, $V_1, V_2, \ldots \ldots \underline{V_n}$

Where, $L_1, L_2 \ldots \ldots L_n$ are the flow rate of solution produced from each effect, kg/h

$V_1, V_2, \ldots \ldots V_n$ are the amounts of vapor produced from each effect, kg/h

2- The amount of heat transferred per hour in the first effect of a multiple effect evaporator with forward feed in Fig. 2 will be:

$$q_1 = U_1 A_1 \Delta T_1 \qquad (8)$$

Where ΔT_1 is the difference between the condensing steam and the boiling point of the liquid, $(Ts - T_1)$. Assuming that the solutions have no boiling point rise and no heat of solution and neglecting the sensible heat necessary to heat the feed to the boiling point, approximately all the latent heat of the condensing steam appears as latent heat in the vapor. This vapor then condenses in the second effect, giving up approximately the same amount of heat.

$$q_2 = U_2 A_2 \Delta T_2 \qquad (9)$$

This same reasoning holds for q_n. Then since $q_1 = q_2 = \ldots \ldots = q_n$, then approximately,

$$U_1 A_1 \Delta T_1 = U_2 A_2 \Delta T_2 = \ldots \ldots = U_n A_n \Delta T_n \qquad (10)$$

Usually, in commercial practice the areas in all effects are equal and

$$q/A = U_1 \Delta T_1 = U_2 \Delta T_2 = \ldots \ldots = U_n \Delta T_n \qquad (11)$$

3- Hence, the temperature drops ΔT in a multiple effect evaporator are approximately inversely proportional to the values of U. Calling $\sum \Delta T$ as follows with the presence of boiling point rise,

$$\sum \Delta T = \Delta T_1 + \Delta T_2 + \ldots \ldots + \Delta T_n$$
$$= T_{sf} - T_5 - [(BPR)_1 + (BPR)_2 + \ldots(BPR)_n] \qquad (12)$$

Where, ΔT_1 = Available temperature difference for first effect $(T_{sf} - T_1)$.

ΔT_2 = Available temperature difference for second effect $(Ts_1 - T_2)$.

ΔT_5 = Available temperature difference for fifth effect $(Ts_{n-1} - T_n)$.

T_1 = Temperature of solution in the first evaporator, °C.

T_2 = Temperature of solution in the second evaporator, °C.

T_n = Temperature of solution in the (n) evaporator, °C.

T_{sf} = inlet temperature of the steam to the first effect, °C.

T_{s1} = inlet temperature of steam to the second effect, °C.

T_{sn} = inlet temperature to the (nth) effect, °C.

Note that,

$$\Delta T_1\ °C = \Delta T\ K, \Delta T_2\ °C = \Delta T_2\ K, \Delta T_3\ °C$$
$$= \Delta T_3\ K, \Delta T_n\ °C = \Delta T_n\ K \qquad (13)$$

Since ΔT_1 is proportional to $1/U_1$, then:

$$\Delta T_1 = \Sigma \Delta T \frac{1/U_1}{1/U_1 + 1/U_2 + \ldots + 1/U_n} \qquad (14)$$

Similar equations can be written for $\Delta T_2, \Delta T_3, \ldots \Delta T_n$

Where, U_1 is the heat transfer coefficient for the first effect, kW/m²K

U2 is the heat transfer coefficient for the second effect, kW/m2K

U3 is the heat transfer coefficient for the third effect, kW/m2K

Un is the heat transfer coefficient for the (n) effect, kW/m2K

Any effect that has an extra heating load, such as a cold feed, requires a proportionally larger ΔT.

4- Using heat and material balances in the first effect, the amount vaporized can be calculated consequently the amount of steam (D_{steam}) entering the first evaporator, the value of q transferred in each effect is estimated.

5- Using the rate equation $q = UA\Delta T$ for each effect, we can calculate the areas $A_1, A_2, ..., A_n$. The average value A_m:

$$A_m = \frac{A_1 + A_2 + \ldots\ldots + A_n}{n} \qquad (15)$$

6- If these areas are reasonably close to each other, the calculations are complete and a second trial is not needed. If these areas are not nearly equal, a second trial should be performed as mentioned above, then obtain new values of $\Delta T_1'$, $\Delta T2'$, ..., $\Delta T_n'$ from these equations:

$$\Delta T_1' = \frac{\Delta T_1 + A_1}{Am} \qquad (16)$$

$$\Delta T_2' = \frac{\Delta T_2 + A_2}{Am} \qquad (17)$$

$$\Delta T_n' = \frac{\Delta T_n + A_n}{Am} \qquad (18)$$

The Sum $\Delta T_1' + \Delta T_2' + \ldots\ldots + \Delta T_n'$ must equal the original $\sum \Delta T$. If not, proportionately readjust all $\Delta T'$ values then calculate the boiling point rise in each effect, and as shown before we can get q_1, q_2, ..., q_n and A_1, A_2,.....A_n using the new values $\Delta T_1' + \Delta T_2' + \ldots\ldots + \Delta T_n'$

3. Cost Estimation

The various cost items can be expressed in terms of the system-operating variables and parameters as follows, [9]:

3.1. Annual Cost of Steam

The steam consumption may be represented according to the following equations:
For single effect evaporator

$$D_{steam} = Q/\lambda \qquad (19)$$

For multiple effect evaporator

$$D_{steam}\lambda_{sf} + G_f C_p T_f = G_1 C_p T_1 + V_1 H_1 \qquad (20)$$

Where, D_{steam} is mass flow rate of steam consumption, kg/s, λ is latent heat of vaporization, kJ/kg, G_f is the flow rate of feed solution, kg/s, G_1 is the flow rate of first effect evaporator, kg/s, T_1 is temperature of solution in the first effect, °C, C_p is specific heat of the solution, kJ/kg.deg., V_1 is the amount of vapor produced from first effect, kg/h., H_1 is the enthalpy of the vapor leaving the evaporator, kJ/kg, Q is the heat load of the evaporator, kJ/s

3.2. Annual Cost of Steam

$$\text{Cost of steam} = D_{steam}\, C_{III} \qquad (21)$$

Where C_{III} is the cost per ton steam, L.E/ton

Annual cost of steam = cost of steam x Annual operating days x operating hours $\qquad (22)$

3.3. Annual Cost of Evaporator

Annual cost of evaporator may be calculated as follows, [9]:

$$\text{Purchased cost} = 75228.2\ (A)^{0.5053} \qquad (23)$$

Annual cost of evaporator
= Purchased cost $(1+0.6)\ 0.15 \qquad (24)$

Where, $(1+0.6)$ is the cost of piping and installation, and 0.15 is the depreciation cost. A is the area of heat transfer of the evaporator, calculated according to the equation:

$$Q = UA\, \Delta T \qquad (25)$$

Where,

ΔT is the driving force = $(T_{steam} - T_{boiling})$, °C $\qquad (26)$

U is the overall heat transfer coefficient, kW/m^2 K

3.4. Annual Total Cost

The annual total cost is calculated by applying the following equations:

Annual total cost = Annual cost of steam + Annual cost of evaporator, L.E/year $\qquad (27)$

4. Results and Discussion

The flow rate and concentration of the feed stream introduced to the evaporator used in this process are fixed for all the cases studied at the following values:
(1) Feed flow rate of Tomato juice 7498.8 kg/h;
(2) Concentration of the Tomato juice feed 5% (wt);
(3) Temperature of the feed solution 30°C;
(4) The product (Tomato paste) leaving at 36%

4.1. Determination of Boiling Point Rise

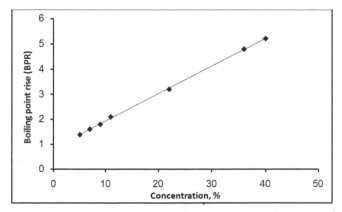

Fig. 3. Boiling Point Rise Of Tomato Juice Solutions

In the majority of cases in evaporation, the thermal properties of the solution being evaporated may differ considerably from those of water. The concentrations of the solutions are high enough so that the heat capacity and boiling point are quite different from that of water. Boiling point rise expresses the difference (usually in °F) between the boiling point of a constant composition solution and the boiling point of pure water at the same pressure, [10].

Table 1 shows the cost items of single and multiple effect evaporators. The results observed that annual cost of steam decrease with increasing number of evaporator effects, while annual cost of evaporators increase with increasing the number of evaporator effects. However the minimum total cost of evaporation will be at the economically optimal number of effects which is double effect evaporator as shown in Fig. 4

Table 1. *Cost Items of Single And Multiple Effect Evaporators*

No. of effects	Annual cost of steam	Annual cost of evaporator	Annual total cost
1	1382480	131651.8973	1514131.897
2	1091426	288676.4304	1380102.43
3	933120	526209.804	1459329.804
4	861280	767122.4444	1628402.444
5	766000	1044717.771	1810717.771
6	593000	1343161.957	1936161.957

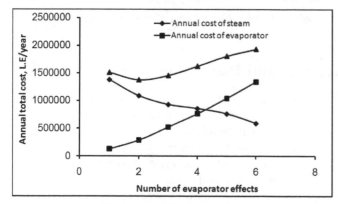

Fig. 4. *Optimal Number of Effects*

5. Sample Calculation

5.1. Double Effect Evaporator

As an example of calculation, a double effect evaporator is schematically represented in (Fig. 5).

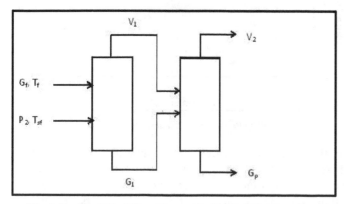

Fig. 5. *Simplified Block Flow Diagram for Double Effect Evaporator*

This sample simulates the double effect schemes of evaporator using the operating conditions (G_p, X_1, X_2, P_1, P_2) of the "Paste and Juice Company at Egypt" aiming at cost calculations with aim of item estimation as mentioned before.

- Material and component balance on first effect and second effect

$$G_1 = 0.4734 \text{ kg/s}$$

$$V_1 = 1.6096 \text{ kg/s}$$

$$V_2 = 0.1841 \text{ kg/s}$$

- $\sum \Delta T = Tsf - T_2 - (BPR_1 + BPR_2)$

From figure 3 BPR_1 and BPR_2 are 3.2, 4.8 respectively

$T_{sf} = 120°C$ at P2 (steam entering the first evaporator)=2 bar

$$T_2 = 50°C \text{ from Company}$$

$$\sum \Delta T = 62°C$$

- From equations (14): U_1 and U_2 are 2.1, 1.04 (respectively from company)

$$\sum \Delta T_1 = 28.5302°C$$

$$\sum \Delta T_2 = 33.4697°C$$

$$T_1 = T_{sf} - \Delta T_1 = 120 - 28.5302 = 91.4698°C$$

$$T_2 = (T_1 - BPR_1) - \Delta T_2 = 54.8°C$$

$$T_{s1} = T_1 - BPR_1 = 88.2698°C$$

- λ_{sf} at (120°C) =2202.3 kJ/kg (from steam table)

- λ_{s1} at (88.2696°C) =2656.82 kJ/kg

C_p for tomato juice = 4.17 kJ/kg.deg, H_1 at Ts_1 = 2656.82 kJ/kg

From equation (20): D_{steam} = 1.9047 kg/h

$$q_1 = D_{steam} \lambda_{sf} = 4194.721 \text{ kJ/s}$$

$$q_2 = V_1 \lambda_{s1} = 3683.52 \text{ kJ/s}$$

- From equation (8):

$$q_1 = U_1 A_1 \Delta T_1, \quad q_1 = U_2 A_2 \Delta T_2$$

$$A_1 = 120.5144 m^2 \text{ and } A_2 = 105.824 m^2$$

From equation (15): $A_m = \dfrac{A_1 + A_2}{2}$

From equation (16-18): $\Delta T_1' = 30.38°C$, $\Delta T_2' = 31.297°C$

- $T_1 = T_{sf} - \Delta T_1' = 120 - 30.38 = 89.62°C$

$T_2 = (T_1 - BPR_1) - \Delta T_2' = (89.62 - 3.2) - 31.297 = 55.123°C$

$$T_{s1} = T_1 - BPR_1 = 86.42°C$$

- λ_{sf} at 120°C =2202.3 kJ/kg (from steam table)

λ_{s1} at 86.42°C =2293.64 kJ/kg

As shown before:

$$q_1' = 4194 \text{kJ/s and } q_2' = 3691.843 \text{ kJ/s}$$

$$A_1 = 113.176 \text{ m}^2 \text{ and } A_2 = 113.427 \text{ m}^2$$

- From equations (21-27) operating hours in the company=7200L.E/year

Annual cost of steam =987033.6 L.E/year

- Cost of first evaporator

Purchased cost = $75228.2(113.176)^{0.5053}$= 820621.21 L.E/year

Annual cost= 820621.21×1.6×0.15= 196949.0908 L.E/year

The same for the second evaporator Annual cost = 196949.09 L.E/year

Annual total cost = 1380931.78L.E/year

The three, four,.....seven effect evaporator can be calculated in the same trend as shown from the previous example of double effect evaporator.

6. Conclusion

A classic example in the chemical industry is the determination of the optimal number of effects in an evaporation system, were the optimum is found when there is an economic balance between energy saving and added investment, this is, a minimization of the total cost. The results observed that as number of effects of evaporators increased, annual cost of steam decreases and annual cost of evaporator increases. The optimum total cost for tomato juice concentration, in which the tomato juice enters at 5% and must be processed to 36% was by using double effect evaporator.

References

[1] P. D. Smith, C.L. E. Swartz and S. T. L. Harrison, Control and Optimization of a Multiple effect evaporator, Proc. S. Afr. Sug. Technol. Ass., 2000, 74, pp. 274-279

[2] H. Nuñez1, R. Simpson, S. Almonacid, and A. Abakarov, A study of different multi effect evaporators designed for Thermal processing of tomato paste: optimization and quality Evaluation", International Conference in Food Innovation, 2010.

[3] D. Kern, Procesos de Transferencia de Calor. Editorial Continental, S.A. México, 1999

[4] J. S. Dhara, G. Bhagchandanc, Design, Modelling and Simulation of Multiple Effect Evaporators. International Journal of Scientific Engineering and Technology, 2012, 1, pp.01-05.

[5] V. Jaishree, Optimization of a multiple Effect Evaporator System, A Thesis submitted to the National Institute of Technology, Rourkela In partial fulfillment of the requirements of Bachelor of Technology (Chemical Engineering) Department of Chemical Engineering National institute of technology, Rourkela, Orissa, 2010, pp. 769, India.

[6] S. Gunajit, D. B. Surajit, Energy Management in Multiple-Effect Evaporator System: A Heat Balance AnalysisApproach, Gen. Math. Notes, 2010, 1, pp. 84-88

[7] G.V. Barbosa-Cánovas, P. Juliano and M. Peleg. Engineering properties of foods, In: Food engineering. Encyclopedia of life support sciences, Barbosa-Cánovas G.V. (Ed.). EOLSS Publishers/UNESCO, Paris, France, 2005.

[8] M. A. Sorour, A comparative study of energy conservation methods in juice concentration by evaporation, Faculty of Eng., Cairo University, 2001, pp. 57–66.

[9] M. S. Peter, K. D. Timmerhaus, Plant Design and Economics for Chemical Engineers, 4th Ed., McGraw Hill, New York, 1991.

[10] B. G. Liptak, Process Control Trends, Control, March, May, and July, 2004.

8

Spectral and Thermal Properties of Biofield Energy Treated Cotton

Mahendra Kumar Trivedi[1], Rama Mohan Tallapragada[1], Alice Branton[1], Dahryn Trivedi[1], Gopal Nayak[1], Rakesh Kumar Mishra[2], Snehasis Jana[2, *]

[1]Trivedi Global Inc., Henderson, USA
[2]Trivedi Science Research Laboratory Pvt. Ltd., Bhopal, Madhya Pradesh, India

Email address:
publication@trivedisrl.com (S. Jana)

Abstract: Cotton has widespread applications in textile industries due its interesting physicochemical properties. The objective of this study was to investigate the influence of biofield energy treatment on the spectral, and thermal properties of the cotton. The study was executed in two groups namely control and treated. The control group persisted as untreated, and the treated group received Mr. Trivedi's biofield energy treatment. The control and treated cotton were characterized by different analytical techniques such as differential scanning calorimetry (DSC), thermogravimetric analysis (TGA), fourier transform infrared (FT-IR) spectroscopy, and CHNSO analysis. DSC analysis showed a substantial increase in exothermic temperature peak of the treated cotton (450 °C) as compared to the control sample (382°C). Additionally, the enthalpy of fusion (ΔH) was significantly increased by 86.47% in treated cotton. The differential thermal analysis (DTA) analysis showed an increase in thermal decomposition temperature of treated cotton (361°C) as compared to the control sample (358°C). The result indicated the increase in thermal stability of the treated cotton in comparison with the control. FT-IR analysis showed an alterations in –OH stretching (3408→3430 cm^{-1}), carbonyl stretching peak (1713-1662 cm^{-1}), C-H bending (1460-1431 cm^{-1}), -OH bending (580-529 cm^{-1}) and –OH out of plane bending (580-529 cm^{-1}) of treated cotton with respect to the control sample. CHNSO elemental analysis showed a substantial increase in the nitrogen percentage by 19.16% and 2.27% increase in oxygen in treated cotton as compared to the control. Overall, the result showed significant changes in spectral and thermal properties of biofield energy treated cotton. It is assumed that biofield energy treated cotton might be interesting for textile applications.

Keywords: Cotton, Biofield Energy Treatment, Thermal Analysis, Fourier Transform Infrared Spectroscopy, CHNSO Analysis

1. Introduction

Cotton is the most popularly used textile fiber due to its easy availability, low cost as well as good mechanical and physical properties. Cotton is mainly derived from a shrub that is native to tropical and subtropical regions around the word, including Africa, USA, and India [1]. The two main product that are derived from the cotton plant is cotton fiber and cottonseed [2-4]. The main component of cotton is cellulose that is the most abundant natural polymer on Earth [5]. It is the renewable biopolymer of outstanding properties and variety of useful applications [5]. Cotton is mainly used as a material for the manufacture of textile fabrics such as towel, robes, jeans, shirts, etc. Textile industries use the cotton fibers for the production of these materials by weaving and knitting process. Cotton has diverse applications in the medical field ranging from a single-thread suture to the composites for the bone replacement [6]. Cotton-based textiles have been used to prevent the growth of microorganisms [7-8]. Moreover, it was shown that cotton has electric and dielectric behavior. Additionally, cotton based fibers have attracted significant attention as a phase change materials (PCMs) [9]. During early 1980's under the National Aeronautics and Space Administration (NASA) research program PCMs capsules were embedded in textile structure to improve their thermal performance [10]. It was used as fabric in the astronaut space suite to provide improved thermal protection against the extreme temperature fluctuations in the outer space [10]. Hence, in order to use cotton-based textile for these applications their thermal and physical stability should be improved. Moreover, the chemical processing of cotton is difficult because this natural polymer is not meltable as well as it is insoluble in most of the available solvents due to strong hydrogen bonds and partially crystalline nature [11]. Many research groups in past few decades have devoted significant attention on modification of cotton for various applications. Yin *et al.* reported the chemical modification of cellulose carbamate using supercritical carbon dioxide and reported that this method has remarkably increased the nitrogen content in the modified polymer [5]. Recently, cotton fabrics were chemically modified with polyamidoamine dendrimer to yield antimicrobial and efficient polymer materials for ink jet printing [12]. Zhao *et al.* reported the carboxymethylation of cotton and elaborated that their high water absorbency maintains its fabric structure makes them potential candidates for wound therapy. El-gandy *et al.* modified the cotton fabrics by grafting with acrylic acid, acrylonitrile under gamma radiation treatment [13]. However, all these methods are not cost effective, hence some alternative strategies should be designed which can modify the physical and thermal properties of cotton. Recently, biofield energy treatment was used as an effective approach for modification of physicochemical properties of metal [14], ceramic [15] and polymers [16]. Hence, authors planned to investigate the impact of biofield treatment on spectral and thermal properties of cotton.

Energy medicine, energy therapy, and energy healing are the divisions of alternative medicine. It is believed that healers can channel the healing into the patients and confer positive results. Moreover, The National Centre for Complementary and Alternative Medicine/National Institute of Health (NCCAM/NIH), has authorized the use of this therapy in health care sector [17]. Biofield energy therapy is known as a treatment method that embraces an improvement in people's health and well-being by effectively interrelating with their biofield [18]. It is believed that good health of a human being entirely depends on the perfect balance of bioenergetic fields [19]. Thus, it is envisaged that human beings have the ability to harness the energy from the surrounding environment/Universe and can transmit into any object (living or non-living) around the Globe. The object(s) always receive the energy and responding in a useful manner that is called biofield energy. Mr. Trivedi is a well-known biofield expert who can transform the characteristics in various research fields such as biotechnology [20] and microbiology [21]. This biofield energy treatment is also known as The Trivedi Effect®.

Hence, by considering the outcomes of unique Mr. Trivedi's biofield energy treatment and excellent properties of cotton, this work was embarked on to investigate the impact of this treatment on spectral and thermal properties

of cotton.

2. Materials and Methods

Cotton was procured from Sigma Aldrich, USA. The sample was divided into two parts; one was kept as the control sample while the other was subjected to Mr. Trivedi's unique biofield energy treatment and coded as treated sample. The treated sample was in sealed pack and handed over to Mr. Trivedi for biofield energy treatment under laboratory condition. Mr. Trivedi gave the energy treatment through his unique energy transmission process to the treated samples without touching it. The control and treated samples were analyzed using various analytical techniques such as differential scanning calorimetry, thermogravimetric analysis, fourier transform infrared spectroscopy, and CHNSO analysis.

2.1. Differential Scanning Calorimetry (DSC)

The control and treated cotton samples were analyzed using Pyris-6 Perkin Elmer DSC at a heating rate of 10°C/min and the air was purged at a flow rate of 5 mL/min. The predetermined amount of sample was kept in an aluminum pan and closed with a lid. A reference sample was prepared using a blank aluminum pan. The percentage change in latent heat of fusion was calculated using following equations:

% change in Latent heat of fusion

$$= \frac{[\Delta H_{Treated} - \Delta H_{Control}]}{\Delta H_{Control}} \times 100 \qquad (1)$$

Where, $\Delta H_{Control}$ and $\Delta H_{Treated}$ are the latent heat of fusion of control and treated samples, respectively.

2.2. Thermogravimetric Analysis-Differential Thermal Analysis (TGA-DTA)

A Mettler Toledo simultaneous TGA and differential thermal analyzer (DTA) was used to investigate the thermal stability of control and treated cotton samples. The heating rate was 5°C/min, and the samples were heated in the range of room temperature to 400°C under air atmosphere.

2.3. FT-IR Spectroscopy

The FT-IR spectra were recorded on Shimadzu's Fourier transform infrared spectrometer (Japan) with the frequency range of 4000-500 cm^{-1}. The predetermined amount of sample was mixed with potassium bromide (KBr), and KBr pellets were prepared by pressing under the hydraulic press.

2.4. CHNSO Analysis

The control and treated cotton were analyzed for their elemental composition (C, H, N, O, S etc.). The powdered polymer samples were subjected to CHNSO Analyser using Model Flash EA 1112 Series, Thermo Finnigan Italy.

3. Results and Discussion

3.1. DSC Characterization

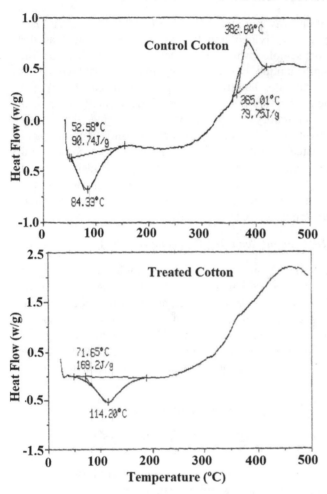

Fig. 1. DSC thermogram of control and treated Cotton.

DSC thermogram of control and treated cotton sample are presented in Figure 1. DSC thermogram of control cotton showed an endothermic peak at 84.33°C which was associated with the elimination of bound water. The DSC thermogram of control sample showed an exothermic peak at 382°C. This exothermic peak was may be due to crystallization or decomposition of cotton leading to the formation of levoglucosan and other volatile products. Potter reported that most of the exothermic events in DSC of textile fibers are due to crystallization and decomposition of the polymer network [22]. Whereas, the biofield treated cotton also exhibited an endothermic peak at 114.20°C that was associated with moisture elimination from the sample. The DSC thermogram showed the exothermic peak at around 450°C. This was again due to decomposition of cotton and volatilization of levoglucosan in the sample [23]. Calamari *et al.* during their work on the cotton ball and plant trash reported a similar exothermic peak in DSC thermogram. They proposed that this exothermic peak was due to β cellulose decomposition in the sample [24]. It is worth noting here that this exothermic peak was significantly increased in the treated sample. This may be inferred as an increase in

thermal stability of the treated sample as compared to the control cotton.

The enthalpy of fusion was obtained from the DSC thermogram of control and treated cotton. The enthalpy change of a material is known as energy absorbed during the phase change of a material from solid to liquid phase. The enthalpy of fusion of control cotton was 90.74 J/g and it was increased up to 169.2 J/g in the treated cotton. The result suggested the 86.47% increase in enthalpy of fusion of the treated cotton as compared to control sample. Recently, from our group it was reported that biofield energy treatment on monoterpenes (thymol and menthol) altered the latent heat of fusion as compared to the control sample [25]. Hence, it is assumed that biofield energy treatment might alter the internal energy that led to increase in enthalpy of fusion of treated cotton.

3.2. TGA-DTA

DTA thermogram of control and treated cotton is presented in Figure 2. DTA thermogram of control sample showed an endothermic transition due to water elimination at around 48°C. DTA thermogram of control sample showed second endothermic transition at 358°C. This was due to thermal decomposition of the cellulosic content present in the cotton sample. However, the DTA thermogram of treated cotton also showed two endothermic peaks at 50°C and 361°C. The former endothermic inflexion was due to dehydration of the moisture, and the later peak was due to thermal decomposition of the cellulose in the sample. Joseph et al. reported that in the case of cotton polymer fabric the first step may be attributed to the thermo-oxidative degradation, and this decomposes during the second step thermal degradation process [26]. The results suggested the increase in thermal decomposition temperature of the treated cotton as compared to control. This demonstrated the high thermal stability of treated cotton after the biofield treatment.

TGA of treated sample was carried out to get further insights about the thermal stability of the sample. TGA thermogram of treated cotton is shown in Figure 3, and it showed one-step thermal degradation pattern. The TGA thermogram of treated cotton showed onset temperature at 242.55°C and this thermal degradation process was terminated at 391.35°C. The result showed around 76% of weight loss during this thermal event. Since, the treated cotton started to degrade thermally only at 242°C that may be associated with good thermal stability of the treated sample. Yue suggested that mercerization process substantially increased the thermal stability of the treated cellulose as compared to control sample. They suggested that due to the strong interaction of the –OH group in cellulose required high energy to start the thermal degradation process [27]. It was reported that thermal properties of textiles are most important and desired features for their applications [28, 29]. For example, thermal insulation determines the elementary function of the garments. Thermal insulation is a key parameter for determining apparel comfort for the user. Hence, it is assumed that biofield energy treated cotton due to good thermal stability could be utilized for fabrication of textile fabrics.

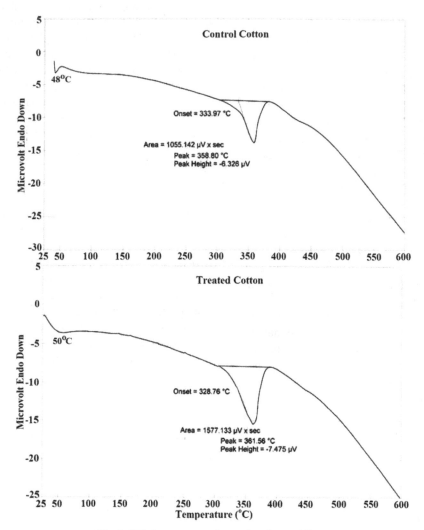

Fig. 2. *DTA thermograms of control and treated Cotton.*

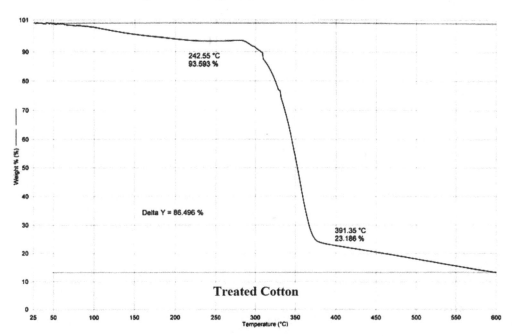

Fig. 3. *TGA thermogram of treated Cotton.*

3.3. FT-IR Spectroscopy

FT-IR spectroscopy of control and treated cotton is presented in Figure 4. The FT-IR spectrum of control cotton showed a characteristic peak at 3408 cm⁻¹ that was due to –OH stretching vibrations. This band is mainly due to polymeric O-H or hydroxyl group associated with hydrogen bonding [30]. The absorption bands at 2910 and 2800 cm⁻¹ were due to C-H stretching vibrations peaks. The vibration peaks at 1460 and 1713 cm⁻¹ were appeared due to carbonyl functional group as well as overlapping involvements of C-H bending vibrations [31]. Boeriu *et al.* and Chung *et al.* reported that FT-IR region below 1500 is the fingerprint region and composed of additional bands that correlate mainly to carbohydrates and other bio-constituents [32, 33]. The small peak at 883 cm⁻¹ in the control cotton was attributed to β glycosidic linkages between the sugar molecules in hemicelluloses and celluloses. The stretching vibration peak at 580 cm⁻¹ was due to –OH out of plane bending vibrations or atmospheric CO_2 (deformation vibration) contamination [34-36].

The FT-IR spectrum of treated cotton is shown in Figure 4. The vibration bands at 3430, 2925 and 2852 cm⁻¹ were mainly due to stretching vibration peaks of –OH and –CH groups in the treated sample. The FT-IR peaks at 1662 and 1431 cm⁻¹ were attributed to the –C=O and –CH bending vibrations. The peak originally present at 883 cm⁻¹ in control sample was disappeared and merged with –OH bending peak in the treated cotton sample. The –OH out of plane bending vibration was appeared at 529 cm⁻¹ in the treated sample. Overall, the FT-IR results showed upward shifting in hydrogen bonded –OH group stretching peak 3408→3430 cm⁻¹ in the treated cotton as compared to the control. It was previously suggested that the frequency (*v*) of the vibrational peak depends on two factors *i.e.* force constant and reduced mass. If the mass is constant, then the frequency is directly proportional to the force constant; therefore, increase in the frequency of any bond suggested a possible enhancement in force constant of respective bond and *vice versa* [37]. Hence, it is assumed that biofield energy treatment might cause an increase in force constant that led to increase in the frequency of the –OH bond. Additionally, the carbonyl stretching peak (1713-1662 cm⁻¹), C-H bending (1460-1431 cm⁻¹), OH bending (580-529 cm⁻¹) and –OH out of plane bending (580-529 cm⁻¹) showed downward shift that could be due to biofield energy treatment. It is assumed here that decrease in frequency of these peaks could be due to decrease in force constant. Hence, it is assumed that biofield energy treatment had caused structural changes in the treated cotton as compared to the control.

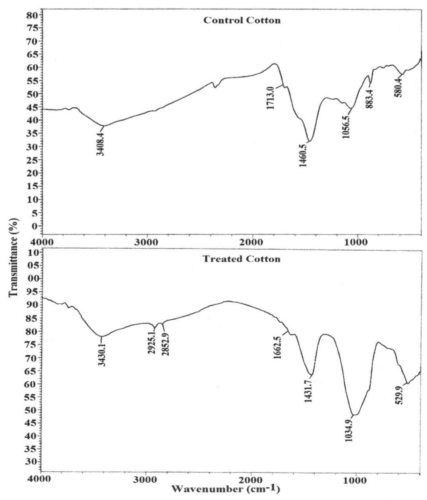

Fig. 4. *FT-IR spectra of control and treated cotton.*

3.4. CHNSO Analysis

CHNSO analysis was conducted to determine the elemental composition of the control and treated cotton. The results are presented in the Table 1. The percentage of nitrogen was 0.43 in the control sample and, it was increased to 0.51 in the treated cotton. The percentage of carbon was almost similar in both control (41.06) and treated cotton (40.94) samples. The hydrogen percentage was 6.06 and 5.97 in the control and treated samples, respectively. Whereas, the percentage of oxygen was 33.02 in the control sample, and it was increased slightly to 33.77 in the treated cotton. The control sample showed 2.15% of Sulphur; while no Sulphur was detected in the treated cotton. The CHNSO result suggested 19.16 and 2.27% increase in percentage nitrogen and oxygen, respectively after biofield treatment. However, the percentage carbon and hydrogen was decreased by 0.29 and 1.50% in treated cotton as compared to the control sample. It was reported that more nitrogen is required for high yield and production of cotton [38]. Hence, it is assumed that biofield treated cotton might of superior quality as compared to the control sample.

Table 1. CHNSO analysis data of control and treated cotton.

Element	Control	Treated	Percent change
Nitrogen	0.43	0.51	18.60
Carbon	41.06	40.94	-0.29
Hydrogen	6.06	5.97	-1.50
Oxygen	33.02	33.77	2.27
Sulphur	2.15	ND	-

4. Conclusions

In summary, the DSC result showed the significant increase in an exothermic peak in treated cotton (450°C) as compared to the control (382°C). DTA analysis showed an increase in thermal decomposition temperature of treated cotton (361°C) with respect to the control (358°C). This indicated the increase in thermal stability of the biofield treated cotton. FT-IR spectroscopic analysis showed an alterations in the frequency of the treated cotton. It is presumed that biofield treatment caused changes in force constant of the treated cotton in comparison with control. Additionally, CHNSO analysis showed the considerable increase in the elemental composition of nitrogen (19.16%) and oxygen (2.27%) in treated cotton. Therefore, the high thermal stability of biofield energy treated cotton might improve its application for preparation of textile fabric.

Abbreviations

DSC: Differential scanning calorimetry;
TGA: Thermogravimetric analysis;
FT-IR: Fourier transform infrared spectroscopy;
PCMs: Phase change materials

Acknowledgments

The authors wish to thank all the laboratory staff of MGV Pharmacy College, Nashik for their kind assistance during handling the various instrument characterizations. The authors would also like to thank Trivedi Science, Trivedi Master Wellness and Trivedi Testimonials for their support during the work.

References

[1] True AC (1896). The cotton plant. Washington: Govt. Printing Office.

[2] Yafa S (2002) Cotton: The Biography of a Revolutionary Fiber. Penguin, USA.

[3] Bailey AE (1948) Cottonseed and cottonseed products. Wiley Interscience, New York.

[4] National Cottonseed Products Association, Inc. http://www.cottonseed.com/publications.

[5] Yin C, Li J, Xu Q, Peng Q, Liu Y, Shen X (2007) Chemical modification of cotton cellulose in supercritical carbon dioxide: Synthesis and characterization of cellulose carbamate. Carbohydrate Polymers 67: 147-154.

[6] Kumar RS (2014) Textiles for Industrial Applications. CRC Press, Taylor & Francis Group, Boca Raton, Florida.

[7] Shahidi S, Ghoranneviss M, Moazzenchi B, Rashidi A, Mirjalili M (2007) Investigation of antibacterial activity on cotton fabrics with cold plasma in the presence of a magnetic field. Plasma Processes and Polymers 4: S1098-S1103.

[8] Windler L, Height M and Nowack B (2013) Comparative evaluation of antimicrobials for textile applications. Environment International 53: 62-73.

[9] Mondal S (2008) Phase change materials for smart textiles-An overview. Applied Thermal Engineering 28: 1536-1550.

[10] Nelson G (2001) Microencapsulation in textile finishing. Review of Progress in Coloration and related Topics 31: 57-64.

[11] Fink HP, Weigel P, Purz HJ, Ganster J (2001) Structure formation of regenerated cellulose materials from NMMO-solutions. Progress in Polymer Science 26: 147-1524.

[12] Gorgani AS, Najafi F, Karami Z (2015) Modification of cotton fabric with a dendrimer to improve ink-jet printing process. Carbohydrate Polymers 131: 168-176.

[13] El Gendy, Eglal HK (2002) Modification of cotton fabrics via radiation graft copolymerization with acrylic acid, acrylonitrile and their mixtures. Indian Journal of Fiber and Textile Research 27: 266-273.

[14] Trivedi MK, Patil S, Tallapragada RM (2013) Effect of biofield treatment on the physical and thermal characteristics of silicon, tin and lead powders. Journal of Material Science & Engineering 2: 125.

[15] Trivedi MK, Patil S, Tallapragada RM (2013) Effect of biofield treatment on the physical and thermal characteristics of vanadium pentoxide powders. Journal of Material Science & Engineering S11, 001.

[16] Trivedi M.K, Nayak G, Patil S, Tallapragada RM, Mishra R (2015) Influence of Biofield Treatment on Physicochemical Properties of Hydroxyethyl Cellulose and Hydroxypropyl Cellulose. Molecular Pharmaceutics & Organic Process Research 3: 126.

[17] Barnes PM, Powell-Griner E, McFann K, Nahin RL (2004) Complementary and alternative medicine use among adults: United States, 2002. Seminars in Integrative Medicine 2: 54-71.

[18] http://www.drfranklipman.com/what-is-biofield-therapy/.

[19] Warber SL, Cornelio D, Straughn J, Kile G (2004) Biofield energy healing from the inside. The Journal of Alternative and Complementary Medicine 10: 1107-1113.

[20] Patil SA, Nayak GB, Barve SS, Tembe RP, Khan RR (2012) Impact of biofield treatment on growth and anatomical characteristics of *Pogostemon cablin* (Benth.). Biotechnology 11: 154-162.

[21] Trivedi MK, Patil S, Shettigar H, Bairwa K, Jana S (2015) Phenotypic and biotypic characterization of *Klebsiella oxytoca:* An impact of biofield treatment. Microbial and Biochemical Technology 7: 202-205.

[22] Potter C (2012) Thermal analysis of textile fibers. AATCC Review 39-45.

[23] Ibrahim SF, El-Amoudy ES, Shady KE (2011) Thermal analysis and characterization of some cellulosic fabrics dyed by a new natural dye and mordanted with different mordants. International Journal of Chemistry 3: 40-54.

[24] Clamari TA, Donaldson DJ, Thibodeaux DP (1990) Distinguishing weathered from un-weathered cotton by thermal analysis. Amer Dyest Rep 79: 42-47.

[25] Trivedi MK, Patil S, Mishra RK, Jana S (2015) Structural and physical properties of biofield treated thymol and menthol. Molecular Pharmaceutics & Organic Process Research 3: 127.

[26] Wanna, JT, Powell JE (1993) Thermal decomposition of cotton cellulose treated with selected salts. Thermochimica Acta 226: 257-263.

[27] Yue Y (2011) A comparative study of cellulose I and II fibers and nanocrystals. (MS Thesis).

[28] Le CV, Ly NG, Postle R (1995) Heat and moisture transfer in textile assemblies. Part I steaming of wool, cotton, nylon and polyester fabric beds. Textile Research Journal 65: 203-212.

[29] Kawabata S (2000) A guide line for manufacturing ideal fabrics. International Journal of Clothing Science and Technology 11: 134-144.

[30] Coates J (2000) Interpretation of infrared spectra, a practical approach. In: Meyers, R, Ed, Encyclopedia of analytical chemistry. John Wiley & Sons Ltd., Chichester, England.

[31] Himmelsbach DS, Akin DE, Kim J, Hardin IR (2003) Chemical structural investigation of the cotton fiber base and associated seed coat: Fourier-transform infrared mapping and histochemistry. Textile Research Journal 73: 218-288.

[32] Boeriu C, Bravo D, Gosselink RJA, van Dam JEJ (2004) Characterization of structure-dependent functional properties of lignin with infrared spectroscopy. Industrial Crops and Products 20: 205-218.

[33] Chung C, Myunghee L, Choe E (2004) Characterization of cotton fabric scouring by FT-IR ATR spectroscopy. Carbohydrate Polymers 58: 417-420.

[34] Rana AK, Basak RK, Mitra BC, Lawther M, Banerjee AN (1997) Studies of acetylation of jute using simplified procedure and its characterization. Journal of Applied Polymer Science 64: 1517-1523.

[35] Grobe A (1989) In: Brandru, J, Immergut EH, Eds, Polymer Handbook, John Wiley, New York.

[36] Günzler H, Gremlich HU (2002) IR Spectroscopy-An Introduction. Wiley-VCH, Weinheim.

[37] Pavia DL, Lampman GM, Kriz GS (2001) Introduction to spectroscopy. (3rdedn), Thomson Learning, Singapore.

[38] Livingston SD, Stichle CR (1995) Correcting Nitrogen Deficiencies in Cotton with Urea-Based Products. The Texas A&M University System, College Station, Texas.

9

Multidisciplinary Engineering for the Utilization of Traditional Automated Storage and Retrieval System (*ASRS*) for Firefighting in Warehouses

Ahmed Farouk Abdel Gawad

Mech. Eng. Dept., Umm Al-Qura Univ., Makkah, Saudi Arabia

Email address:

afaroukg@yahoo.com

Abstract: *ASRS* is usually installed in a factory or warehouse to increase storage capacity, floor space utilization, labor productivity in storage operations and stock rotations *etc*. *ASRS* systems are custom-planned for each individual application, and they range in complexity from relatively small mechanized systems that are controlled manually to very large computer-controlled systems that are fully integrated with factory and warehouse operations. The purpose of the present paper is to gain experience in multidiscipline engineering through the design, fabrication and implementation of *ASRS* model in a capstone (graduation) project. The scope of the project was extended by considering the fire fighting in *ASRS* warehouses. In the present project, the students designed, manufactured and implemented an *ASRS* model. The model was automatically controlled by a microcontroller-based electronic circuit. The students succeeded in designing, building and testing their model. Based on their practical experience, they proposed a new technique for firefighting in *ASRS* warehouses.

Keywords: Multidisciplinary Engineering, *ASRS*, Warehouse, Firefighting

1. Introduction

1.1. Multidisciplinary Engineering

As new skills are required of engineering graduates to respond to ever-changing societal needs, research on engineering learning systems is required to effectively adapt and respond to those needs. Multidisciplinary, team-taught, project-based instruction has shown effectiveness in teaching teamwork, communication, and life-long learning skills, and appreciation for other disciplines. Unfortunately, this instruction mode has not been widely adopted, largely due to its resource-intensiveness [1].

Another important objective is the development of professional skills of students. Developing professional skills to improve students' awareness of the engineering practice enhances the overall competence of students and provides lifelong learning skills including proficiency in communication, engineering ethics, professional presentations, professional etiquette, and project management [2].

The multidisciplinary approach was considered by many researchers [3-10] for both under- and post-graduate engineering studies.

The present paper covers an engineering graduation project as an adaptation of the multidisciplinary learning technique. The students developed a model of an automated storage and retrieval system (*ASRS*). The multidisciplinary topics cover problem-definition and analysis, mechanical design, material selection, manufacturing, selection of electrical motors, and design, programming and implementation of electronic control circuits and sensors. Also, multidisciplinary engineering may be extended to firefighting in *ASRS* warehouses in emergency situations.

1.2. What is ASRS

An automated storage and retrieval system (*ASRS* or *AS/RS*) consists of a variety of computer-controlled systems for automatically placing and retrieving loads from defined storage locations. Automated storage and retrieval systems (*ASRS*) are typically used in applications where: (*i*) There is a very high volume of loads being moved into and out of storage. (*ii*) Storage density is important because of space constraints; no value adding content is present in this process.

(*iii*) Accuracy is critical because of potential expensive damages to the load. *ASRS* can be used with standard loads as well as nonstandard loads [11]. Automated storage and retrieval systems have been widely used in distribution and production environments since their introduction in the 1950s [12]. More details about *ASRS* can be found in Refs. [13-21]. Figure 1 shows some examples of real-life *SARS*.

The utilization of *ASRS* has many benefits that may be listed as: (*i*) Improving storage capacity and operator efficiency, (*ii*) Increasing accuracy of stock handling, (*iii*) Maximizing the benefits of available storage space, (*iv*) Reducing labor costs, (*v*) Reducing stock damage and waste costs, (*vi*) Improving customer service, (*vii*) Real-time inventory control, (*viii*) Preventing personnel from harsh working conditions, *e.g.*, working in cold food-storage utilities.

(a) [22]

(b) [23]

(c) [24]

Figure 1. Some examples of real-life SARS.

1.3. Problems of Conventional Storage Systems

The performance of any manufacturing industry depends mostly on its material handling and storage system [25]. Thus, the conventional storage systems have many practical problems that can be listed as: (*i*) Lost or damage of stored products, (*ii*) Difficulties of inventory access, (*iii*) Much time is usually spent for stock searching, (*iv*) Delaying the customer's delivery as orders spend much time in the factory, (*v*) Inaccurate inventory records, (*vi*) Wasting of considerably much space, (*vii*) Workers may be exposed to dangerous situations, (*viii*) Excess quantities of inventory.

1.4. Types of ASRS

Several important categories of *ASRS* can be distinguished based on certain features and applications. The following are the principle types [26]:

1.4.1. Unit-Load ASRS

The unit load *ASRS* is used to store and retrieve loads that are palletized or stored in standard-sized containers. The system is computer controlled. The *SR* (storage and retrieval) machines are automated and designed to handle the unit load containers. Usually, a mechanical clamp mechanism on the *SR* machine handles the load. However, there are other mechanisms such as a vacuum or a magnet-based mechanism for handling sheet metal. The loads are generally over 500 *lb per unit*. The unit-load system is the generic *ASRS*.

1.4.2. Mini-Load ASRS

This system is designed to handle small loads such as individual parts, tools, and supplies that are contained in bins or drawers in the storage system. Such a system is applicable where the availability of space is limited. It also finds its use where the volume is too low for a full-scale unit load system and too high for a manual system. A mini-load *ASRS* is generally smaller than a unit-load *ASRS* and is often enclosed for security of items stored.

1.4.3. Deep-Lane ASRS

This is a high-density unit load storage system that is appropriate for storing large quantities of stock. The items are stored in multi deep storage with up to 10 items in a single rack, one load behind the next. Each rack is designed for flow-through, with input and output on the opposite side. Machine is used on the entry side of the rack for input load and loads are retrieved from other side by an *SR*-type machine. The *SR* machines are similar to unit-load *SR* machine except that it has specialized functions such as controlling rack-entry vehicles.

1.4.4. Man-on-Board ASRS

This system allows storage of items in less than unit-load quantities. Human operator rides on the carriage of the *SR* machine to pick up individual items from a bin or drawer. The system permits individual items to be picked directly at their storage locations. This provides an opportunity to increase system throughput. The operator can select the items and place them in a module. It is then carried by the *SR* machine to the end of the aisle or to a conveyor to reach its destination.

1.4.5. Automated Item Retrieval System

This system is designed for retrieval of individual items or small product cartoons. The items are stored in lanes rather than bins or drawers. When an item is retrieved from the front by use of a rear-mounted pusher bar, it is delivered to the pickup station by pushing it from its lane and dropping onto a conveyor. The supply of items in each lane is periodically replenished and thus permitting first-in/first-out inventory rotation. After moving itself to the correct lane, the picking head activates the pusher mechanism to release the required number of units from storage.

2. Present *ASRS* Model

2.1. Objectives

The main objectives of the present project can be summarized as:

(*i*) Design a model, with a suitable size, of automated storage and retrieval system (*ASRS*) for small loads.

(*ii*) Fabrication, assembling, and testing the model.

(*iii*) Integrating simple control system (microcontroller-based) on the storage/retrieval (*SR*) machine.

(*iv*) Proposing a new firefighting technique for small-load units.

(*v*) Covering the learning skills that are mentioned in *Sec.* 1.1.

2.2. Project Scope

The present project focuses on design and fabrication of the *ASRS* model. The design was mainly developed using "Autodesk Inventor Professional" software. Programmable autonomously storage/retrieve (*SR*) machine was used in conjunction to store and retrieve the small-load units. Furthermore, students designed and implemented an electronic circuit to control the movement of the *SR* machine. The movement of this system is limited to store and retrieve small-load units into and from storage structure (cupboards). Also, the students came up with a new technique for the firefighting of fire that may ignite in such small-load units. This technique is based on the practical experience of the present project.

3. Design and Fabrication of the Present *ASRS* Model

This section covers the details of design and fabrication of the present *ASRS* model. The different components of the *ASRS* will be explained with design drawings and pictures. Figure 2 shows an overall picture of the *ASRS* model including movement mechanism and storage structure (cupboard). The overall dimensions of *ASRS* model can be stated as: length 1.35 *m* (*x*-axis), height 0.65 *m* (*y*-axis), width 0.25 *m* (*z*-axis).

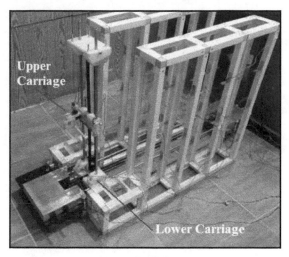

(a) An overall picture.

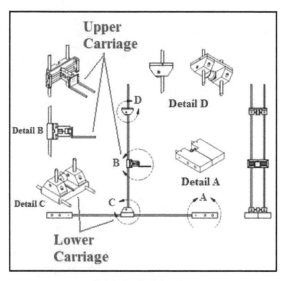

(b) Mechanical drawing.

(c) Main axes of motion.

***Figure 2.** Present ASRS model.*

Some parts of the *ASRS* model were fabricated in suitable workshops. Table 1 shows the details of these parts (components) including material selection, manufacturing technique, and quantity. Other parts (components) of the *ASRS* model are ready-made that can be bought directly from the market. Table 2 shows the details of these ready-made parts.

Table 1. *Details of fabricated parts (components) in workshops.*

No.	Part (Component)	Material Selection	Manufacturing Technique	Quantity
1	Rigid Base	Aluminium		4
2	Moving Base	PTFE		1
3	Guide Base	PTFE		4
4	Vertical Slider	PTFE		1
5	Guide Head Plate	PTFE	Water Jet & Traditional Machining	1
6	Holder Plate	PTFE		1
7	Holder Arm	PTFE		2
8	Fork plate	PTFE		1
9	Fork arm	PTFE		2
10	Back holder	PTFE		1
11	H-Holder	PTFE		2
12	Guide rods	Stainless steel	Sawing	6
13	Pulley	PTFE	Turning	1
14	Timing pulley	PTFE		1

Table 2. *Details of ready-made parts (components).*

No.	Part (Component)	Type	Quantity
1	Chain	Roller	1
2	Belt	Timing flat	1
3	Gears	Spur	6
		Chain gears	2
4	Bearings	Ball bearings	8
5	coupling	Rigid coupling	1
6	Springs	Compression & extension	1
7	Bolts	Tap bolts	As needed
8	Nuts	Hexagonal nuts	As needed

3.1. Details of Fabricated Parts

This section gives details of the parts (components) that were fabricated specifically for the present project in workshops. These parts are listed in Table 1.

3.1.1. Rigid Base

The rigid base is the base that carries the whole mechanism of *ASRS*. The rigid base, Fig. 3, is consisted of four symmetrical parts that are made of aluminum, Table 1. These four parts constitute two main side bases; one on each side of the mechanism. The rigid base holds the two guide rods (10 *mm*-diameter) that guide motion in *x*-axis. Figure 3 shows the mechanical drawings (Isometric, Elevation, Plan, Side view) of one of the four parts of the rigid base.

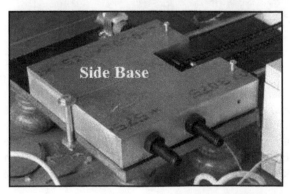

Figure 3. *One of the two side bases of the rigid base of ASRS model.*

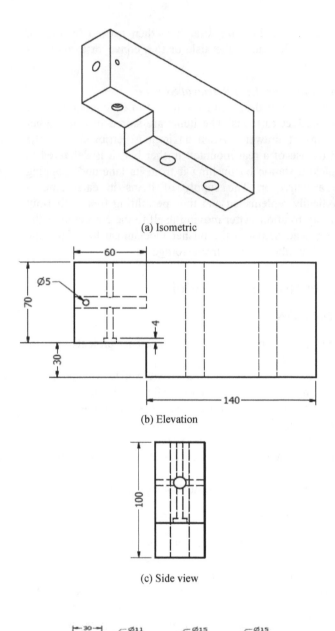

(a) Isometric

(b) Elevation

(c) Side view

(d) Plan

Figure 4. *Mechanical drawings of one of the four parts of the rigid base.*

3.1.2. Moving Base

The moving base, which is made of *PTFE*, was designed to facilitate the movement of *ASRS* components in the horizontal direction, Fig. 5. It is moved in the either directions of *x*-axis (right or left) by a chain. The chain is driven be an electric motor. The chain is fixed to the moving base. The moving base is supported and guided by two metal rods. These two solid rods have a diameter of 10 *mm* and are made of stainless steel. Figure 6 shows the mechanical drawings (Isometric, Elevation, Plan, Side view) of the moving base.

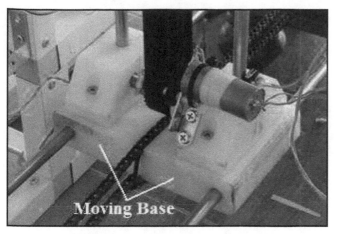

Figure 5. *Moving base installed in ASRS model.*

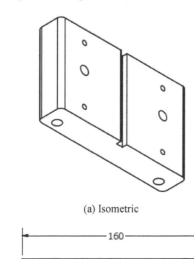

(a) Isometric

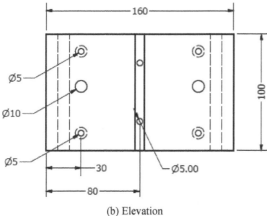

(b) Elevation

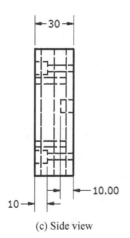

(c) Side view

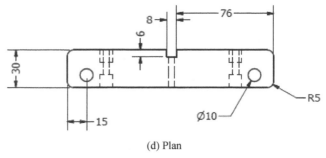

(d) Plan

Figure 6. *Mechanical drawings of moving base.*

3.1.3. Guide Bases

There are four guide bases that are designed to carry the vertical timing belt as well as the two guide rods (guides) in the vertical direction (y-axis), Fig. 6. The guide bases are made of *PTFE*. Two of them are fixed on either side of the lower moving base, Fig. 4. The other two are fixed to an upper plate (Guide Head Plate), Fig. 7. The two vertical solid guide rods have a diameter of 10 *mm* and are made of stainless steel. The two vertical guide rods guide the motion along the y-axis. Figure 8 shows the mechanical drawings (Isometric, Elevation, Plan, Side view) of one of the guide bases.

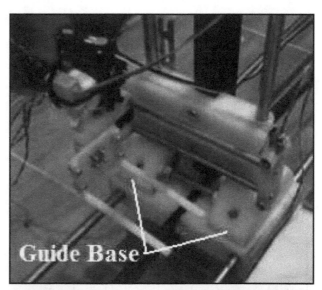

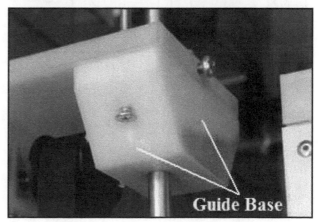

Figure 7. *Guide base installed in ASRS model.*

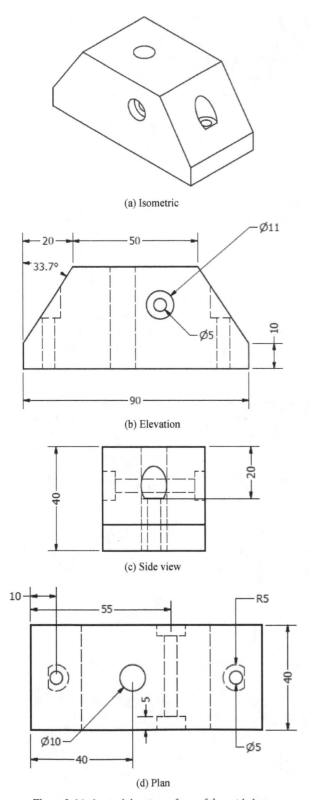

(a) Isometric

(b) Elevation

(c) Side view

(d) Plan

Figure 8. Mechanical drawings of one of the guide bases.

3.1.4. Vertical Slider

The vertical slider is designed to carry the holder plate, Fig. 9. The vertical slider is made of *PTFE*. The vertical slider is fixed to the vertical belt by two bolts, Fig. 9. Thus, it moves strictly with the vertical belt. It slides on the two vertical solid guide rods. Figure 10 shows the mechanical drawings (Isometric, Elevation, Plan, Side view) of the vertical slider.

Figure 9. Vertical slider installed in ASRS model.

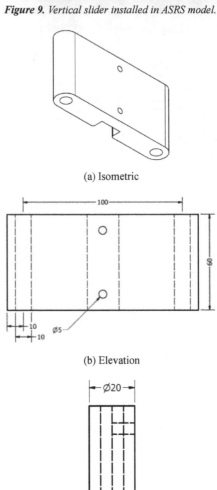

(a) Isometric

(b) Elevation

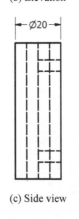

(c) Side view

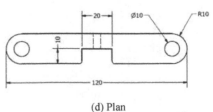

(d) Plan

Figure 10. Mechanical drawings of the vertical slider.

3.1.5. Guide Head Plate

The guide head plate is an upper plate that is fixed to two of the guide bases, Fig. 11. It is made of *PTFE*. Also, it holds the two vertical solid guide rods into place. Figure 12 shows the mechanical drawings (Isometric, Elevation, Plan, Side view) of one of the guide head plate.

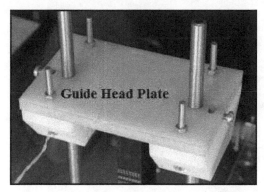

Figure 11. *Guide head plate installed in ASRS model.*

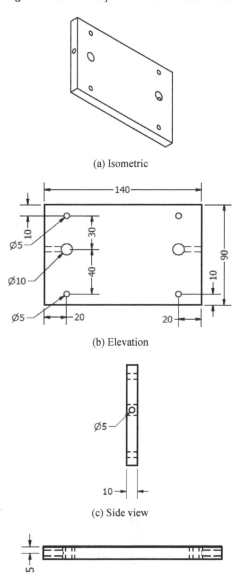

(a) Isometric

(b) Elevation

(c) Side view

(d) Plan

Figure 12. *Mechanical drawings of the guide head plate.*

3.1.6. Holder Plate

The holder plate is fixed to the vertical slider by screws. It is designed to carry two horizontal solid guide rods that are used to guide the movement of the upper fork arms, Fig. 13. The holder plate is made of *PTFE*. The middle section of the holder plate is machined to present a toothed rack to enable the movement of the fork plate and fork arms in the *z*-axis. The holder plate is fixed to the belt that moves in the vertical direction (*y*-axis) by two small bolts. Figure 14 shows the mechanical drawings (Isometric, Elevation, Side view) of the holder plate.

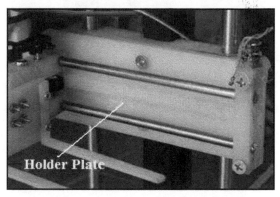

Figure 13. *Holder plate installed in ASRS model.*

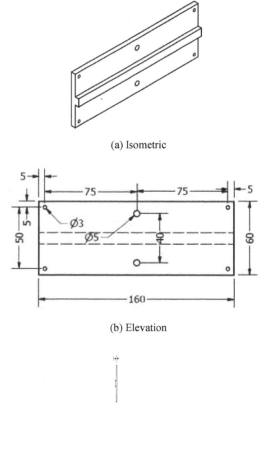

(a) Isometric

(b) Elevation

(c) Side view

Figure 14. *Mechanical drawings of the holder plate.*

3.1.7. Holder Arm

Two holder arms are designed to fix the two upper horizontal solid guide rods (5 *mm*-diameter) that guide motion along *z*-axis, Fig. 15. Each holder arm is fixed to one of the two sides of the holder plate by a pair of screws. The two holder arms are made of *PTFE*. Figure 16 shows the mechanical drawings (Isometric, Elevation, Plan, Side view) of the holder arm.

3.1.8. Fork Plate

The fork plate is designed to carry the two fork arms, Fig. 17. The fork plate is made of *PTFE*. It is fixed into place by four bolts. Figure 18 shows the mechanical drawings (Isometric, Elevation, Side view) of the fork plate.

Figure 15. One of the two holder arms installed in ASRS model.

Figure 17. Fork plate installed in ASRS model.

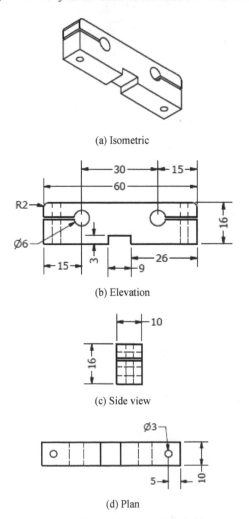

(a) Isometric

(b) Elevation

(c) Side view

(d) Plan

Figure 16. Mechanical drawings of the holder arm.

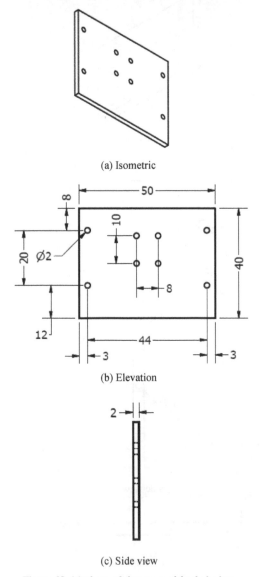

(a) Isometric

(b) Elevation

(c) Side view

Figure 18. Mechanical drawings of the fork plate.

3.1.9. Fork Arms

The two fork arms are designed to carry the target stock (unit-load), Fig. 19. They are made of *PTFE*. They are fixed to fork plate by four screws. Figure 20 shows the mechanical drawings (Isometric, Elevation, Side view) of one of the fork arms.

3.1.10. Back Holder

The back holder was designed to fix the fork plate to the H-holder, Fig. 21. The back holder is made of *PTFE*. It is fixed into place by four bolts. Figure 22 shows the mechanical drawings (Isometric, Elevation, Plan, Side view) of the back holder.

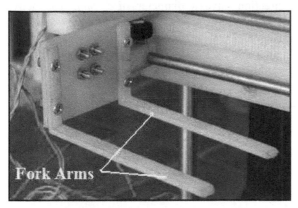

Figure 19. *Fork arms installed in ASRS model.*

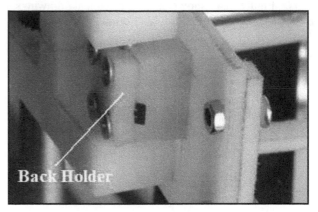

Figure 21. *Back holder installed in ASRS model.*

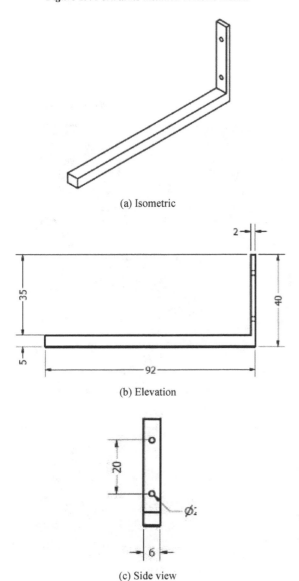

(a) Isometric

(b) Elevation

(c) Side view

Figure 20. *Mechanical drawings of the fork arms.*

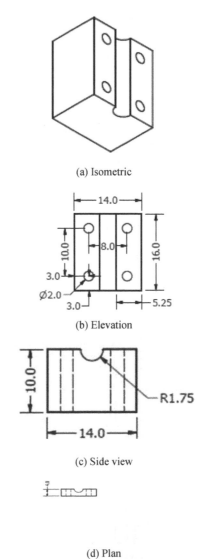

(a) Isometric

(b) Elevation

(c) Side view

(d) Plan

Figure 22. *Mechanical drawings of the back holder.*

3.1.11. H-holder

The H-holder was designed to carry the fork plate, fork arms, and transmission gears. It is consisted of two similar parts that have T-shape, Fig. 23. The two parts are made of *PTFE*. They are fixed to each other by a long bolt. Also, the two transmission gears are fixed in place to the H-holder by two bolts. Figure 24 shows the mechanical drawings (Isometric, Elevation, Plan, Side view) of one of the T-shape parts.

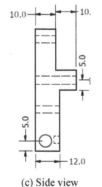

Figure 23. H-holder installed in ASRS model.

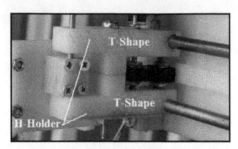

(a) Isometric

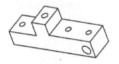

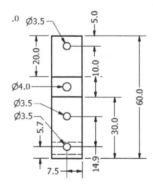

(b) Elevation

(c) Side view

(d) Plan

Figure 24. Mechanical drawings of one of the T-shape parts.

3.1.12. Solid Guide Rods

There are six giude rods that guide the movement in the three main directions, Fig. 25. A pair of similar solid guide rods, which are made of stailess steel are used in every one of the three diections. The two rods have a diameter of 10 *mm* in both the *x*- and *y*-dierction. Whereas, the two rods of the *z*-direction have a diameter of 5 *mm*. The active lengthes of rods are 95, 55, 15 *cm* in *x*-, *y*-, and *z*-direction, respectively.

Figures 26-28 show two pictures for each of pair of rods along the three main directions; *x*-, *y*-, and *z*-axis, respectively.

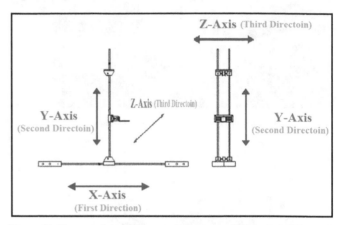

Figure 25. Three main axes (directions) of motion of the present ASRS model.

(a) Overall view.

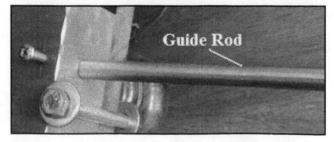

(b) Closer view.

Figure 26. The two guide rods along x-axis.

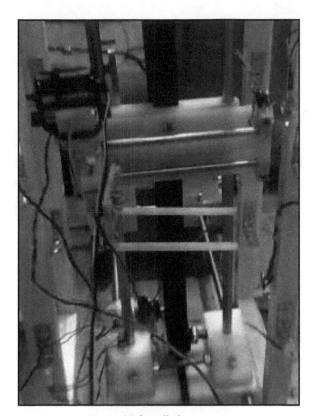

(a) Overall view.

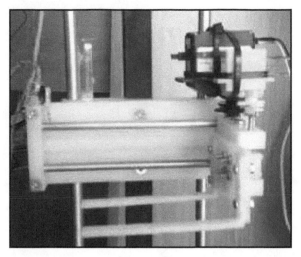

(a) Overall view.

(b) Closer view.

Figure 28. The two guide rods along z-axis.

3.1.13. Pulley

A pulley was designed to carry the vertical belt that moves the holder plate (*Sec.* 3.1.6, Fig. 13). The pulley is made of *PTFE*, Fig. 29. The pulley is carried by a solid stainless steel rod that has a diameter of 4 *mm*. The two ends of the rod are supported by two small bearings that have housing in the two upper guide bases (*Sec.* 3.1.3, Fig. 7). Figure 30 shows the mechanical drawings (Isometric, Elevation, Side view) of the pulley.

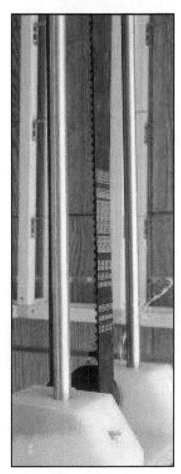

(b) Closer view.

Figure 27. The two guide rods along y-axis.

Figure 29. Pulley installed in ASRS model.

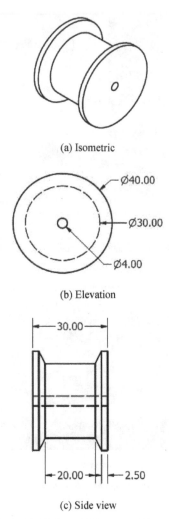

(a) Isometric

(b) Elevation

(c) Side view

Figure 30. *Mechanical drawings of the pulley.*

3.1.14. Timing Pulley

A timing pulley was designed to move the vertical belt that moves the holder plate (*Sec.* 3.1.6, Fig. 13). The pulley is made of *PTFE*, Fig. 31. The pulley is carried by a solid stainless steel rod that has a diameter of 4 *mm*. The two ends of the rod are supported by two small bearings that have housing in the two lower guide bases (*Sec.* 3.1.3, Fig. 7). Figure 32 shows the mechanical drawings (Isometric, Elevation) of the timing pulley.

Figure 31. *Timing pulley installed in ASRS model.*

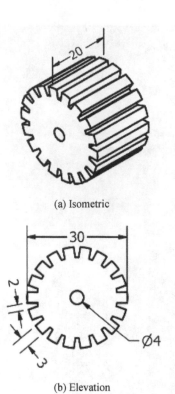

(a) Isometric

(b) Elevation

Figure 32. *Mechanical drawings of the timing pulley.*

3.2. Details of Ready-Made Parts

This section gives details of the ready-made parts (components) that were purchased directly from the market. These parts are listed in Table 2.

3.2.1. Chain

A suitable chain, Fig. 33, was used to move the moving base (*Sec.* 3.1.2, Fig. 5) that facilitates the movement of *ASRS* components in the *x*-axis (first direction). The chain is connected to two chain gears that rest on the rigid base, Fig. 3. Each of the two chain gears is carried by a solid stainless steel rod that has a diameter of 4 *mm*. The two ends of the rod are supported by two small bearings that have housing in the two sides of the rigid base. One of these chain gears, which is driven by a suitable electrical motor, is the driving gear of the chain.

Figure 33. *Chain (x-axis) installed in ASRS model.*

3.2.2. Timing Belt

The timing belt, Fig. 34, is used to move the holder plate (*Sec.* 3.1.6, Fig. 13) vertically in the *y*-axis (second direction). The timing belt is driven by a timing pulley (*Sec.* 3.1.14, Fig. 31). A flat pulley (*Sec.* 3.1.13, Fig. 29) supports the other side of the timing belt.

Figure 34. Timing belt (y-axis) installed in ASRS model.

3.2.3. Gears

A group of seven gears was used in the present project. Two of them are chain gears Fig. 35a. They were used to drive the chain (*Sec.* 3.2.1, Fig. 33). Five spur gears were also used, Fig. 35b. Two of the spur gears are housed in the H-holder (*Sec.* 3.1.11, Fig. 23). These two spur gears are used to move the fork plate (*Sec.* 3.1.8, Fig. 17) and fork arms (*Sec.* 3.1.9, Fig. 19). Three other spur gears, Fig. 35c, were used to move the timing belt (*Sec.* 3.2.2, Fig. 34) in the y-axis (second direction).

(a) Chain gear.

(b) Two supr gears of fork arms.

(c) Four spur gears of the timing belt.

Figure 35. Gears installed in ASRS model.

3.2.4. Bearings

Eight roller bearings with internal diamter of 4 *mm* were used in the *ASRS* model, Fig. 36. Four of them were used to support the chain (*Sec.* 3.2.1, Fig. 33) in the x-axis (first directon). They rest in the two side parts of the rigid base (*Sec.* 3.1.1, Fig. 3). The other four bearings support the timing belt (*Sec.* 3.2.2, Fig. 34) in the y-axis (second direction). They rest in the four guide bases (*Sec.* 3.1.3, Fig. 7).

Figure 36. Bearings with 4 mm-internal diameter.

3.2.5. Coupling

A rigid coupling, Fig. 37, was used to connect an electrical motor to the chain gear (*Sec.* 3.2.3, Fig. 35a) that drives the chain (*Sec.* 3.2.1, Fig. 33) in the x-axis (first direction). The connection goes through one of the parts of the rigid base (*Sec.* 3.1.1, Fig. 3).

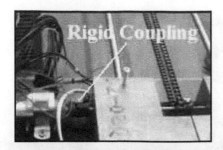

Figure 39. An overall view of the cupboard.

3.3.2. Components of Cupboard

The cupboard may be divided into two types of components as follows, Fig. 40:

(*i*) Pick-and-deposit station: It is where loads are transferred into and out of the *ASRS*. It is generally located at the end of the aisles for access the external handling system that brings loads to the *ASRS* and takes loads away.

(*ii*) Compartments: They are the unit-load containers of the stored material.

Figure 37. Rigid coupling installed in ASRS model.

3.2.6. Spring

A spring, Fig. 38, was used to ensure the engagement of the gears of the gear train that transfers motion from an electrical motor to the timing pulley (*Sec.* 3.1.14, Fig. 31). The timing pulley drives the *ASRS* in the *y*-axis (second direction).

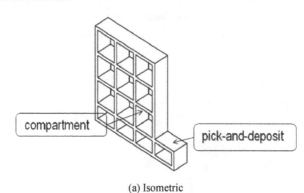

(a) Isometric

(a) Overall view.   (b) Detailed view.

Figure 38. Spring installed in ASRS model.

3.3. Cupboards (Pallets)

3.3.1. Description

Cupboard was design to simulate the shelves (pallets) that hold the stock pieces, Fig. 39. It was fabricated of aluminum rods and plastic connections. Inventor software was used to carry out the design process. Two similar cupboards were used to simulate a working track of *ASRS* between them.

(b) Elevation

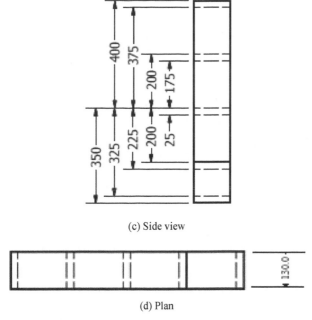

(c) Side view

(d) Plan

Figure 40. Mechanical drawings of one of the two cupboards.

3.3.3. Cupboard Fabrication

The cupboard was fabricated from ready-made parts that are listed in table 3.

Table 3. Ready-made parts.

No.	Component	Material
1	Sheet	Plastic
2	Joint	Plastic
3	Column	Aluminium
4	Nail	Steel

4. Electrical and Electronic Elements

4.1. Electrical Motors

In the present project, there are two types of motors. The first type is the *DC* motors and the second is the servomotors.

4.1.1. DC Motors

DC motor is an electric motor that runs on direct current (*DC*) electricity. Today, *DC* motors are still found in applications as small as toys and disk drives, or in large sizes to operate steel rolling mills and paper machines. Modern *DC* motors are nearly always operated in conjunction with power electronic devices [27].

Two suitable *DC* motors were used in the present project. The first *DC* motor was used to drive the *ASRS* components in the *x*-axis (first direction), Fig. 41a. The motor is connected to the driving chain gear (*Sec.* 3.2.3, Fig. 35a) through a rigid coupling (*Sec.* 3.2.5, Fig. 37). The second *DC* motor was used to drive the *ASRS* components in the *y*-axis (second direction), Fig. 41b. The motor is connected to the timing pulley (*Sec.* 3.1.14, Fig. 31) through a gear train (*Sec.* 3.2.3, Fig. 35c).

(a) *DC* Motor for driving in the *x*-axis.

(b) *DC* Motor for driving in the *y*-axis.

Figure 41. DC motors installed in ASRS model.

4.1.2. Servomotors

A servomotor is a motor which forms part of a servomechanism. The servomotor is paired with some type of encoder to provide position/speed feedback. This feedback loop is used to provide precise control of the mechanical degree of freedom driven by the motor. Servomotors have a range of 0°-180°. For the serious designer of heavy duty and speed orientated equipment, servos offer the best solution [28].

In the present project two servomotors were used, Fig. 42. A servomotor was used to drive the *ASRS* components in the *z*-axis (third direction), Fig. 43 (black motor). The servomotor rests on the H-holder (*Sec.* 3.1.11, Fig. 23). Another servomotor, Fig. 42 (orange motor), controls the movement of the fork plate (*Sec.* 3.1.8, Fig. 17) and the fork arms (*Sec.* 3.1.9, Fig. 19) in a complete 180° as can be seen in Fig. 44.

Figure 42. Servomotors installed in ASRS model for motion in z-axis (third direction).

Figure 43. Forks moves in z-direction, right and left.

Figure 44. Forks turn from 0° to 180°.

4.2. Sensors

In the present project, two types of sensors were used. A Light Dependent Resistor (*LDR*) sensor was used to detect the pointing light. The other one is micro-switch sensor for cutting the circuit at the end of guide rod.

4.2.1. Light Dependent Resistor (LDR)

Light Dependent Resistor (*LDR*), Fig. 45, is also known as the general purpose photoconductive cell. It is a type of semiconductor and its conductivity changes with proportional change in the intensity of light. A photo resistor or light dependent resistor (*LDR*) is a resistor whose resistance decreases with increasing incident light intensity; in other words, it exhibits photoconductivity. *LDR*s are very useful especially in light/dark sensor circuits [30].

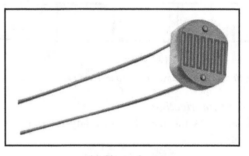

(a) Closer view.

(b) *LDR* installed in *ASRS*.

Figure 45. Light dependent resistor (LDR) [29].

A number of 6 *LDR*s were used in the present project. A beam of red light is emitted from the *ASRS* upper carriage. The carriage stops at the appropriate location when the light beam hits the corresponding *LDR*.

Figure 46 explains how the *LDR* is utilized in the present project. At the beginning, the upper carriage stops at the "Pick-and-deposit station" when the light beam hits the corresponding *LDR*, Fig. 46a. Then, the carriage moves horizontally between the two cupboards until it reaches the following corresponding *LDR* which is fixed on a column next to the target compartment, Fig. 46b. At that position, the carriage moves vertically looking for another *LDR* at a point next to the target compartment, Fig. 46c. At that point, the carriage stops and the fork plate moves horizontally to deliver the load in the target compartment, Fig. 46d.

(a) *LDR* at "Pick-and-deposit station".

(b) *LDR* at a column next to the target compartment.

(c) *LDR* at a point next to the target compartment.

(d) Carriage unloading at the target compartment.

Figure 46. *Steps of operation of SARS using LDRs.*

4.2.2. Micro-Switch

It is also called miniature snap-action switch, Fig. 47. It is an electric switch that is actuated by very little physical force, through the use of a tipping-point mechanism, sometimes called an "over-center" mechanism. Switching happens reliably at specific and repeatable positions of the actuator, which is not necessarily true of other mechanisms. They are very common due to their low cost and durability, greater than 1 million cycles and up to 10 million cycles for heavy duty models. This durability is a natural consequence of the design. The defining feature of micro switches is that a relatively small movement at the actuator button produces a relatively large movement at the electrical contacts, which occurs at high speed (regardless of the speed of actuation). Most successful designs also exhibit hysteresis, meaning that a small reversal of the actuator is insufficient to reverse the contacts; there must be a significant movement in the opposite direction. Both of these characteristics help to achieve a clean and reliable interruption to the switched circuit [32].

(a) Closer view.

(b) Micro-switch installed in *ASRS*.

Figure 47. *Micro-switch [31].*

In the present project, 6 micro-switches were used to limit the movement of the upper and lower carriages in all the three directions (*x*-, *y*-, and *z*-axis).

4.3. Electronic Control Circuit

An electronic circuit was designed and implemented based on experience of others [33]. The circuit consists of main components and other auxiliary components. The main components, including microcontroller, relays, and H-bridges, are illustrated in the following section. Auxiliary components include *IC*s, switches, fuses, wires, *etc.*

4.3.1. Main Components

(i) PIC Microcontroller

The *PIC* microcontroller, Fig. 48, is a family of modified Harvard architecture microcontrollers made by Microchip Technology, derived from the *PIC1650* originally developed by General Instrument's Microelectronics Division. The name *PIC* initially referred to "Peripheral Interface Controller". *PIC*s are popular with both industrial developers and hobbyists alike due to their low cost, wide availability, large user base, extensive collection of application notes, availability of low cost or free development tools, and serial programming (and re-programming with flash memory) capability [35].

In the present circuit, the microcontroller was used as the main controller for the *SARS* machine. It receives the input from sensors and sends the output to the motors depending on the input signals from the sensors.

Figure 48. *PIC Microcontroller [34].*

(ii) Relay

A relay is an electrically operated switch. Many relays use an electromagnet to operate a switching mechanism mechanically, but other operating principles are also used, Fig. 49. Relays are used where it is necessary to control a circuit by a low-power signal (with complete electrical isolation between control and controlled circuits), or where several circuits must be controlled by one signal [37].

In the present circuit, a group of six similar relays were used, Fig. 49.

Figure 49. *Relay [36].*

(iii) H-bridge

An H-bridge, Fig. 50, is an electronic circuit that enables a voltage to be applied across a load in either direction. These circuits are often used in robotics and other applications to allow *DC* motors to run forwards and backwards. H-bridges are available as integrated circuits, or can be built from discrete components [39].

In the present project, H-bridges were used to switch the rotational direction of the motors. Thus, moves the *ASRS* carriages in either direction.

Figure 50. *H-bridge [38].*

4.3.2. Programming Language

Assembly language is a low-level programming language for computers, microprocessors, microcontrollers, and other integrated circuits. It implements a symbolic representation of the binary machine codes and other constants needed to program a given *CPU* architecture. An assembly language is thus specific to certain physical (or virtual) computer architecture. This is in contrast to most high-level programming languages, which, ideally, are portable. A utility program called an assembler is used to translate assembly language statements into the target computer's machine code. The assembler performs a more or less isomorphic translation (one-to-one mapping) from mnemonic statements into machine instructions and data. This is in contrast with high-level languages, in which a single statement generally results in many machine instructions. Many sophisticated assemblers offer additional mechanisms to facilitate program development, control the assembly process, and aid debugging. In particular, most modern assemblers include a macro facility, and are called macro assemblers [40].

In the present project, the assembly language was used to program the microcontroller (*Sec.* 4.3.1.1, Fig. 48).

4.3.3. Power Supply

A power supply, Fig. 51, was used for electrical supply of the circuit. The input to the power supply is *AC* 220 *V*. The output of the power supply is *DC* that ranges from 3 to 12 *V*. This output is utilized by the different components of the circuit and the four motors.

Figure 51. *Power supply.*

5. Actual Operation of *ASRS*

5.1. Test Case

The present control circuit was implemented in conjunction with the *ASRS* model to ensure that the design, manufacturing, programming were correct. Figure 52 shows the control circuit in operation. The objective is to move the load from the "Pick-and-deposit station" to the target compartment. To ensure that the *ASRS* model operates correctly, the control circuit was programmed to move the load (marked in "Blue") to a certain compartment, on one of the two cupboards, that is marked in "Red" as shown in Fig. 53.

When the microcontroller, which is connected to the sensors and push-button switches, receives the operation order, it sends an appropriate signal to the corresponding relay that in turn controls an H-bridge. Thus, the H-bridge controls the related motor to rotate either in clockwise or anti-clockwise directions. The operation continues sequentially as programmed until the packer delivers the unit-load into place.

The test was carried out very successfully as was shown in step-by-step sequence in *Sec.* 4.2.1 and illustrated in Fig. 46.

(a) Overall View.

(b) Closer view.

Figure 52. *Electronic control circuit of ASRS model.*

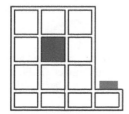

Figure 53. *Locations of load (blue) and target compartment (red).*

5.2. Faced Problems

During the design and manufacturing phases of this project, the students faced some practical problems that may be listed as follows:

1. Deciding the most suitable design as there is a big variety of *ASRS*.
2. Material selection as there is somehow limitations in the local market.
3. Shortage in workshops that provide accurate machine tooling.
4. Standard gears with small sizes are not available in the local market. They had to be specifically manufactured.
5. Fixing the motors on *ASRS* model due to their small sizes.
6. Arranging the big number of connecting and control wires.

6. Proposed Firefighting Utilizing *ASRS*

6.1. Firefighting in Warehouses

The traditional method for firefighting (suppression) in *ASRS* warehouses is the sprinklers. Sprinkler systems are widely known and they have the advantage of being relatively easy to supply and construct in site. However, using them in *ASRS* warehouses have some disadvantages [41-43]: (*i*) Huge amount of water is needed to supply the sprinkler systems. (*ii*) Water destroys the stock if it is consisted of cartooned items. (*iii*) *ASRS* warehouses are usually so dense that firefighters' accessibility is noticeably low. (*iv*) Dense (high-storage capacity) warehouses prevent the sprinkler water from reaching the lower levels of pallets.

Thus, an alternative technique may be used instead of sprinklers or besides them. Based on the practical experience of this project, a proposed method is explained in the following section.

6.2. Proposed Firefighting Technique

The proposed technique depends on using the *ASRS* to deliver a package (unit-load) of dry chemical powder. Thus, immediate fire-extinction is expected by chemical reaction in an instant. Usually, dry powder is used in situations where water would have a negative impact on fires or substances stored near a fire source. It is preferable that the powder is suitable for a wide range of fire classes (A, B, C, D) [44-46].

The proposed technique is more suitable to the case when the warehouse contains valuable/precious stock pieces. Examples of such warehouses are *ASRS* university libraries [47-50], libraries of motion pictures and television programs [51], *etc.*

The idea of the technique can be summarized in the following points:

1. Suitable set of fire detectors [52-54] are used such that a detector is placed in each compartment of the warehouse.
2. Dry powder packages are prepared in certain section of

the warehouse near the "Pick-and-deposit station".

3. The size of the powder packages fits the compartment size.

4. The material of the powder package is to be chosen to tear out immediately when exposed to fire heat/flame.

5. All the fire detectors are connected to the electronic control circuit (*Sec.* 4.3).

6. When a detector of certain compartment sends a fire signal to the electronic circuit, a fire alarm operates. Thus, operators put one of the powder packages in "Pick-and-deposit station".

7. Then, the microcontroller gives the pre-programmed orders to the *ASRS* to carry the powder package from the "Pick-and-deposit station" to the target compartment (on-fire compartment).

8. This procedure may continue several times till the fire is completely extinct.

Unfortunately, this technique was not tested experimentally in this project due to some practical difficulties. However, it is planned to be implemented as a further extension of the present project.

7. Conclusions

Based on the above illustrations and test observations, the following points can be stated:

1. The present *ASRS* model operated successfully and performed exactly the target task.

2. The students gained a really big experience in design, material selection, and manufacturing of such complicated mechanical system.

3. Although the electronic circuit was simple, it was implemented successfully.

4. There are some practical problems that were faced by the students. These problems should be put into consideration when considering similar types of projects.

5. The strategy of using the *ASRS* packer for firefighting is an interesting trend of thinking out of the box.

6. Economic feasibility study may be needed when considering the manufacturing of a full-scale prototype.

7. Although, the proposed technique of firefighting was not tested experimentally in this project due to some practical difficulties, it seems very promising.

Acknowledgement

The author would like to acknowledge Engs. M. A. Alhazmi, A. A. Imam, M. M. Olaqi, A. Y. Bawzer, and W. M. Sindi as being members of the team of the B.Sc. Graduation project of the present work under the author's supervision.

Nomenclature

AC	Alternating Current
ASRS	Automated Storage and Retrieval System (*AS/RS*)
CPU	Central Processing Unit
DC	Direct Current
IC	Integrated circuit
LDR	Light Dependent Resistor
SR	Storage and Retrieval
PIC	Peripheral Interface Controller
PTFE	Polytetrafluoroethylene (Teflon)

References

[1] J. Rhee, C. Oyamot, D. Parent, L. Speer, A. Basu, and L. Gerston, "A Case Study of a Co-Instructed Multidisciplinary Senior Capstone Project in Sustainability", Advances in Engineering Education, Vol. 4, No. 2, 2014.

[2] https://eeic.osu.edu/multidisciplinary-engineering-capstone-design

[3] P. K. Imbrie, K. Haghighi, P. Wankat, and W. Oakes, "Creating a Model Multidisciplinary Engineering Program", Proceedings of the 2005 ASEE/AaeE 4th Global Colloquium, 2005.

[4] K. Wolff, and K. Luckett, "Integrating Multidisciplinary Engineering Knowledge", Teaching in Higher Education, Vol. 18, No. 1, pp. 78-92, 2013.

[5] K. Craig, and P. Voglewede, "Multidisciplinary Engineering Systems Graduate Education: Master of Engineering in Mechatronics", Transforming Engineering Education: Creating Interdisciplinary Skills for Complex Global Environments, 2010 IEEE , Dublin, 6-9 April 2010, pp. 1-14.

[6] C. Telenko, B. Camburn, and K. Wood, "Designettes: New Approaches to Multidisciplinary Engineering Design Education", Proceedings of the ASME 2014 International Design Engineering Technical Conferences & Computers and Information in Engineering Conference (IDETC/CIE 2014), Buffalo, New York, USA, August 17-20, 2014, DETC2014-35137.

[7] M. Parten, D. Vines, J. Jones, and A. Ertas, "Program for Multidisciplinary Engineering Projects", Proceedings of Frontiers in Education Conference, 1996, FIE '96. 26th Annual Conference, Salt Lake City, UT, 6-9 Nov., 1996, Vol. 3, pp. 1309-1312.

[8] F. Sahin, and W. Walter, "Multidisciplinary Microrobotics Teaching Activities in Engineering Education", Proceedings of the 2003 American Society for Engineering Education Annual Conference & Exposition 2003.

[9] J. L. Cezeaux, E. W. Haffner, and T. Keyser, "The Evolution of a Collaborative Multidisciplinary Engineering Design Experience", Proceedings of The International Conference on Engineering & Technology (The 2008 IAJC-IJME International Conference, ISBN 978-1-60643-379-9), Music City Sheraton, Nashville, TN, USA, November 17-19, 2008, Paper #094, ENG 107.

[10] P. J. Robbie, I. Baker, W. Lotko, J. P. Collier, "A Multidisciplinary Approach to Introductory Engineering Design", 38th ASEE/IEEE Frontiers in Education Conference, Saratoga Springs, NY, USA, October 22-25, 2008.

[11] http://racksandrollers.com/automated-storage-retrieval-systems-warehouse

[12] K. J. Roodbergen, and I. F. A. Vis, "A Survey of Literature on Automated Storage and Retrieval Systems", European Journal of Operational Research, Vol. 194, No. 2, pp. 343-362, 16 April 2009, DOI: 10.1016/j.ejor.2008.01.038.

[13] M. Dotoli, M. P. Fanti, and G. Iacobellis, "Comparing Deadlock Detection and Avoidance Policies in Automated Storage and Retrieval Systems", Systems, Man and Cybernetics, 2004 IEEE International Conference, 10-13 Oct. 2004, pp. 1607-1612, Vol. 2, DOI: 10.1109/ICSMC.2004.1399861

[14] Y.-H. Hu, S. Y. Huang, C. Chen, W.-J. Hsu, A. C. Toh, C. K. Loh, and T. Song, "Travel Time Analysis of a New Automated storage and Retrieval System", Computers & Operations Research, Vol. 32, pp. 1515-1544, 2005.

[15] M. Dotoli, and M. P. Fanti, "Deadlock Detection and Avoidance Strategies for Automated Storage and Retrieval Systems", Systems, Man, and Cybernetics, Part C: Applications and Reviews, IEEE Transactions, Vol. 37, No. 4, pp. 541-552, July 2007, DOI: 10.1109/TSMCC.2007.897690

[16] C. Kator, "Automated Storage and Retrieval System (AS/RS) Basics", Modern Materials Handling, August 1, 2007.

[17] G. Moon, and G.-P. Kim, "Effects of Relocation to AS/RS Storage Location Policy with Production Quantity Variation", Computers & Industrial Engineering, Vol. 40, pp. 1-13, 2001.

[18] Automated Storage & Retrieval System: Material Handling Machines, Application Bulletin, No. 0108AB9707, Square D Company, 1997.

[19] http://www.mhia.org/industrygroups/as-rs

[20] http://www.atlastechnologies.com/pdfs/ASRS.pdf

[21] http://www.westfaliausa.com/products/ASRS/documents/ASRS_Innovations_WhitePaper_2009-Web.pdf

[22] http://library.csun.edu/About/ASRS

[23] http://www.vrhandling.ch/En/LinkClick.aspx?fileticket=aVBPt8g0kGA%3d&tabid=2839&language=de-CH

[24] http://www.directindustry.com/prod/lodige/storage-systems-19118-1161081.html

[25] A. H. B. Hadzir, Development of Automated Storage and Retrieval System (AS/RS) Prototype", B.Sc. Mechanical Engineering, Faculty of Mechanical Engineering, Universiti Malaysia Pahang, December 2010: http://umpir.ump.edu.my/1751/1/Abdul_Halim_Hadzir_(_CD_5008_).pdf

[26] http://www.ignou.ac.in/upload/Unit4-55.pdf

[27] S. Shrivastava, J. Rawat, and A. Agrawal, "Controlling DC Motor Using Microcontroller (PIC16F72) with PWM", International Journal of Engineering Research, Vol. 1, No. 2, pp. 45-47, 1 Dec. 2012.

[28] http://unirobotics.co.za/store/22-servo-motors-controllers

[29] http://www.raspberrypi-spy.co.uk/2012/08/reading-analogue-sensors-with-one-gpio-pin/

[30] D.V.P. Latha, "Simulation of PLC based Smart Street Lighting Control using LDR", International Journal of Latest Trends in Engineering and Technology (IJLTET), Vol. 2, No. 4, pp. 113-121, July 2013.

[31] http://www.futureelectronics.com/en/technologies/electromechanical/switches/snap-acting/Pages/6907612-DG13B1LA.aspx

[32] http://www.langir.com/htm/cm-micro-switch.htm

[33] M. M. A. Zaki, H. M. Monier, and R. F. Rageh, "Automatic Storage Machine", B. Sc. Graduation Project, Supervisor: M. Abd-Elqawy, Computer & Systems Dept., Faculty of Eng., Zagazig Univ., Egypt, 2011.

[34] http://www.societyofrobots.com/microcontroller_tutorial.shtml

[35] http://bitstream24.com/embedded-c-programming-with-the-microchip-pic-microcontroller

[36] http://ram-e-shop.com/oscmax/catalog/index.php?cPath=29&osCsid=3319a177028ed9b332352a465b26dbcf

[37] https://www.princeton.edu/~achaney/tmve/wiki100k/docs/Relay.html

[38] http://www.electronics-lab.com/blog/?tag=h-bridge

[39] http://www.ez-robot.com/Tutorials/Hardware.aspx?id=25

[40] http://www.princeton.edu/~achaney/tmve/wiki100k/docs/Assembly_language.html

[41] D. T. Gottuk, and J. Dinaburg, Fire Detection in Warehouse Facilities, Springer Science & Business Media, 2012. ISBN: 978-1-4614-8115-7. DOI: 10.1007/978-1-4614-8115-7

[42] S. Martorano, "Sprinkler Protection for High Bay and Automated Storage in Warehouse-Type Storage Facilities", Technical Article, February, 2010: http://www.vikinggroupinc.com/techarticles/Sprinkler%20Protection%20for%20High%20Bay%20and%20Automated%20Storage%20in%20Warehouse%20Type%20Storage%20Facilities.pdf

[43] http://www.klausbruckner.com/blog/global-research-update-high-challenge-storage-protection/

[44] http://www.ruehl-ag.de/en/fire-extinguishing-agents/products/fire-fighting-dry-chemical-powders.html

[45] http://www.protech-i.jp/fire/drypowder_eg.html

[46] http://www.rosenbauer.com/en/rosenbauer-world/products/industrial-vehicles/dry-powder-fire-trucks.html?tx_rosenbauersprachmenu_pi2%5Bcountry%5D=253&tx_rosenbauersprachmenu_pi2%5Baction%5D=showCountry&tx_rosenbauersprachmenu_pi2%5Bcontroller%5D=Land&cHash=8df2dd36d587602f9facdf3917bbda18

[47] Helen Heinrich, and Eric Willis ,"Automated storage and retrieval system: a time-tested innovation", Library Management, Vol. 35, No. 6/7, pp.444 – 453, 2014. DOI: http://dx.doi.org/10.1108/LM-09-2013-0086

[48] http://library.csun.edu/About/ASRS

[49] http://library.umkc.edu/newmnl/about-robot

[50] http://library.sonoma.edu/about/building/ars/

[51] http://www.elecompack.com/casestudies/asrsparamountcs.pdf

[52] "Selection and Location of Fire Detectors". http://www.fire.org.nz/Business-Fire-Safety/UnwantedAlarms/Documents/69efc74f885837062a01ae02b0e57a39.pdf

[53] Sierra monitor corporation, Flame Detector Selection Guide, Rev. A1-11/10, 2010. http://www.sierramonitor.com/docs/pdf/flame_detector_selection.pdf

[54] L. Grosse, J. DeJong, and J. Murphy, "Risk Analysis of Residential Fire Detector Performance", August 1995. http://hrrc.arch.tamu.edu/media/cms_page_media/558/95-01R.pdf

10

Forced Convection Heat Transfer Analysis through Dimpled Surfaces with Different Arrangements

Hasibur Rahman Sardar[1], Abdul Razak Kaladgi[2, *]

[1]Department of Electronics & Communication Engineering, P.A College of Engineering, Karnataka, India
[2]Department of Mechanical Engineering, P.A College of Engineering, Karnataka, India

Email address:
hasibpace@gmail.com (H. R. Sardar), abdulkaladgi@gmail.com (A. R. Kaladgi)

Abstract: Dimples play a very important role in heat transfer enhancement of electronic cooling systems, heat exchangers etc. This work mainly deals with the experimental investigation of forced convection heat transfer over circular shaped dimples of different diameters on a flat copper plate under external laminar flow conditions. Experimental measurements on heat transfer characteristics of air (with various inlet flow rates) on a flat plate with dimples were conducted. From the obtained results, it was observed that the heat transfer coefficient and Nusselt number were high for the copper plate in which the diameter of dimples increases centrally in the direction of flow (case c) as compared to the other two cases.

Keywords: Forced Convection, Dimples, Heat Transfer, Passive Techniques

1. Introduction

There are various heat transfer applications where the use of fluid-to-gas heat exchanger is important. The issues like accurate heat transfer rate analysis, estimations of pressure drops, long-term performance and economic aspect of the equipment make the design of heat exchangers quite complicated. Also higher performance, higher heat transfer rate with minimum pumping power requirements are some of the main challenges of the heat exchanger design. Therefore, improving the heat exchanger efficiency through the enhancement techniques resulting in a considerable reduction in cost is one of main task faced by the engineers [1].Various heat transfer enhancement techniques are developed and used for heat exchanger applications over the past couple of years. Several attempts are also made to reduce the size and cost of the heat exchangers. Among these the passive techniques can be considered important one because of its wide variety of applications like in electronic cooling (heat sinks), process industries, cooling and heating in evaporators, solar air heaters, turbine airfoil cooling etc [2]. The main principle of heat transfer enhancement in passive techniques is the surface modifications such as protrusions, pin fins, and dimples. Among these, the dimples (concavities) can be considered special one as they not only enhance the heat transfer rate but also produce minimum pressure drop

penalties [3]. The dimple produces vortex pairs, induces flow separation & creates reattachment zones to increase the heat transfer. And as they do not protrude into the flow so they contribute less to the foam drag, to produce minimum pressure drop penalties [4]. Another added advantage in dimple manufacture is the removal of material which reduces cost and weight of the equipment.

Kuethe [5] can be considered as the first person to make dimples on flat surfaces to increase the heat transfer rate. According to him the dimples are expected to promote turbulent mixing in the flow, acting as vortex generator & hence increase the heat transfer rate. Afanasyev et al [6] carried an experimental to study the heat transfer characteristics of flow over a flat plate having spherical dimples and reported an increment of 30-40% in the heat transfer rate with a minimum pressure drop. Chyu et al [7] conducted an experiment to study local heat transfer coefficient distribution in a channel having dimples of spherical & tear drop type. They observed a considerable increase in the distribution of local heat transfer coefficient everywhere on these dimple surfaces as compared to flat surface. Mahmood et al [8], experimentally investigated the effect of dimples on heat transfer augmentation .They used the flow visualization techniques and concluded that the periodic nature of shedding off of vortices is the main cause of enhancement of heat transfer and is much more pronounced at the downstream rims of the dimples. Xie et al

[9] numerically investigated the heat transfer and fluid flow characteristics of teardrop dimple along with teardrop protrusion having different eccentricities. They used the K-Ɛ model to capture the turbulence effects. They concluded that the heat transfer enhancement along with energy savings are more in teardrop dimples as compared to flat surfaces.Farhad sangtarash & hosseinshokuhmand [10] conducted experimental & numerical investigation on inline & staggered arrangement of dimples on multilouvered fins to study the heat transfer and pressure drop characteristics of air through these multilouvered fin banks at varying Reynolds number. They concluded that the augumentation of heat transfer was more in staggered arrangements as compared to online arrangement.

From the literature above, it is abundantly clear that dimples or vortex generators and the vortex heat transfer enhancement (VHTE) techniques have a high potential to increase the heat transfer rate along with the production of lower pressure drop penalties. The other advantages are:

a. Fouling rate reduction b. Cost reduction c. weight reduction etc [11], however, much of the research work either experimental or numerical is on spherical dimples of uniform diameter [7, 12]. It is also seen that most of the research is confined to flow in the channel i.e. Internal flow, with a very few studies on external flow [12]. So the main focus of this experimental work is to study the effect of circular dimples of various diameters under external laminar flow conditions

2. Experimental Setup

The prime objective of the present work was to study experimentally the heat transfer enhancement through dimple surfaces of different diameters on a flat plate using force convection technique. For this to be possible we required a forced convection setup which was fabricated as required. The fabricated setup is shown below.

Figure 1. Experimental setup.

Figure 2. Schematic representation.

The main components of the test apparatus are a test plates of dimensions 100x100x2 mm, a calibrated orifice flow meter, Strip plate heater with capacity of 100 watts, Dimmer stat, Digital temperature, voltmeter, and ammeter with J type thermocouple, a gate valve, and a centrifugal blower. In this work Strip plate heater was fabricated to provide heat input

to the test surface. The provision was made to fix the heater at the base of the each test plate in a rectangular channel connect to the blower through a orifice plate with pipe at inlet and to the atmosphere at outlet. U-TUBE manometer was connected across the orifice plate to indicate the pressure difference in terms of centimeters of water column difference. A PVC pipe was used to connect the blower outlet to the rectangular duct. Next to the blower outlet, flow regulating valve was connected to the pipe to regulate the air flow.

Table 1. Components and Specifications.

Components	Specification
Test plate	10x10x2 cm copper plates
Blower	110W, 0.4BHP, 280rpm
Heater	100W, 4"x4"
Dimmer stat	6A,230V
Digital Temperature Indicator	6 channel,1200^{0}C, 230V
Orifice plate	12mm dia.
Manometer	"U-tube" glass manometer
Casing	A wooden casing of size of 8"x8" and 2feet long.
Thermocouple	K-Type, 300^{0}C, 1m long.
Digital Multi-meter	Voltmeter, Ammeter

3. Results and Discussion

Experiments were conducted on copper test plates with circular dimples of different diameters. The dimples were arranged in a staggered fashion with different arrangements like:

Case a. Gradual Increase in the diameter of dimples in the left & right columns of the plate in the direction flow.

Case b. Gradual Decrease in the diameter of dimples in the left & right columns of the plate in the directionflow (reverse case).

Case c. centrallyincreasing the diameter of dimples in the direction flow& maintaining the left & right columnwith constant diameter dimples.

The data obtained were used to find heat transfer parameters like Nusselt number, heat transfer coefficient, and heat transfer rate. And the experimental findings have been plotted in the form of graphs, mainly
- Nusselt number(Nu) vs Reynolds number(Re)
- Heat transfer coefficient(h) vs Reynolds number(Re)
- Heat transfer rate Q vs Reynolds number(Re)

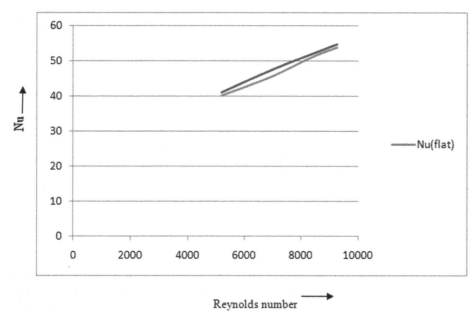

Figure 3. *Variation of Nusselt number with Reynolds number(case.a).*

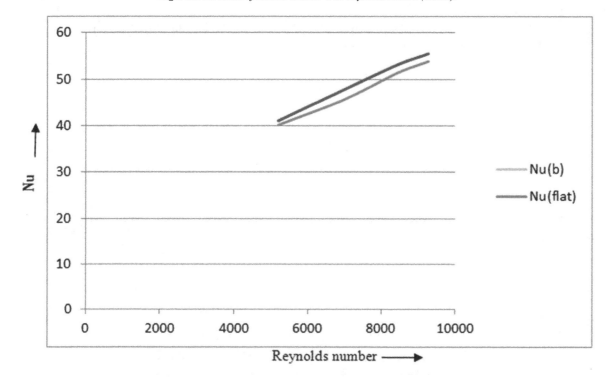

Figure 4. *Variation of Nusselt number with Reynolds number(case.b.).*

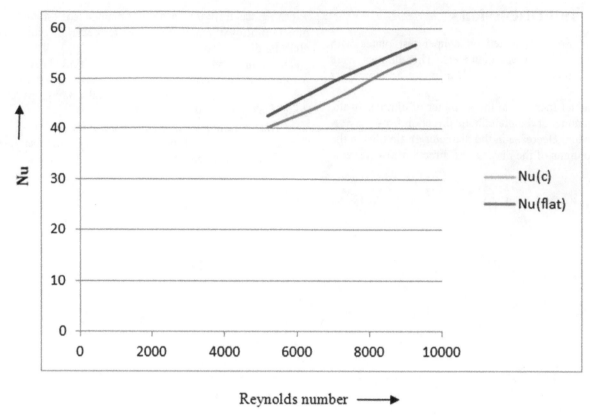

Figure 5. Variation of Nusselt number with Reynolds number(case.c.).

Figure 3, 4, 5 shows variation of Nusselt number 'Nu' with Reynolds number for the three cases considered. It is obvious that the 'Nu' increases as Reynolds number increases due to direct flow impingement on the downstream boundary and strengthened flow mixing by vortices at the downstream [3,13]. The formation of vortex pairs which are periodically shed off from the dimples, a large up wash regions with somefluids coming out from the central regions of the dimples, from vortex pairs & near dimple diagonals are the main causes of enhancement of Nusselt number & is more pronounced near the downstream rims of the dimples [8].It can also be seen that the variation in the Nusselt number is gradual with Reynolds number as expected [14, 15].

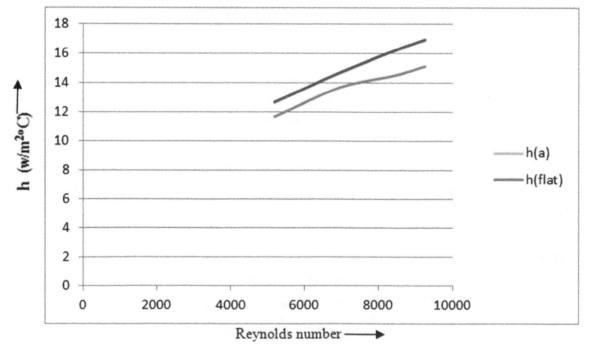

Figure 6. Variation of Heat transfer coefficient with Reynolds number(case.a.).

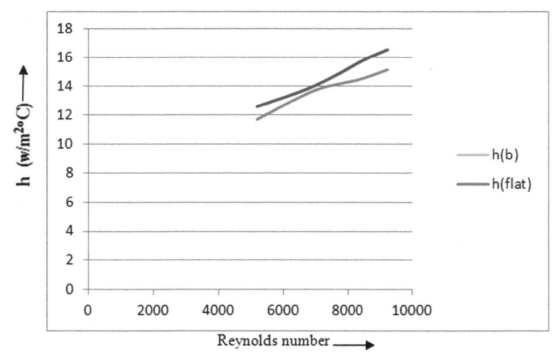

Figure 7. Variation of Heat transfer coefficient with Reynolds number(case.b.).

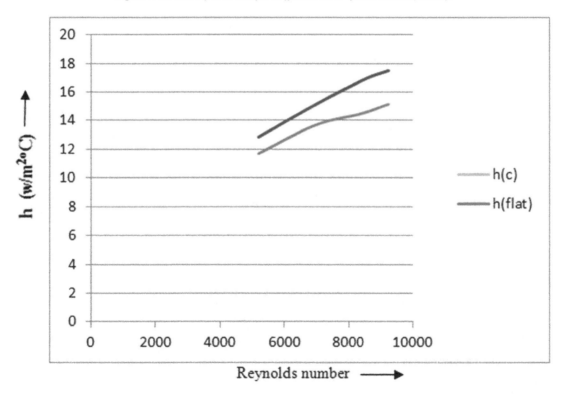

Figure 8. Variation of Heat transfer coefficient with Reynolds number(case.c.).

Figure 6, 7, 8 shows the variation of heat transfer coefficient 'h' with Reynolds number 'Re' for the various cases considered. It is obvious that 'h' increases with 'Re' as expected because the development of the thermal boundary layer is delayed or disrupted & hence enhances the local heat transfer in the reattachment region and wake region and increases the heat transfer coefficient [3].

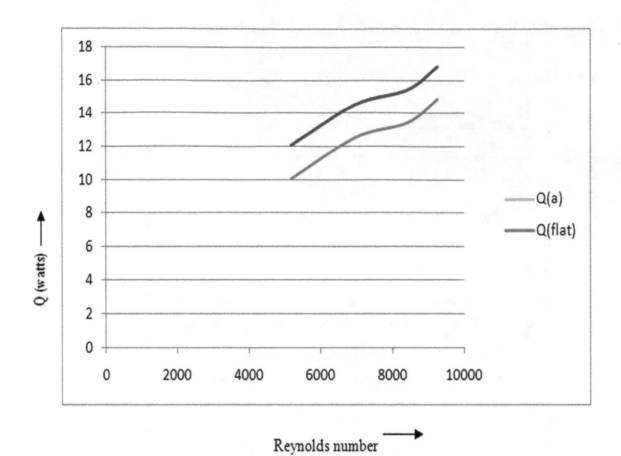

Figure 9. *Variation of Heat transfer rate with Reynolds number(case.a.).*

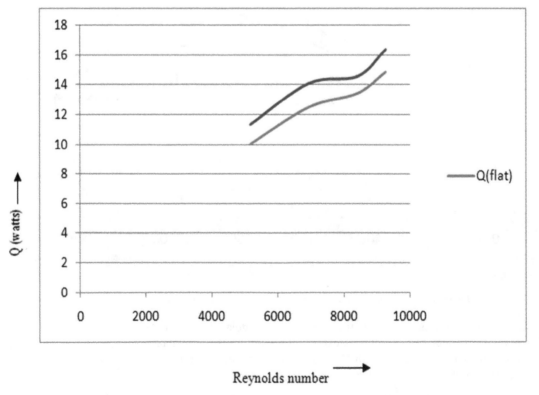

Figure 10. *Variation of Heat transfer rate with Reynolds number(case.b.).*

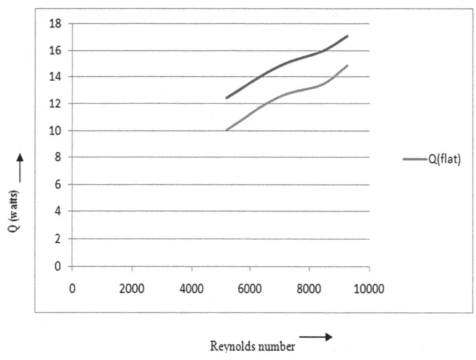

Figure 11. Variation of Heat transfer rate with Reynolds number(case.c.).

Figure 9, 10, 11 shows variation of Heat transfer rate 'Q' with Reynolds number 'Re' for the various cases considered. It can be seen that again 'Q' increases as 'Re' increases in all the three cases. It can also be seen that 'Q' is very much higher for case c(dimples diameter decreasing centrally) because of increased flow area as compared to the other two cases. It can also be seen that for high Reynolds number the 'Q' curve for case c is higher than the curve of case b & case c. Hence it can be concluded that case c helps in better enhancing the heat transfer compared to case a & case b.

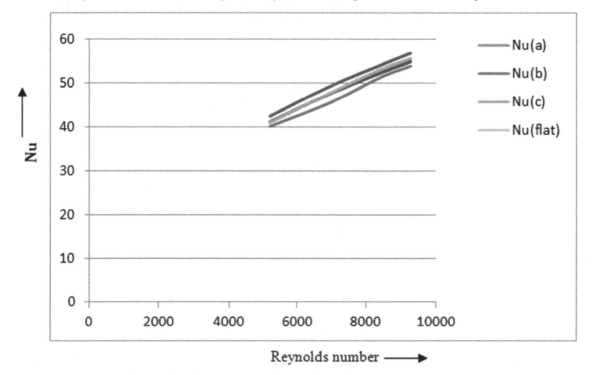

Figure 12. Variation of Nusselt number with Reynolds number.

Figure 12 shows comparison of Nusselt number 'Nu' with Reynolds number 'Re' for the all the three cases considered. It can be seen that 'Nu' increases as 'Re' increases in all the three cases. It can also be seen that the variation for the first two cases is very less compared to the third case may be due the fact that the dimple diameter is not increased or decreased centrally where the pronounce effect of heat transfer will   occur so the variation is negligibly small.

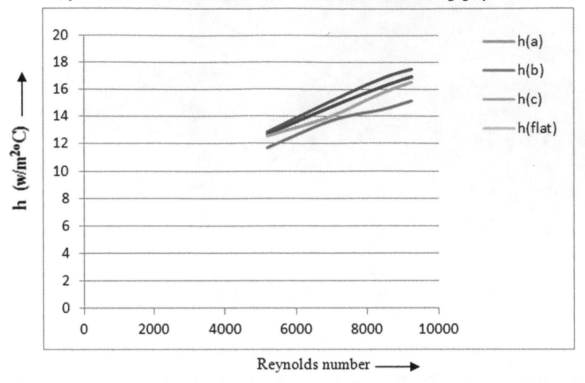

Figure 13. Variation of Heat transfer coefficient with Reynolds number.

Figure 13 shows the comparison of the variation of heat transfer coefficient 'h' with Reynolds number 'Re' for the various cases considered. It is obvious that 'h' increases with 'Re' as expected and it is also observed that heat transfer coefficient is high for the third case (case of dimple diameter decreasing centrally) due to higher heat transfer rate occurring at the central region where the fluid flow rate is highest compared to other two cases.

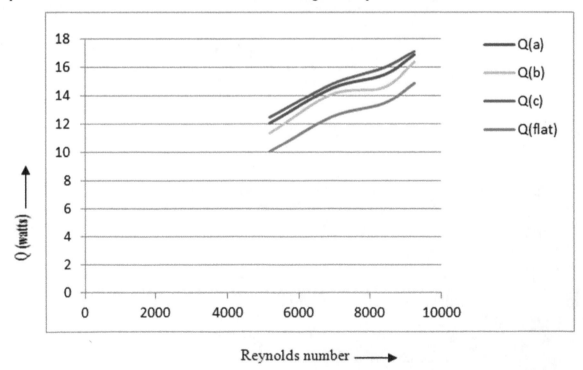

Figure 14. Variation of Heat transfer rate with Reynolds number.

Figure 14 showsthe variation of Heat transfer rate 'Q' with Reynolds number 'Re' for the various cases considered. It can be seen that again 'Q' increases as 'Re' increases in all the three cases. It can also be seen that 'Q' is very much lower for case b & is highest for case c because of increased flow rate as compared to the other two cases. It can also be seen that for high Reynolds number the 'Q' curve for case c is higher than the curve of case b & case c. Hence it can be concluded that case c helps in better enhancing the heat transfer compared to case a & case b.

4. Conclusion

In this experimental work an investigation of the effect of air flow over a flat plate with different diameter dimples on the flat plate is carried out. The main conclusions of the work were:

- Nusselt number increases with Reynolds number for all the three cases of dimple arrangement considered due to direct flow impingement on the downstream boundary and strengthened flow mixing by the vortices at the downstream.
- Case 'c' dimple arrangement has highest Nusselt number because of the strong flow impingement on the upstream side of these dimples. Case a & b dimple arrangement gives nearly the same value of Nusselt number.
- Heat transfer coefficient increases with Reynolds number for all the three cases of dimples arrangement considered due to the disruption of the thermal boundary layer development & hence enhance the local heat transfer in the reattachment and wake regions.
- Case 'c' dimple arrangement gives slightly higher value of heat transfer coefficient as compared to case 'a' & 'b' dimples.
- Case 'c' dimple arrangement has better heat transfer enhancing capacity as compared to case 'a' & 'b' because of the increased flow rate in this type of dimple arrangement. However, the Augmentation depends on the configuration [12].

References

[1] Dewan, A., Mahanta, P., Raju,K.S., and Kumar.P.S.,Review of passive heat transfer augmentation techniques. Proc. Instn Mech. Engrs, Part A: J. Power and Energy, Vol. 218, pp. 509–527, 2004.

[2] Liu ,S, and Sakr ,M.A., comprehensive review on passive heat transfer enhancements in pipe exchangers. Renewable and Sustainable Energy Reviews, Vol.19 pp. 64–81, 2013.

[3] Zhang, D., Zheng, L., Xie, G., and Xie, Y.,An Experimental Study on Heat Transfer enhancement of Non-Newtonian Fluid in a Rectangular Channel with Dimples/Protrusions, Transactions of the ASME, Vol. 136, pp.021005-10, 2014.

[4] Beves, C.C., Barber, T.J., and Leonardi,E., An Investigation of Flow over Two-Dimensional Circular Cavity. In 15th Australasian Fluid Mechanics Conference, the University of Sydney, Australia, pp.13-17, 2004.

[5] Kuethe A. M., Boundary Layer Control of Flow Separation and Heat Exchange. US Patent No. 1191, 1970.

[6] Afanasyev, V. N., Chudnovsky, Y. P., Leontiev, A. I., and Roganov, P.S., Turbulent flow friction and heat transfer characteristics for spherical cavities on a flat plate. Experimental Thermal Fluid Science, Vol. 7, Issue 1, pp. 1–8, 1993.

[7] Chyu, M. K., Yu, Y., Ding, H., Downs, J. P., and Soechting, F.O., Concavity enhanced heat transfer in an internal cooling passage. In Orlando international Gs Turbine & Aero engine Congress & Exhibition, Proceedings of the 1997(ASME paper 97-GT-437), 1997.

[8] Mahmood, G. I., Hill, M. L., Nelson, D. L., Ligrani, P. M., Moon, H.K., and Glezer, B., Local heat transfer and flow structure on and above a dimpled surface in a channel. J Turbomach, Vol.123, Issue 1, pp: 115–23, 2001.

[9] Yonghui Xie, Huancheng Qu ,Di Zhang ,Numerical investigation of flow and heat transfer in rectangular channel with teardrop dimple/protrusion. International Journal of Heat and Mass Transfer ,Vol 84 pp.486–496,2015

[10] Farhad sangtarash & hosseinshokuhmand ,Experimental and numerical investigation of the heat transfer augmentation and pressure drop in simple, dimpled and perforated dimpled louver fin banks with an in-line or staggered arrangement ,Applied Thermal Engineering Vol 82 ,pp194-205, 2015.

[11] Gadhave, G., and Kumar. P., Enhancement of forced Convection Heat Transfer over Dimple Surface-Review. International Multidisciplinary e - Journal .Vol-1, Issue-2, pp.51-57, 2012

[12] Katkhaw, N., Vorayos, N., Kiatsiriroat, T., Khunatorn, Y., Bunturat, D., and Nuntaphan., A. Heat transfer behavior of flat plate having 450 ellipsoidal dimpled surfaces. Case Studies in Thermal Engineering, vol.2, pp. 67–74, 2014.

[13] Patel,I.H ., and Borse ,S.H. Experimental investigation of heat transfer enhancement over the dimpled surface. International Journal of Engineering Science and Technology, Vol.4, Issue 6, pp.3666–3672, 2012.

[14] Faheem Akhtar, Abdul Razak R Kaladgi and Mohammed Samee, Heat transfer augmentation using dimples in forced convection -an experimental approach. Int. J. Mech. Eng. & Rob. Res. Vol 4,Issue 1 ,pp 150-153,2015

[15] Faheem Akhtar, Abdul Razak R Kaladgi and Mohammed Samee, Heat transfer enhancement using dimple surfaces under natural convection—an experimental study, Int. J. Mech. Eng. & Rob. Res. Vol 4,Issue 1 ,pp 173-175,2015

11

Single and multicomponent droplet models for spray applications

Shah Shahood Alam[1], Ahtisham Ahmad Nizami[1], Tariq Aziz[2]

[1]Pollution and Combustion Engineering Lab. Department of Mechanical Engineering, Aligarh Muslim University, Aligarh-202002, U.P. India
[2]Department of Applied Mathematics, Aligarh Muslim University, Aligarh-202002, U.P. India

Email address:

sshahood2004@yahoo.co.in (S. S. Alam)

Abstract: An unsteady, spherically symmetric, single component, diffusion controlled gas phase droplet combustion model was developed first, by solving numerically the time dependent equations of energy and species. Results indicated that flame to droplet diameter ratio (flame standoff ratio) increased throughout the droplet burning period, its value being much smaller than that of the quasi-steady case, where it assumes a large constant value. Effects of fuels on important combustion characteristics suggested that combustion parameters were influenced primarily by the fuel boiling point. Droplet mass burning rate variation was smallest for ethanol in comparison with methyl linoleate (biodiesel) and n-heptane. Also, effects of fuels on CO, NO, CO_2, and H_2O concentrations were determined from the point of view of getting a qualitative trend.For multicomponent spherical combustion of a heptane-dodecane droplet, it was observed that the mass fraction of heptane decreased abruptly to a minimum value as the droplet surface was approached. For a $200 \mu m$ hexane-decane droplet (at its boiling point), vaporising in conditions of 1 atm and 1000K with $Le_f = 10$, it was observed that mixing of air and fuel vapour resulted in a higher concentration of hexane at the droplet surface at the end of droplet lifetime thereby altering the vaporisation behaviour. Other conditions remaining same, an increase in Lewis number resulted in a higher mass fraction of hexane being present at the droplet surface. A detailed multicomponent (MC) droplet vaporisation model (diffusion limit model with convection and no internal liquid circulation) was also evolved by numerically solving the transient-diffusive equations of species and energy for a $280 \mu m$ (heptane-dodecane) droplet vaporising at 1 atm and 1000 K wi*th* $Re_\sigma = 100$, and $Le_f = 10$. The present MC model was compared with other existing models and was found to be simpler and quite accurate. The submodels developed in the present work can be implemented in spray analysis.

Keywords: Single and Multi-Component Droplet Models, Numerical Technique, Simplified Approach, Different Fuels, Spray Combustion Application

1. Introduction

The subject of 'Liquid Droplet Combustion' serves as the basic step for understanding the complex process of spray combustion prevalent in diesel engines, industrial boilers and furnaces, liquid rockets and gas turbines. In these devices, liquid fuels are atomised into fine sprays, in order to increase the surface area per volume so as to promote evaporation.

An isolated droplet combustion under microgravity (near zero gravity) condition is an ideal situation for studying liquid droplet combustion phenomenon. It leads to a simplified, one dimensional solution approach of the droplet combustion model' (a spherical liquid fuel droplet surrounded by a concentric, spherically symmetric flame).

For the present work, different droplet evaporation/combustion models are evolved with respect to single/pure component and multicomponent fuels. At first, a comprehensive, single component, spherically symmetric gas phase droplet combustion model is developed and tested for important combustion and emission characteristics. After that, simplified droplet vaporisation and combustion models are developed followed by a more detailed multicomponent droplet vaporisation model. The results obtained are compared with the existing experimental/modelling results of other authors for the same conditions.

Some noteworthy contributions related to single component, spherically symmetric, liquid droplet combustion are mentioned below. Kumagai and co-workers [1,2,3] were pioneers in conducting spherically symmetric droplet combustion experiments in microgravity conditions through drop towers, capturing the flame movement and further showed that F/D (also called flame standoff ratio) increases throughout the droplet burning history, thereby signifying that droplet combustion is transient phenomenon.

Waldman [4] and Ulzama and Specht [5] used analytical procedure whereas Puri and Libby [6] and King [7] employed numerical techniques in developing spherically symmetric droplet combustion models. The results of these authors were mainly confined to the observations that unlike quasi-steady case, flame is not stationary and flame to droplet diameter ratio increases throughout the droplet burning period.

A study on convection was carried out by Yang and Wong [8] who investigated the effect of heat conduction through the support fibre on a droplet vaporising in a weak convective field. Another aspect related to convection is the presence of internal circulation within the droplet. Law, C.K [9] introduced the 'infinite diffusivity' or 'batch distillation' model which assumed internal circulation within the droplet. Droplet temperature and concentrations were assumed spatially uniform but temporally varying. It was suggested that the more volatile substance will vaporise from the droplet surface leaving only the less volatile material to vaporise slowly.

In the absence of internal circulation, the infinite diffusivity model was found to be inappropriate. For such conditions Landis and Mills [10] carried out numerical analysis to solve the coupled heat and mass transfer problem for a vaporising spherically symmetric, miscible bicomponent droplet. This model in literature is termed as 'diffusion limit' model.

Law, C.K [11] generalised the formulation of Landis and Mills and suggested that regressing droplet surface problems are only amenable to numerical solutions. Tong and Sirignano [12] devised a simplified vortex model which required less computing time than the more detailed model of Lara-Urbaneja and Sirignano [13].

In an experimental investigation, Aldred et al. [14] used the steady state burning of n-heptane wetted ceramic spheres for measuring the flame structure and composition profiles for the flame corresponding to 9.2 mm diameter sphere. Their results indicated that oxidizer from the ambient atmosphere and fuel vapour from the droplet surface diffuse towards each other to form the flame where the products of combustion are formed and the flame temperature is highest.

Marchese and Dryer [15] considered detailed chemical kinetic modelling for the time dependent burning of isolated, spherically symmetric liquid droplets of methanol and water using a finite element chemically reacting flow model. It was noted that a favourable comparison occurred with microgravity droplet tower experiment if internal liquid circulation was included which could be caused by droplet generation/deployment techniques. However, significant deviations from the quasi-steady d^2-law were observed.

A liquid composed of a multitude of chemical species is called a multicomponent (MC) liquid. The overwhelming majority of liquids used for power production are MC liquids; these include gasoline, diesel fuel and kerosene.

Much of the earlier studies on droplet combustion used pure fuels. Multicomponent effects were not considered to be serious for the reason that the requirements of combustor efficiency and emission were generally not stringent. A review of the existing literature reveals that multicomponent droplet studies constitute only a relatively small fraction of the available literature.

However, recent developments in engine design and fuel formulation indicate that multicomponent effects will become progressively more important in the utilisation of liquid fuels. Combustion processes within engine will be more tightly controlled to further improve efficiency and reduce emissions.

The impetus for continued research in this field comes from the search for alternative sources of fuel like vegetable oils and better ways of fuel utilisation in the face of increasing demand and dwindling oil reserves. Also of major concern are the problems of combustion related pollution and the use of combustion for disposal of hazardous wastes. In the process, fuels developed are blends of several components.

There also exists considerable interest in the utilisation of such hybrid fuels as water/oil emulsions, alcohol/oil solutions and emulsions, and coal/oil mixtures. The widely different physical and chemical properties of the constituents of these hybrid fuels necessitate consideration of multicomponent effects in an essential way.

To understand heterogeneous multicomponent fuel combustion either as a droplet or in some other form (e.g. pool burning), the following factors have to be considered:

The relative concentrations and volatility of the liquid constituents, as would be expected. The miscibility of the liquid constituents. This controls the phase change characteristics. The internal circulation which influences the rate with which the liquid components can be brought to the surface where vaporisation takes place.

Liquid phase mass diffusion is much slower than liquid phase heat diffusion so that thin diffusion layers can occur near the surface, especially at high ambient temperatures at which the surface regression rate is large. The more volatile substances tend to vaporise faster at first until their surface concentration values are diminished and further vaporisation of those quantities becomes liquid phase mass diffusion controlled [16].

It is then obvious that not only the fuel vaporisation process, but also strongly kinetically dependent gas phase combustion phenomena such as ignition, extinction and pollution formation will depend sensitively on composition of the liquid fuel and how its vaporisation is modelled [17].

These differences have been attributed to transient liquid mass transport in the droplet interior, volatility differential between the constituent fuels, phase equilibrium at the droplet surface, and thermo transport properties that are functions of mixture compositions, temperature and pressure [18].

Here, different components vaporise at different rates, creating concentration gradients in the liquid phase and causing liquid phase mass diffusion. The theory requires coupled solutions of liquid phase species continuity equations, multicomponent phase-equilibrium relations, typically Roult's law and gas phase multicomponent energy and species continuity equations [19].

Feeling a need for an analytical solution, Law and Law [17] derived a simplified approximate analytical solution for quasi-steady, spherically symmetric, liquid phase mass diffusion controlled vaporisation and combustion of multicomponent fuel droplets for the case where liquid phase species diffusion was slow compared to droplet surface regression rate. An ideal solution behaviour was assumed.

A unique feature of the diffusion dominated droplet vaporisation mechanism was the possible attainment of approximately steady state temperature and concentration profiles within the droplet, which then led to a steady state vaporisation rate. Based on this concept, Law and Law [17] formulated a d^2-law model for multicomponent droplet vaporisation and combustion. It was noted that, the mass flux fraction or the fractional vaporisation rate ε_m was propotional to the initial liquid phase mass fraction of that species prior to vaporisation.

Their solution allowed direct evaluation of all combustion properties of interest, including the liquid phase composition profiles, once the droplet surface temperature was determined iteratively. Therefore utilisation of the their multicomponent d^2-law is almost as simple as the classical pure component d^2-law.

Tong and Sirignano [12] analysed the problem of transient vaporisation of a multicomponent droplet in a hot convective environment. The model accounted for the liquid phase internal circulation and quasi-steady, axisymmetric gas phase convection. Essentially it was called the simplified vortex model for the liquid phase (which was basically a diffusion limit model with axisymmetry rather than spherical symmetry) and a simplified, quasi-steady, axisymmetric convective model for the gas phase.

The objective of the study were (i) to develop an algorithm for multicomponent droplet vaporisation simple enough to be feasibly incorporated into a complete spray combustion analysis and yet be accountable for important physics, (ii) comparison of the developed model with existing models, and (iii) to compare the different models with the available experimental data.

Lerner et al. [20] conducted experiments for measuring overall vaporisation rates, droplet composition and droplet trajectories for free, isolated, bicomponent paraffin droplets subjected to large relative gas-droplet velocities. The experimental results for bicomponent fuel droplets of heptane and dodecane vaporising at atmospheric pressure were used for comparison with other theoretical models of Tong and Sirignano [12].

Aggarwal et al. compared different evaporation models for their feasibility in spray calculations [21].

Shaw, B.D [22] investigated spherically symmetric combustion of miscible droplets for the case where liquid phase species transport was slow relative to droplet surface regression rates. Attention was focussed on later periods of combustion, following decay of initial transients, when droplet species profiles change slowly relative to droplet size changes and d^2-law combustion closely holds.

Spherical combustion of heptane-dodecane droplet was considered at one atmosphere and $300\ K$. Asymptotic analysis was employed. The gas phase was assumed to remain quasi-steady. Properties were not calculated as a function of temperature. A concentration boundary layer where species profile changed sharply in the radial coordinate was shown to be present at the droplet surface.

Mawid and Aggarwal [23] numerically analysed transient combustion of a spherically symmetric 50-50 by mass heptane-decane liquid fuel droplet. The unsteady effects caused by the liquid and gas phase processes were considered and were divided into three main periods.

An important aspect of multicomponent droplet combustion is the *combustion of chlorinated hydrocarbons, dealing with the effects of chlorination and blending.* Direct incineration is a promising technology for the disposal of hazardous wastes with the potential of complete detoxification. Many hazardous wastes are chlorinated hydrocarbons (CHC$_S$) which are incineration resistant

A comprehensive experimental investigation was conducted [24] to quantify the combustion characteristics of pure CHC$_S$ as well as their mixtures with regular hydrocarbon fuels, with the specific interest of enhancing the incinerability of CHC$_S$ through judicious blending with hydrocarbon fuels.

Mixtures of TECA (1,1,2,2,-tetrachloroethane) and various alkanes were studied to determine the role of volatility differentials in the burning of TECA.

It was observed that the incineration of a heavily chlorinated hydrocarbon could be promoted through the addition of a small quantity of a less volatile regular hydrocarbon fuel, and emphasized the importance of developing rational blending strategies in the incineration of hazardous wastes.

Another important aspect of multicomponent droplet burning is vaporisation *of alcohols with respect to the water vapour condensation phenomenon.* Some previous studies on the light alcohols, namely methanol and ethanol, have suggested that the droplet vaporisation rate can be substantially enhanced through condensation of water vapour from the environment.

That is because the saturation temperatures of ethanol and methanol are lower than that of water and because they are also completely water miscible, water vapour from a humid environment could condense onto and subsequently dissolve into the relatively cool alcohol droplet. The condensation heat release could be used by alcohol for its own vaporisation, thereby facilitating its evaporation rate [25].

An interesting phenomenon accompanying multi-component droplet combustion is *microexplosion*. Microexplosion (frag-mentation of liquid droplets due to violent internal vaporisation) has potential in improving engine performance since it can be used to promote the atomisation of heavy fuels by adding certain amounts of light fuels.

Zeng and Lee [26] presented a numerical model of microexplosion for multicomponent droplets. The first part of the model addressed the mass and temperature distribution inside the droplet and the bubble growth within the droplet. The bubble generation was described by a homogeneous nucleation theory, and the subsequent bubble growth led to the final explosion i.e. break up.

The second part of the model determined when and how the break up process proceeded. Unlike adhoc/empirical approaches reported in the literature, the size and velocity of sibling droplets (secondary droplets) were determined by a linear instability analysis.

The vaporisation behaviour of an oxygenate diesel blend was analysed at the end. It was found that microexplosion was possible under typical diesel engine environments for this type of fuel. Occurrence of microexplosion shortened the droplet lifetime, and this effect was stronger for droplets with larger sizes or a near 50/50 composition.

2. Selection of Fuels

The selected fuels included alkanes, alcohols and biodiesels. In addition to pure fuels, like n- heptane, ethanol and biodiese fuel methyl linoliate, mixtures of n-heptane-n-dodecane and n-hexane-n-decane were considered in the multi-component droplet analysis.

These family of fuels have different thermophysical and transport properties and structure. Therefore variation in their properties should affect combustion characteristics like flame temperature, flame radius, transfer number, burning constant, burning rate, combustion lifetime, flame to droplet diameter ratio as well as emission characteristics. n-heptane is used as a test fuel in many experimental and theoretical studies and therefore lot of experimental and modelling data are available for comparison purposes. Some experimental and theoretical studies used n-heptane for sooting , since it is a sooting fuel.

For the combustion of liquid fuels, it is convenient to express the composition in terms of a single hydrocarbon, even though it is a mixture of many hydrocarbons (multicomponent fuel) [27].

Thus gasoline is usually considered to be octane and diesel fuel is considered to be dodecane. n-octane and n-dodecane are important fuels for experimental and modelling studies. Methanol has relatively simple gas chemistry making it more suitable to theoretical treatment, it is also a non sooting fuel.

Ethanol is regaining its popularity for use in practical applications. The Clean Air Act Amendments of 1990 mandated the use of oxygenated fuels such as ethanol and methanol in regions of country experiencing high levels of CO [28]. In India, Ministry of Petroleum and Natural Gas launched a nationwide "Ethanol Blended Petrol Program" in 2006 as a step towards gradually reducing the dependence on fossil fuels and increasing the use of alternative fuels that are renewable in nature.

Blending of ethanol with diesel fuels helps in reducing the amount of aromatic precursors such as acetylene that lead to a drastic suppression of polycyclic aromatic hydrocarbons (PAHs) / soot formation [28].

Biodiesel is a substitute for petroleum diesel and can be used in diesel engines with minor modifications. Indian government has set up a biofuel board to promote its cultivation on a large scale. Biodiesel is methyl or ethyl ester of fatty acid made from virgin or used vegetable oils (both edible and non edible) and animal fat. The main resources for biodiesel production can be non edible oils obtained from plant species such as Jatropha curcas (Ratanjyot), Pongamia pinnata (Karanj), Calophyllum inophyllum (Nagchampa) etc. [29].

As evident from the available literature on biodiesels regarding their structure, nearly all biodiesels are made up of a combination (in different percentages) of either Linoleic acid (Methyl linoleate), Oleic acid (Methyl oleate) or Stearic acid (Methyl stearate). Out of these, Safflower oil contains about 78.0 % Methyl linoleate; Sunflower oil contains about 74 % methyl linoleate; karanj 44.5-71.3 % Methyl oleate and 10.8-18.3 % Methyl Linoleate; Rice-bran 26.4-35.1 % Methyl linoleate, 39.2-43.7 % Methyl oleate; Neem 49.1-61.9 % Methyl oleate, 2.3-25.8 % Methyl linoleate, 14.4-24.1 % Methyl stearate [30]; while Jatropha about 34.3 % Methyl oleate, 43.12 % Methyl linoleate and 7.46 % Methyl stearate. For high percentage of a particular methyl ester in a biodiesel, the behaviour of that biodiesel with regards to combustion and emission characteristics can be predicted by considering it as a pure or single component fuel made up of that particular methyl ester.

It should be added that at present relatively less amount of thermophysical and transport data of biodiesel fuels exists in the literature. Also, combustion and emission data of biodiesels with respect to single droplet is limited. One of the motivation factor for the present work was to overcome this deficiency.

The succeeding paragraph describes methods of estimating properties of biodiesel fuel based on their chemical composition and structure at specified temperature and pressure, used in combustion modelling for the present work. Biodiesel contains six or seven fatty acid esters. It is possible to estimate the properties of each pure component and then compute the mixture properties based on some mixing rules [31]. Further, in the formulation of biodiesel fuel, there is an advantage over petroleum that the chemistry of the components is well defined. The fuel is an ester of fatty acids derived from natural sources. As a result fatty acids present in significant quantity are palmitic acid, stearic acid, oleic acid, linoleic acid, linolenic acid and erucic acid. Relative composition change by sources, but the components do not [32].

Once properties of one of the major components is estimated, the same methodology can be applied for other components.

2.1. Properties Etimation for Biodiesel Fuel (Methyl Linoleate)

The critical properties are an important starting point as they are used to estimate other important thermodynamic properties. For the present work, critical properties of fatty acid methyl esters of pure vegetable oils were estimated by two widely used methods: Joback modification of Lydersen's method, and Ambrose's method, since these two methods were previously shown to be able to provide reasonably accurate estimates for most compounds. Thermophysical properties of chosen fuels are given in Table 1.

On comparing with reported values [31,33], slight discrepancy (10-12%) was found with normal boiling point temperature as calculated in the present work. The reason may be that we have used chemical structure as the basis for estimation whereas Yuan et al.[31] have used different correlations for determing the normal boiling point. For comparison of critical data with the reported results [31], it was observed that there was a very good agreement with respect to critical pressure and critical molar volume (Table 2), however critical temperature values differ since for the present work, the calculation for the critical temperature involves the normal boiling point, hence the variation. Once these properties were known, they were used in turn for determining combustion parameters like transfer number B_T, burning constant k_b, droplet lifetime t_d and thermal diffusivity α_g.

3. Problem Formulation

3.1. Development of Spherically Symmetric, Single Component, Droplet Combustion Model for the Gas Phase

Following assumptions are invoked:

Spherical liquid fuel droplet is made up of single chemical species and is assumed to be at its boiling point temperature surrounded by a spherically symmetric flame, in a quiescent, infinite oxidising medium with phase equilibrium at the liquid-vapour interface expressed by the Clausius-Clapeyron equation.

Droplet processes are diffusion controlled (ordinary diffusion is considered, thermal and pressure diffusion effects are neglected). Fuel and oxidiser react instantaneously in stoichiometric proportions at the flame. Chemical kinetics is infinitely fast resulting in flame being represented as an infinitesimally thin sheet.

Ambient pressure is subcritical and uniform. Conduction is the only mode of heat transport, radiation heat transfer is neglected. Soret and Dufour effects are absent.

Thermo physical and transport properties are evaluated as a function of pressure, temperature and composition. Ideal gas behaviour is assumed. Specific enthalpy 'h' is a function of temperature only. The product of density and diffusivity is taken as constant. Gas phase Lewis number Le_g is assumed as unity.

The overall mass conservation and species conservation equations are given respectively as:

$$\frac{\partial \rho}{\partial t}+\frac{1}{r^2}\frac{\partial}{\partial r}\left(r^2 \rho v_r\right)=0 \tag{1}$$

$$\frac{\partial(\rho Y)}{\partial t}+\frac{1}{r^2}\frac{\partial}{\partial r}r^2\left[\rho v_r Y-\rho D\frac{\partial Y}{\partial r}\right]=0 \tag{2}$$

Where;
t is the instantaneous time
r is the radial distance from the droplet center
ρ is the density
v_r is the radial velocity of the fuel vapour
D is the mass diffusivity
Y is the mass fraction of the species
equations (1) and (2) are combined to give species concentration or species diffusion equation for the gas phase as follows

$$\frac{\partial Y}{\partial t}=D_g\frac{\partial^2 Y}{\partial r^2}+\frac{2D_g}{r}\frac{\partial Y}{\partial r}-v_r\left(\frac{\partial Y}{\partial r}\right) \tag{3}$$

D_g is the gas phase mass diffusivity

The relation for energy conservation can be written in the following form

$$\frac{\partial(\rho h)}{\partial t}+\frac{1}{r^2}\frac{\partial}{\partial r}\left[r^2\rho\left(v_r\cdot h-D.\,C_p\frac{\partial T}{\partial r}\right)\right]=0 \tag{4}$$

The energy or heat diffusion equation for the gas phase, equation (5) can be derived with the help of overall mass conservation equation (1) and equation (4), as:

$$\frac{\partial T}{\partial t}=\alpha_g\frac{\partial^2 T}{\partial r^2}+\frac{2\alpha_g}{r}\frac{\partial T}{\partial r}-v_r\left(\frac{\partial T}{\partial r}\right) \tag{5}$$

T is the temperature, α_g is gas phase thermal diffusivity. Neglecting radial velocity of fuel vapour v_r for the present model, equations (3) and (5) reduce to a set of linear, second order partial differential equations (equations 6 and 7).

$$\frac{\partial Y}{\partial t}=D_g\frac{\partial^2 Y}{\partial r^2}+\frac{2D_g}{r}\frac{\partial Y}{\partial r} \tag{6}$$

$$\frac{\partial T}{\partial t}=\alpha_g\frac{\partial^2 T}{\partial r^2}+\frac{2\alpha_g}{r}\frac{\partial T}{\partial r} \tag{7}$$

These equations can be accurately solved with finite difference technique using appropriate boundary conditions and provide the solution in terms of species concentration profiles (fuel vapour and oxidiser) and temperature profile for the inflame and post flame zones respectively.

The boundary and initial conditions based on this combustion model are as follows

$$at\ r=r_f;\ T=T_f,\ Y_{o,f}=0,\ Y_{F,f}=0$$

$$at\ r=r_\infty;\ T=T_\infty,\ Y_{o,\infty}=0.232,$$

at $t = 0$; $r = r_{lo}$, $T = T_b$, $Y_{F,S} = 1.0$

where $r_l = r_{lo}(1 - t/t_d)^{1/2}$ for $(0 \le t \le t_d)$, is the moving boundary condition coming out from the d^2-law. Here, T_f and T_∞ are temperatures at the flame and ambient atmosphere respectively.

$Y_{F,s}$ and $Y_{F,f}$ are fuel mass fractions respectively at the droplet surface and flame. $Y_{o,\infty}$ and $Y_{o,f}$ are oxidiser concentrations in the ambience and at the flame respectively, t is the instantaneous time, t_d is the combustion lifetime of the droplet, r_{lo} is the original or initial droplet radius and r_l is the instantaneous droplet radius at time "t".

The location where the maximum temperature $T = T_f$ or the corresponding concentrations $Y_{F,f} = 0$ and $Y_{o,f} = 0$ occur, was taken as the flame radius r_f. Instantaneous time "t" was obtained from the computer results whereas the combustion lifetime "t_d" was determined from the relationship coming out from the d^2-law.

Other parameters like instantaneous flame to droplet diameter ratio (F/D), flame standoff distance $(F\text{-}D)/2$, dimensionless flame diameter (F/D_0) etc, were then calculated as a function of time. Products species concentrations (CO, NO, CO2 and H2O) were estimated using Olikara and Borman code [34] with gas phase code of the present study.

3.2. Solution Technique

Equations (6) and (7) are a set of linear, second order, parabolic partial differential equations with variable coefficients. They become quite similar when thermal and mass diffusivities are made equal for unity Lewis number. They can be solved by any one of the following methods such as weighted residual methods, method of descritisation in one variable, variables separable method or by finite difference technique. In weighted residual methods, the solution is approximate and continuous, but these methods are not convenient in present case since one of the boundary condition is time dependent.

The method of descritisation in one variable is not suitable since it leads to solving a large system of ordinary differential equations at each step and therefore time consuming. Variables separable method provides exact and continuous solution but cannot handle complicated boundary conditions.

Keeping in view the limitations offered by other methods, "finite difference technique" is chosen (which can deal efficiently with moving boundary conditions, as is the case in present work) and successfully utilised in solving equations (6) and (7). The approach is simple, fairly accurate and numerically efficient [35]. Here the mesh size in radial direction is chosen as h and in time direction as k.

Using finite difference approximations, equations (6) and (7) can be descritised employing three point central difference expressions for second and first space derivatives, and the time derivative is approximated by a forward difference approximation resulting in a two level, explicit scheme (eqn 8), which is implemented on a computer.

$$T_m^{n+1} = \alpha\lambda_1(1 - p_m)T_{m-1}^n + (1 - 2\alpha\lambda_1)T_m^n + \alpha\lambda_1(1 + p_m)T_{m+1}^n \quad (8)$$

Here, $\lambda_1(mesh\ ratio) = k/h^2$, $p_m = h/r_m$,

$$r_m = r_{lo} + mh' \quad Nh = r_\infty - r_{lo}, \quad m = 0,1,2....N.$$

The solution scheme is stable as long as the stability condition $\lambda_1 < 1/2$ is satisfied. Equations (6) and (7) can also be descritised in Crank-Nicholson fashion, which results in a six point, two level implicit scheme. From a knowledge of solution at the nth time step, we can calculate the solution at the $(n+1)^{th}$ time step by solving a system of $(N-1)$ tridiagonal equations. Although the resulting scheme is more accurate, the cost of computation is fairly high.

3.3. Multicomponent Droplet Vaporisation and Combustion Models

3.3.1. Droplet Vaporisation Model

Multicomponent droplet vaporisation and combustion models for the case when d^2-law is followed are formulated and effects of air-fuel vapour mixing and Lewis number are obtained on droplet vaporisation behaviour.

After that, a more realistic MC diffusion limit model with convection is also evolved and compared with other models.

For pure vaporisation/evaporation case, the evaporation constant k_{ev} of the multicomponent mixture is given as [17]:

$$k_{ev} = \frac{8\bar{\lambda}_g}{\bar{C}_{pg}\bar{\rho}_l}\ln\left[1 + \frac{\bar{C}_{pg}(T_\infty - \bar{T}_{b_l})}{\sum y_{ilo}L_i}\right] \quad (9)$$

where, $\bar{\lambda}_g, \bar{C}_{pg}$ are respectively the gas mixture thermal conductivity and specific heat.

$\bar{T}_{b_l}, T_\infty$ are liquid mixture boiling point temperature and ambient temperature respectively.

L_i is the latent heat of vaporisation of a particular species, and y_{ilo} is the initial liquid phase mass fraction of a species, $\bar{\rho}_l$ is the liquid mixture density.

With the assumption that surface temperature is equal to its boiling point $(T_s = T_b)$, d^2-law can be followed, where the instantaneous droplet radius r_l can be determined from

$$r_l = r_{lo}(1 - t/t_{ev})^{1/2} \quad (10)$$

here, r_{lo} is the initial or original droplet radius and t_{ev} is the droplet lifetime of multicomponent droplet (for pure vaporization case)

That is:

$$t_{ev} = 4r_{lo}^2 / k_{ev} \tag{11}$$

Once t_{ev} is calculated, it is used as an input to the code developed for solving the species diffusion equation in the liquid phase. Other variables needed as input are liquid phase thermal diffusivity of the species, gas and liquid phase densities and diffusion coefficients, liquid phase Lewis number and initial droplet radius.

Apart from these, certain parameters related to droplet evaporation need to be determined which are discussed below.

Generally, initial liquid side mass fractions of the bicomponent fuel droplet are known or chosen, then the corresponding liquid side mole fractions can be determined using

$$x_{mls} = \frac{y_{mls}/W_m}{\sum_i (y_i/W_i)} \tag{12}$$

is the species liquid side mole fraction at the droplet surface, y_{mls} is the species liquid side mass fraction at the droplet surface, W_m is the species molecular weight.

The gas side mole fraction can be calculated by the relation

$$\frac{x_{mgs}}{x_{mls}} = \frac{1}{P} \exp\left(\frac{L_m}{R_m} \left(\frac{1}{T_{b_m}} - \frac{1}{T_s} \right) \right) \tag{13}$$

here, T_s is the surface temperature, $P = 1 \text{atm}$, L_m, R_m and T_{b_m} are latent heat, specific gas constant and boiling point of the species respectively. Then y_{mgs} (species gas side mass fraction at the droplet surface) can be obtained from the relation

$$y_{mgs} = \frac{x_{mgs}W_m}{\sum_i x_i W_i} \tag{14}$$

also, $Y_{F,S} = \sum_i y_{igs} = \sum_i y_{mgs} \tag{15}$

and fractional vaporisation rate of the species, ε_m is given as:

$$\varepsilon_m = y_{mgs} + y_{mgs}\left(1 - Y_{F,S}\right)/Y_{F,S} \tag{16}$$

For a multicomponent droplet vaporisation model, the transfer number B_m for a particular species is given by the relation:

$$B_m = \frac{y_{mgs}}{\varepsilon_m - y_{mgs}} \tag{17}$$

y_{mgs} and B_m are used as input to the code.

The solution of the species diffusion equation gives the

variation of species concentration or species mass fraction from centre of droplet to the droplet surface as a function of time. One can also calculate the species vaporisation rate (m_i) using the relation

$$m_i = 4\pi\bar{\rho}_g \bar{D}_g r_i \varepsilon_i \ln(1 + B_m) \tag{18}$$

$\bar{\rho}_g, \bar{D}_g$ are gas phase mixture density and diffusion coefficient respectively and r_l is the instantaneous droplet radius which can be determined as a function of time from the d^2-law, for a known initial droplet size.

3.3.2. Droplet Combustion Model

Another expression for mixture burning constant is derived [17] when considering combustion

$$k_b = \frac{8\bar{\lambda}_g}{\bar{C}_{pg}\bar{\rho}_l} \ln\left[1 + \frac{\bar{C}_{pg}(T_f - \bar{T}_{b_i}) + [(Y_{o,\infty}/\sum y_i v_i)](\sum y_i Q_i)}{\sum y_i L_i} \right] \tag{19}$$

T_f is the flame temperature of the bicomponent liquid fuel mixture determined from the stoichiometric combustion reaction, Q_i is the heat of combustion of a particular species. The transfer number for combustion is given for a particular species as:

$$B_m = \frac{v_i Y_{o,\infty} + y_{igs}}{\varepsilon_i - y_{igs}} \tag{20}$$

Here V_i is defined as a ratio of mass of fuel to mass of oxygen, obtained from the stoichiometric combustion reaction of the particular species. The computer programme then gives variation of species liquid mass fraction with dimensionless radius at different times of droplet burning.

3.3.3. Multicomponent Diffusion Limit Vaporisation Model with Convection (Variable Surface Temperature)

An attempt has been made to develop a more realistic bicomponent droplet vaporisation model by relaxing the assumption of Ts = Tb. Once this assumption is dropped, the problem of bicomponent droplet vaporisation becomes involved, since one has to solve the energy equation (equation 21), in the liquid phase in addition to the liquid phase species mass diffusion equation (equation 43).

The solution of energy equation gives the variation of droplet surface and center temperatures as a function of dimensionless radius at different times of droplet burning or droplet vaporisation, as the case may be. The variation of droplet surface temperature with time then becomes an important input parameter for the solution of liquid phase mass diffusion equation, in addition d^2-law does not hold for this particular case.

Mass diffusion in the liquid phase is of primary importance in the vaporisation process for a multicomponent fuel. Therefore for studying multicomponent droplet evaporation/combustion, a detailed liquid phase analysis is a prior step.

3.4. Liquid Phase Analysis

3.4.1. Liquid Phase Energy Equation

For the development of liquid phase model, "conduction limit" approach is followed where conduction is the only mode of heat transport from the energy arriving at the droplet surface to its interior. Here, the droplet temperature is varying both spatially and temporally. The system is spherically symmetric with no internal liquid motion. Gas phase is assumed quasi-steady. Clausius-Clapeyron relation describes the phase equilibrium at the liquid-vapour interface. The liquid phase heat diffusion equation is given as:

$$\frac{\partial T_l}{\partial t} = \frac{\alpha_l}{r^2} \frac{\partial}{\partial r}\left(r^2 \frac{\partial T_l}{\partial r}\right) = \alpha_l \left[\frac{\partial^2 T_l}{\partial r^2} + \frac{2}{r}\frac{\partial T_l}{\partial r}\right] \quad (21)$$

where, T_l and α_l are liquid phase temperature and thermal diffusivity respectively.

Initial and boundary conditions being

$$T(r,0) = T_0(r) \quad (22)$$

$T_0(r)$ is the initial temperature distribution

$$\left(\frac{\partial T}{\partial r}\right)_{r=0} = 0 \quad (23)$$

$$mH = mL + \left(4\pi^2 \lambda_l \frac{\partial T}{\partial r}\right)_{r_s} \quad (24)$$

Where r_s is the radius at the droplet surface, λ_l is liquid thermal conductivity.

The above equation represents the energy conservation at the interface, L and H are specific and effective latent heat of vaporisation respectively.

The complete solution of equation (21) requires certain quasi-steady gas phase relations (with respect to combustion) in terms of droplet surface temperature T_s given as:

$$\hat{Q}_1 = \left(1 - Y_{F,S}\right)\left[\left(\hat{T}_\infty - \hat{T}_s + vY_{o,\infty}\hat{Q}_2\right)\right]/Y_{F,S} + vY_{o,\infty} \quad (25)$$

$$\hat{m} = \ln\{1 + (\hat{T}_\infty - \hat{T}_s + vY_{o,\infty}\hat{Q}_2)/\hat{Q}_1\} \quad (26)$$

where, $\hat{Q}_1$ is a dimensionless parameter, $\hat{m}$ is dimensionless mass evaporation rate,

$\hat{Q}_2 = Q/L$, Q is the heat of combustion of fuel, v is stoichiometric fuel-oxygen ratio on mass basis, $Y_{F,S}$ is the fuel vapour mass fraction at the droplet surface, which can be determined from Clausius-Clapeyron relation (equation 27) and $Y_{o,\infty}$ is oxidiser mass fraction in ambient atmosphere = 0.232.

Ambient temperature T_∞, droplet surface temperature T_s and boiling point temperature T_b are non-dimensionalised by

defining $\hat{T}_\infty = \frac{T_\infty C_{pg}}{L}$, $\hat{T}_s = \frac{T_s C_{pg}}{L}$ and $\hat{T}_b = \frac{T_b C_{pg}}{L}$ where C_{pg} is the gas phase specific heat. The Clausius-Clapeyron relation can be written as [36]:

$$Y_{F,S} = \{(1 + W_g/W_F(P_\infty \exp([C_{pg}/R][1/\hat{T}_s - 1/\hat{T}_b]) - 1)\}^{-1} \quad (27)$$

W_g is the average molecular weight of all gas phase species except fuel, at the surface, W_F is the molecular weight of the fuel. R is the specific gas constant of the fuel, P_∞ is equal to one atmosphere.

3.4.2. Solution Technique

A convenient method of solving problems with moving boundary is to change the moving boundary to a fixed boundary using coordinate transformation.

Equation (21) is a linear, second order partial differential equation which can be solved together with initial and boundary conditions using coordinate transformation method. Therefore, we have

$$\hat{r} = r/r_s(t), \quad \hat{t} = \alpha_l \int_0^t (r_s(t))^{-2} dt, \quad \hat{T} = C_{pg}T/L \quad (28)$$

Where $\hat{r}$, $\hat{t}$ and $\hat{T}$ are the transformed radial coordinate, time and temperature respectively. $r_s(t)$ is the instantaneous droplet radius at the surface at time t. The transformed equations are descritised using finite difference technique employing three point central difference expressions for second space derivative and central difference expression for first space derivatives while the time derivative is approximated by a forward difference expression resulting in a two level, explicit scheme.

The heat diffusion equation (equation 21) governing liquid phase temperature variation takes the form

$$\frac{\partial \hat{T}}{\partial \hat{t}} = \frac{1}{\hat{r}^2} \frac{\partial}{\partial \hat{r}}\left(\hat{r}^2 \frac{\partial \hat{T}}{\partial \hat{r}}\right) - \hat{m}K\hat{r}\left(\frac{\partial \hat{T}}{\partial \hat{r}}\right) \quad (29)$$

and the corresponding initial and boundary conditions are

$$\hat{T}(\hat{r},0) = \hat{T}_0(\hat{r}) \quad (30)$$

$$\left(\frac{\partial \hat{T}}{\partial \hat{r}}\right)_{\hat{r}=0} = 0 \quad (31)$$

$$\left(\frac{\partial \hat{T}}{\partial \hat{r}}\right)_{\hat{r}=1} = \hat{m}K'(\hat{H} - 1) \quad (32)$$

K' and K are dimensionless parameters defined as: $K' = \lambda_g/\lambda_l$ and $K = (\lambda_g/\lambda_l)/(C_{pg}/C_{pl})$. For simplicity we have assumed gas and liquid phase specific heats and thermal conductivities as equal, hence K and K' are equal. For combustion under subcritical conditions, K is of the order of 0.1 to 0.5. We now impose a mesh:

$$[(\hat{r}_i,\hat{t}_j):\hat{r}_i = ih,\ i = 0,1,2,...,N;\ \hat{t}_j = (jk),\ j = 0,1,2,3,..],$$

here h is the mesh size for $\hat{r}$ and k is the time step. We then use the following finite difference approximations

$$\left(\frac{\partial \hat{T}}{\partial \hat{t}}\right)_{(\hat{r}_i,\hat{t}_j)} \cong \frac{\hat{T}_{i,j+1}-\hat{T}_{i,j}}{k} \tag{33}$$

$$\left(\frac{\partial \hat{T}}{\partial \hat{r}}\right)_{(\hat{r}_i,\hat{t}_j)} \cong \frac{\hat{T}_{i+1,j}-\hat{T}_{i-1,j}}{2h} \tag{34}$$

$$\left(\frac{\partial^2 \hat{T}}{\partial \hat{r}^2}\right)_{(\hat{r}_i,\hat{t}_j)} \cong \frac{\hat{T}_{i+1,j}-2\hat{T}_{i,j}+\hat{T}_{i-1,j}}{h^2} \tag{35}$$

$$\text{Where, } \hat{T}_{i,j} = \hat{T}(\hat{r}_i,\hat{t}_j) \tag{36}$$

In general

$$\left(\frac{\partial \hat{T}}{\partial \hat{r}}\right)_{(\hat{r}_i,\hat{t}_j)} \cong \frac{1}{h}\left(C_2\hat{T}_{i+1,j}+C_1\hat{T}_{i,j}+C_0\hat{T}_{i-1,j}\right) \tag{37}$$

Special cases being
Central Difference Formula
($C_2 = 1/2$, $C_1 = 0$, $C_0 = -1/2$)
Forward Difference Formula
($C_2 = 1$, $C_1 = -1$, $C_0 = 0$)
Backward Difference Formula
($C_2 = 0$, $C_1 = 1$, $C_0 = -1$)
Substituting in equation (29) and simplifying, we get the two level explicit scheme given by equation (38):

$$\hat{T}_{i,j+1} = \lambda_1\hat{T}_{i+1,j}\left(1+C_2hf_{i,j}\right)+\hat{T}_{i,j}\left(1-2\lambda_1+C_1h\lambda_1 f_{i,j}\right)$$
$$+\lambda_1\hat{T}_{i-1,j}\left(1+C_0hf_{i,j}\right), \tag{38}$$

where λ_1 (mesh ratio) = k/h^2

$$\begin{aligned}&i = 1,2,3,......,N-1\\&j = 0,1,2,3,......\\&\text{and}\\&f_{i,j} = (2/\hat{r}_i)-\hat{m}_j K\hat{r}_i\end{aligned} \tag{39}$$

Further, initial and boundary conditions may be descritised to obtain following additional equations:

$$\hat{T}_{0,i} = (\hat{T}_0)_i \tag{40}$$

$$(\hat{T}_0)_j = (\hat{T}_1)_j \tag{41}$$

$$(\hat{T}_N)_j = (\hat{T}_{N-1})_j + hK'\hat{m}_j(\hat{H}_j-1) \tag{42}$$

Solution of heat diffusion equation (21) provides the radial variation of droplet temperature at different times of burning. Prior solution of equation (21) is necessary for determining the species concentration or mass fraction distribution within a multicomponent liquid fuel droplet.

3.4.3. Liquid Phase Mass Diffusion Equation

Spherically symmetric, multicomponent liquid phase mass diffusion with no internal circulation is written as:

$$\frac{\partial Y_{l,m}}{\partial t} = D_l\left(\frac{\partial^2 Y_{l,m}}{\partial r^2}+\frac{2}{r}\frac{\partial Y_{l,m}}{\partial r}\right) \tag{43}$$

where, $Y_{l,m}$ is the concentration or mass fraction of the *mth* species in the liquid, and D_l is liquid mass diffusivity which in this case is much smaller than liquid phase thermal diffusivity α_l, hence liquid phase Lewis number, $Le_l \gg 1$.

The boundary condition at the liquid side of the droplet surface is

$$\left.\frac{\partial Y_{l,m}}{\partial r}\right]_s = \frac{\rho_g D_g}{\rho_l D_l r_l}\ln[1+B]\left[Y_{l,ms}-\varepsilon_m\right] \tag{44}$$

m denotes particular species

r_l is the instantaneous droplet radius at time " t "

ε_m is *species* fractional mass vaporisation rate

B is *the* transfer number D_l and D_g are liquid and gas phase mass diffusivities respectively

ρ_l and ρ_g are respectively the liquid and gas phase densities

At the center of the droplet, symmetry yields

$$\left.\frac{\partial Y_{l,m}}{\partial r}\right]_{r=0} = 0 \tag{45}$$

A uniform initial liquid phase composition for the droplet is chosen, given as:

$$Y_{l,m}(0,r) = Y_{l,mo} \tag{46}$$

Equation (43) with boundary conditions must be applied concurrently with phase equilibrium conditions and heat diffusion equation (21) to obtain the complete solution.

For a single component fuel, phase equilibrium is expressed by the Clausius-Clapeyron equation and ε_m is unity. For a multicomponent fuel, Raoult's law provides for the phase equilibrium and $\varepsilon_m \neq 1$.

3.4.4. Solution Technique

Equation (43) is a linear, second order partial differential equation. Using coordinate transformation by defining certain dimensionless variables, we have

$$\zeta = \frac{r}{R(t)}, \tau = \frac{\alpha_l t}{r_{lo}^2}, r_s(\tau) = \frac{R(t)}{r_{lo}}, \beta = \frac{1}{2}\frac{d}{d\tau}(r_s^2),$$

$$\overline{Y} = \frac{(Y_{lm} - Y_{lm0})}{Y_{lm0}} \tag{47}$$

here, r is spherical radial coordinate, r_{lo} is original or initial droplet radius, t is instantaneous time, $R(t)$ is instantaneous droplet radius at time t, ζ is dimensionless radial coordinate, τ is non-dimensional time, $r_s(\tau)$ is instantaneous non-dimensional droplet radius, $\overline{Y}$ is dimensionless mass fraction.

After transformation, liquid phase mass diffusion equation (equation 43) for the mth species can be written as:

$$\frac{1}{\gamma}\left(r_s^2\frac{\partial \overline{Y}}{\partial \tau} - \beta\zeta\frac{\partial \overline{Y}}{\partial \zeta}\right) = \frac{\chi_d}{\zeta^2}\frac{\partial}{\partial \zeta}\left(\zeta^2\frac{\partial \overline{Y}}{\partial \zeta}\right) \tag{48}$$

where, γ is taken as reciprocal of Lewis number "Le_l", and χ_d is a dimensionless parameter defined as a ratio of effective mass diffusivity to mass diffusivity (equal to unity for the present case).

Boundary condition at the liquid side of the droplet surface (equation 44) becomes

$$\frac{\partial \overline{Y}}{\partial \zeta}\bigg|_{\zeta=1} = \frac{\rho_g D_g}{\rho_l D_l}\ln[1+B]\left[\overline{Y}_s - \varepsilon_m\right] \tag{49}$$

Boundary condition at the droplet center (equation 45) is written as:

$$\frac{\partial \overline{Y}}{\partial \zeta}\bigg|_{\zeta=0} = 0 \tag{50}$$

Equation (46) representing initially chosen composition takes the form

$$\overline{Y}(0,\zeta) = \overline{Y}_0 \tag{51}$$

Equation (48) can be rewritten as:

$$r_s^2\frac{\partial \overline{Y}}{\partial \tau} - \beta\zeta\frac{\partial \overline{Y}}{\partial \zeta} = \chi_d\left(\frac{\partial^2 \overline{Y}}{\partial \zeta^2} + \frac{2}{\zeta}\frac{\partial \overline{Y}}{\partial \zeta}\right) \tag{52}$$

Equation (52), along with equations (49) to (51) are then descritised using forward difference formula for first derivatives with respect to τ and ζ and central difference formula for the second derivative with respect to ζ. After simplification, final descritised form of equation (52) is obtained as a two level explicit scheme:

$$\overline{Y}_j^{n+1} = \frac{\lambda_1 \overline{Y}_{j+1}^n}{(1-k\tau_n)}\left[\beta\zeta_j(\Delta\zeta) + \chi_d + \frac{2\chi_d(\Delta\zeta)}{\zeta_j}\right] - \frac{\lambda_1 \overline{Y}_j^n}{(1-k\tau_n)}$$

$$\left[\beta\zeta_j(\Delta\zeta) + 2\chi_d + \frac{2\chi_d(\Delta\zeta)}{\zeta_j}\right] + \overline{Y}_j^n + \frac{\chi_d\lambda_1\overline{Y}_{j-1}^n}{(1-k\tau_n)} \tag{53}$$

where; $\overline{Y}_j^n = \overline{Y}(\tau_n, \zeta_j)$, $\lambda_1 = \frac{\Delta\tau}{(\Delta\zeta^2)}$ is the mesh ratio kept $< \frac{1}{2}$ for a stable solution, k is time step, $\Delta\tau$ is dimensionless time step, $\Delta\zeta$ is mesh size in ζ direction, τ_n is non-dimensional time $= n\,\Delta\tau$ and $\zeta_j = j\Delta\zeta$, $j = 1,2,....,M-1$

Equations (49) to (51) after descritisation take the form

$$\overline{Y}_M^n - \overline{Y}_{M-1}^n = \Delta\zeta\frac{\rho_g D_g}{\rho_l D_l}\ln[1+B][\overline{Y}_s - \varepsilon_m] \tag{54}$$

$$\overline{Y}_1^n - \overline{Y}_0^n = 0, \text{ for } n = 0,1,2, \tag{55}$$

$$\overline{Y}_j^0 = \overline{Y}(0), \text{ for } j = 0,1,2, \text{ , M} \tag{56}$$

The solution of equation (43) gives the variation of species mass fraction within the multicomponent liquid droplet as a function of droplet radius at different times of droplet burning.

Once tested for a pure component (n-octane) fuel droplet undergoing spherically symmetric combustion, a heptane-dodecane bicomponent droplet vaporisation model ($D_0 = 280\mu m$) with variable surface temperature was developed for the case where $Re_g = 100$ (convective situation). The gas phase processes were considered to be quasi-steady. The bicomponent liquid fuel droplet contained an initial liquid mass fraction of (50-50)% heptane-dodecane mixture with conditions of ambient pressure $P_\infty = 1$ atmosphere, ambient temperature $T_\infty = 1000\,K$, $Le_l = 10$ and gas phase Prandtl number $Pr_g = 1$ (since quasi-steady gas phase).

Further, equations (25) and (26) for $\hat{Q}_l$ and $\hat{m}$ were modified as equations (57) and (58) respectively for pure vaporisation case of a bicomponent fuel droplet made up of (n-heptane + n-dodecane) mixture at the given conditions of ambient pressure and temperature of one atmosphere and $1000\,K$ respectively

$$\hat{Q}_l = \left(1 - Y_{F,S_{mix}}\right)\left[\hat{T}_\infty - \hat{T}_s\right]/Y_{F,S_{mix}} \tag{57}$$

and $\hat{m} = \ln\left\{1 + \left[\hat{T}_\infty - \hat{T}_s\right]/\hat{Q}_l\right\} \tag{58}$

After determining $\hat{Q}_l$ and $\hat{m}$, plots of $\hat{Q}_l$ and $\hat{m}$ with $\hat{T}_s$ were obtained. The next step involved the calculation of vaporisation/evaporation lifetime t_{ev} of the bicomponent droplet vaporising in a convective environment using the relation:

$$t_{ev} = \frac{\rho_l D_0{}^2}{8\overline{\lambda}_g/\overline{C}_{pg}\ln(1+B)\left(1+0.3Re_g^{0.5}Pr_g^{0.33}\right)} \qquad (59)$$

For this particular case where $T_s \neq T_b$, the bicomponent droplet transfer number (or B number), fractional vaporisation rate (ε), mass fraction of the fuel mixture at the droplet surface $Y_{F,S}$ and dimensionless parameter β will not be constant but will vary with droplet surface temperature and time. The plots of B, ε, $Y_{F,S}$ and β against dimensionless time τ_n were then obtained by curve fitting and the curve fit equations were incorporated in the computer code for solving the multicomponent liquid phase mass diffusion equation (equation 43).

4. Results and Discussion

Three different categories of fuels namely, alkanes, alcohols and biodiesels were selected to isolate the effects of various thermophysical properties on important combustion and emission characteristics. It was observed that dimensionless flame diameter F/D_0 is influenced primarily by the fuel boiling point. The same conclusion could be drawn for the variation of F/D ratio and flame standoff distance with time. Also, the F/D ratio was found to increase throughout the droplet burning period, suggesting transient burning. From $(D/D_0)^2$ versus time plot, it was observed that for steady state burning, droplet lifetime was highest for the fuel having smallest burning constant (from the relationship coming out from the d^2-law and burning rate variation was low for ethanol in comparison with methyl linoleate and n-heptane.

Important species concentrations in terms of NO, CO, CO_2 and H_2O were quantified for pure fuels and their blends. Multicomponent droplet combustion model was evolved and compared with model of Shaw [22] for the same fuel and burning conditions suggesting that present model was accurate and simpler. For a hexane-decane droplet vaporising at 1 atm and $1000\,K$, substantial effect of mixing of air and fuel vapours was observed on the vaporisation behaviour and should be accounted for in MC droplet modelling. Practical liquids have high Lewis number (of the order of 30). Profound effect on vaporisation characteristics was noted when Lewis number was varied.

In addition, a more detailed multicomponent vaporisation model was developed and compared with various models available in literature. It was felt that the present model could prove out to be more effective for application in spray analysis due to its simplicity.

4.1. Effect of Pure Fuels on Dimensionless Flame Diameter, F/D Ratio, Flame Standoff Distance, Square of Dimensionless Droplet Diameter and Mass Burning Rate

A plot of dimensionless flame diameter with dimensionless time for a $2000\,\mu m$ droplet of different fuels burning under standard conditions is shown in Fig 1. Three

different categories of fuels namely alkanes, alcohols and biodiesels were chosen to isolate the effect of their thermophysical and transport properties on combustion characteristics. It was observed that F/D_0 increased with an increase in the boiling point of the fuel. As a result, F/D_0 values were highest for biodiesels, followed by alkanes and alcohols. For each fuel checked, F/D_0 maximised in the 20–25 % lifetime range.

In Figure 2, F/D ratio also known as flame standoff ratio was plotted against dimensionless time for different fuels such as methyl linoleate, n-heptane and ethanol for a droplet diameter of 100 microns. This ratio is important in droplet combustion since it shows whether the burning is quasi-steady (where this ratio assumes a large constant value) or non steady as evident from experimental observations where the ratio increases throughout the droplet burning period. The F/D ratio for the present work increases throughout the droplet burning period suggesting transient droplet burning irrespective of the type of fuel.

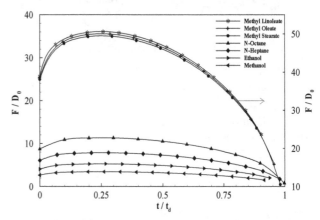

Figure 1. *Effect of fuel properties on the variation of dimensionless flame diameter F/D0 with dimensionless time t/td*

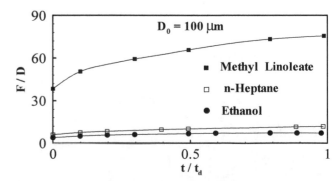

Figure 2. *Effect of fuels on F/D ratio behavior with dimensionless time.*

This is consistent with experimental observations. It is further noted that F/D values increase with an increase in the boiling point, being highest for biodiesel fuel (methyl linoleate), followed by alkane (n-heptane) and alcohol (ethanol).

When flame standoff distance was plotted against dimensionless time for n-heptane, ethanol and methyl linoleate (Fig 3), it was observed that flame standoff distance increased with an increase in boiling point of fuel.

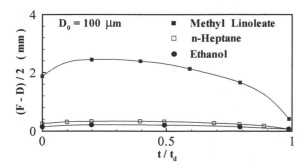

Figure 3. *Flame standoff distance versus time*

Variation of dimensionless droplet diameter squared $(D/D_0)^2$ with time was obtained for n-heptane, ethanol and methyl linoleate as shown in Fig 4. The trend showed a linear variation of $(D/D_0)^2$ with t/t_d. It was noted that droplet lifetime was highest for ethanol followed by methyl linoleate and n-heptane which could be justified from the inverse relationship between lifetime and burning constant coming out from the d^2-law i.e $t_{d_{ss}} = 4r_{lo}^2 / k_{b_{ss}}$.

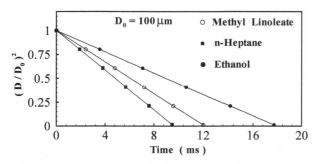

Figure 4. *Dimensionless droplet diameter squared against time*

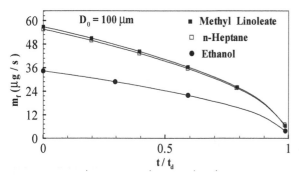

Figure 5. *Droplet mass burning rate*

Mass burning rate variation with dimensionless time t/t_d for methyl linoleate, n-heptane and methanol droplets is shown in Fig 5. Results suggest that for a 100 micron droplet, m_f values are highest at the start of combustion for methyl linoleate followed by n-heptane and ethanol. The values for different fuels were calculated using equation 60, given below, which is a function of liquid density and burning constant

$$m_f = \pi r_l \rho_l k_b / 2 \qquad (60)$$

Effect of fuels on emission characteristics of a spherical combusting droplet

Most experiments related to droplet burning have been concerned primarily with combustion aspects such as measurement of droplet and flame temperatures, flame movement, droplet lifetimes, burning rates etc. As a result, relatively less experimental data related to emission characteristics is available in literature. Such information is needed to examine the formation and destruction of pollutants such as soot, unburned hydrocarbons, NO_X, CO and CO_2 and subsequently to establish design criteria for efficient and stable combustors.

For the determination of species concentration profiles around the burning spherical droplet, the following procedure was adopted.

Adiabatic flame temperature determined earlier, using first law of thermodynamics together with other input data was used to solve numerically, the gas phase energy equation (7) with the help of a computer programme. The solution obtained was in the form of temperature profile at each time step starting from the droplet surface, reaching the maxima at the flame and then reducing to the ambient temperature value. The radius at which the maximum temperature occured was taken as the flame radius.

Temperature values in the vicinity of the flame were then chosen from the computed results, which were designated as number of steps (NOS). The solution of species diffusion equation (6) in the gas phase, gave the species profile (fuel and oxidiser). It was observed that the fuel vapour concentration was maximum at the droplet surface and decreased gradually to the minimum value a the flame front, whereas, oxygen diffusing from the ambient atmosphere became minimum at the flame. Now for the same time step, which was used in the solution of energy equation, values or NOS of fuel mass fraction Y_F corresponding to the temperature values were taken from the computed results of the species diffusion equation.

Then for a particular fuel, $FARS$ (fuel to air ratio stoichiometric) on mass basis was calculated for the stoichiometric reaction of fuel and air at atmospheric conditions of temperature $T_\infty = 298K$, pressure $P_\infty = 1\ atmosphere$ and ambient oxidiser concentration $Y_{o,\infty} = 0.232$.

Fuel to air ratio actual $(FARA)$ on mass basis was calculated next for each step where, $FARA = Y_F / 1 - Y_F$, and finally equivalence ratio, $\phi = FARA / FARS$. The values of flame temperature, equivalence ratio and ambient pressure were then used as input data for the Olikara and Borman Code [34], (modified for single droplet burning case) to obtain concentrations of important combustion products species.

Moreover, since most of the practical combustion systems operate with air as an oxidiser therefore formation of NO is a significant parameter which was another reason for our preference for the Olikara and Borman programme [34].

The behaviour of species concentrations as a function of dimensionless flame radius r/r_f across the flame zone with finite thickness is shown in Figs 6-.9. The general mechanism of formation of products involves the conversion of the reactants to CO and H_2 on the fuel rich side. The CO and H_2 further diffuse towards the thin reaction zone, which is the flame sheet and react with increasing amount of O_2 and and get oxidised to form CO_2 and H_2O. The instantaneous temperature at the given point may affect the mechanism and the rate of these reactions. It is observed that NO concentration is maximum at the flame zone and is found to be very temperature sensitive. Results of the present study show higher than expected concentrations. However, the present work aims at providing a qualitative trend rather than quantitative using a simplified approach.

To isolate the effect of fuels on emission data, three pure fuels: n-heptane, ethanol and methyl linoleate (ml), three practical multicomponent fuels: diesel oil (DF2), gasoline and aviation gas turbine fuel (JP5) and two fuel blends: gasohol (70% gasoline+30% ethanol) and diesohol (80%dieseloil+20%ethanol) were considered for a droplet size of 100 micron burning at 1 atmosphere and 298 K, (Table 4).

Among pure fuels, (Figs 6-.9), minimum concentration of CO was 0.268 % in the flame zone contributed by ethanol (CO was minimum where CO_2 was maximum (9.635 %) at dimensionless flame radius $r/r_f = 0.9$), 1.475 % was for ml and 3.522 % for n-heptane. Minimum concentration of CO_2 was contributed by ethanol and then by n-heptane and ml in the increasing order. NO concentration was lowest for ethanol followed by n-heptane and ml in increasing manner, whereas H_2O concentration was maximum for ethanol followed by n-heptane and ml in decreasing order.

For practical fuels, DF2 accounted for lowest CO emission followed in the increasing order by JP5 and gasoline, further, DF2 contributed maximum CO_2 followed by JP5 and gasoline (lowest). NO concentrations in the increasing order were contributed by DF2, JP5 and gasoline respectively.

From the results, it was generally observed that fuels having higher percentage of fuel carbon and fuel hydrogen led to more CO_2 and H_2O. However, higher flame temperature might lead to dissociation thus slightly reducing CO_2 and H_2O concentrations as seen for methyl linoleate.

NO concentrations were a strong function of flame temperature, and as a result they were lowest for ethanol among pure fuels, lowest for DF2 among practical fuels and minimum for gasohol in case of blended fuels. The variation of CO, CO_2, and NO with respect to different fuels for the present work, showed similar trend when compared with experimental data obtained from engine analysis [29].

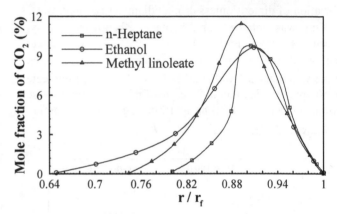

Figure 7. *CO_2 concentration*

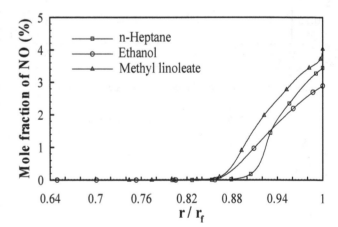

Figure 8. *NO concentration*

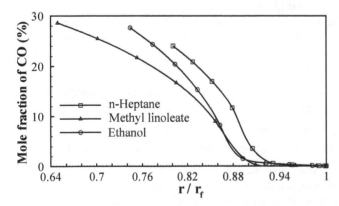

Figure 6. *Variation of CO concentration with dimensionless flame radius for a 100 micron fuel droplet burning in standard atmosphere*

It was also observed that gasohol blend gave 83.9 % less emissions with regards CO, 42 % less emissions with regards NO and 23.16 % more emissions regarding CO_2 when compared with gasoline. Whereas diesohol blend contributed 39.56% less emissions with respect to CO, 31.94 % less emissions wrt NO and 21.96 % more emissions wrt CO_2 when compared with diesel oil (DF2).

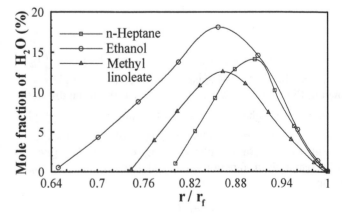

Figure 9. *H_2O concentration*

4.2. Vaporisation and Combustion Behaviour of Multicomponent Fuel Droplets

Various multicomponent droplet vaporisation and combustion models exist in literature. Some of these models seem to be oversimplified, while some are too complex to be incorporated in a spray model. The objective of the present study was to develop multicomponent droplet vaporisation and combustion models which are simple and accurate to be feasibly incorporated in spray analysis.

Law and Law [17] have shown that for the case where liquid phase species diffusion rates are slow relative to droplet surface regression rates, in other words when droplet surface temperature is equal to its boiling point temperature ($T_s = T_b$) and the droplet is vaporising vigorously following d^2-law after an initial transient period, then a similar expression for burning constant can be developed analogous to the single component case.

This analysis comprehensively reduces the complexity of the problem. A conclusion of this work is that the fractional vaporisation rate of any species ε_m at the droplet surface is equal to the initial liquid phase mass fraction of that species prior to vaporisation/combustion.

4.3. Calculation Procedure

The burning conditions for heptane-dodecane droplet were ambient pressure $P_\infty = 1\,atm$, ambient temperature $T_\infty = 300K$, ambient oxidiser mass fraction $Y_{o,\infty} = 0.232$. The combustion reaction was assumed stoichiometric, with no dissociation effects. Initially assumed liquid phase mass fraction y_{ilo} was 0.4 for n-heptane. Then using equation 12, the corresponding liquid side mole fractions were determined for n-heptane and n-dodecane. The adiabatic flame temperature T_f for the mixture was determined as a next step using a computer code (developed in the present study) as 2475.47K. Following properties, $\overline{C}_{pg}$, $\overline{\rho}_l$, $\overline{\lambda}_g$, $\overline{T}_{b_i}$, $\sum_i (y_i v_i)$, $\sum_i (y_i L_i)$ and $\sum_i (Q_i y_i)$;(where a bar above symbols indicates mixture values) were then evaluated at a proper reference temperature and composition using relations provided elsewhere [37]. The mixture burning constant k_b was calculated from equation 19 as $5.67 \times 10^{-7} m^2 / s$ (0.567 mm^2 / s).

The gas side mole fraction for n-heptane x_{mgs} was calculated using equation 13 and the gas side mass fraction y_{mgs} was calculated using equation 14. Similarly x_{mgs}, y_{mgs} for n-dodecane were evaluated and $Y_{F,S}$ was determined using equation 15. Fractional vaporisation rate ε_m for n-heptane was obtained from equation 16 and finally the transfer number B_m (combustion) for n-heptane was determined using equation 20. Other properties such as liquid phase thermal diffusivity α_l, ρ_l, ρ_g, D_l and D_g for

n-heptane were evaluated and used as input parameters to the computer code. The results are depicted in Figs 10-11.

4.4. Plots of Heptane Mass Fraction Versus Dimensionless Radius and Time

The solution of equation 43 was obtained as a variation of n-heptane mass fraction with dimensionless radius r/R at each instant of time. Here r is the radial coordinate and R or r_l is the instantaneous droplet radius. The variation of n-heptane mass fraction as a function of dimensionless radius is depicted in Fig 10 for the present study and compared with the work of Shaw [22] for the same conditions. For both models, droplet was assumed to be at its boiling point. Both results showed the same trend. From an initial chosen value of n-heptane liquid phase mass fraction $y_{ilo} = 0.4$, the mass fraction of n-heptane remained nearly constant from the center of the droplet till the value of dimensionless radius r/R was about 0.76, then as the droplet surface was approached, n-heptane began to vaporise and its mass fraction or concentration decreased abruptly.

For the present work the minimum n-heptane concentration at the surface was about 0.155 whereas for Shaw [22] it was about 0.05. The reason for the concentration of n-heptane remaining nearly constant from the center of the droplet till dimensionless radius was about 0.76 can be attributed to the fact that liquid phase mass diffusion is very small as compared to liquid phase thermal diffusion. Hence higher the value of liquid phase Lewis number Le_l, higher will be the diffusional resistance offered by the liquid and smaller will be the vaporisation. Appropriate relations for determining thermophysical and transport properties were employed for the present work. For the given conditions, $Le_l = 10$ for the present model, whereas for Shaw [22], $Le_l = 7.8$, this difference in the values of Lewis number might be the factor responsible for the higher n-heptane fraction left at the surface for the present work.

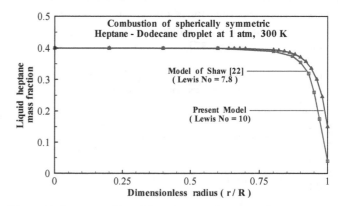

Figure 10. *Variation of liquid heptane mass fraction with dimensionless radius at a particular time*

Figure 11 shows the variation of n-heptane liquid surface mass fraction with dimensionless time for the present model, other conditions remaining same. The plot shows the liquid surface mass fraction of n-heptane varying from the initial chosen value of $y_{ilo} = 0.4$ to a value of about 0.155. The surface mass fraction of heptane initially decreases steadily with time suggesting preferential vaporisation of the more volatile species (which is n-heptane) in the earlier stages of droplet burning and hence high vaporisation rates resembling batch distillation type behaviour.

Then the rate of decrease becomes small and continues till the end of droplet lifetime, which may be because the bicomponent droplet is assumed to be at its boiling point ($426.895K$) throughout its burning period. This high boiling point temperature will lead to the vaporisation of less volatile (higher boiling point) component from the surface and hence less droplet vaporisation rate towards the end of droplet lifetime. Authors like Tong and Sirignano [12] and Aggarwal and Mongia [18] have also suggested the same explanation.

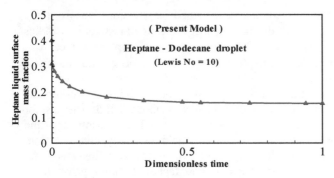

Figure 11. *Heptane liquid surface mass fraction variation with dimensionless time.*

4.5. Effect of Mixing of Air and Fuel Vapours on Vaporisation

A comparison of the variation of hexane liquid surface mass fraction with dimensionless time is made for two situations where mixing of air and fuel vapours is taken into account and the later case where mixing is not considered (Fig 12). The droplet was assumed to be vaporising at its boiling point in ambient conditions of 1 atm and 1000 K. Thermodynamic properties for hexane-decane mixture were calculated as a function of temperature and composition using appropriate relations [37].

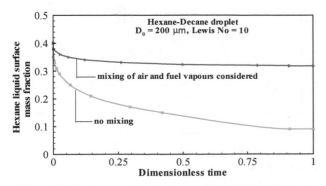

Figure 12. *Vaporisation of hexane-decane droplet (effect of mixing)*

It is observed that when mixing was neglected, a small amount of hexane fraction (about 0.09) was left at the surface at the end of droplet lifetime compared to approx

0.32 when mixing was considered. Further, the decrease in heptane mass fraction was gradual for the non-mixing case, while for the mixing case, apart from initial small decrease, the hexane mass fraction was nearly constant from dimensionless time value of about 0.5 onwards. This difference in vaporisation behaviour for the two cases could be attributed to the use of appropriate mixing rules accounting for air and fuel vapour mixing. The present result could be appreciated if experimental data was available for the same vaporising conditions, fuel and droplet size.

4.6. Effect of Lewis Number

Another important result (Fig 13) compares the effect of Lewis number on the variation of hexane liquid surface mass fraction with dimensionless time for a hexane-decane droplet, other conditions remaining same.

It was observed that the mass fraction of hexane at the droplet surface reduces gradually from an initial chosen value of 0.4 to about 0.09 at the end of lifetime, when $Le_l = 10$. Whereas for $Le_l = 30$, hexane mass fraction drops initially from 0.4 to 0.18 at dimensionless time equal to 0.16, after that becoming more or less constant till the end of droplet lifetime, where it is 0.175. The reason for more amount of hexane left at the droplet surface might be due to the high diffusional resistance offered to hexane (due to high Lewis number), discouraging it to come to the surface rapidly and hence its slow evaporation from the droplet surface. The trend in the result of the present study is in conformity with the results of other authors [19].

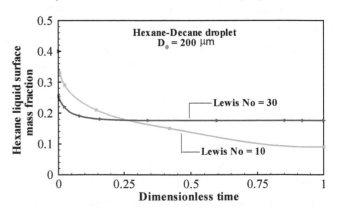

Figure 13. *Effect of Lewis number.*

4.7. Comparison between Different Vaporisation Models

As next part of the present work, different multicomponent droplet models (Figs 14-15) were compared for a heptane-dodecane droplet vaporising at 1 atmosphere and 1000 K. For all models, the initial liquid surface mass fraction was taken as 0.5. The present diffusion limit model (with convection) was evolved by solving transient-diffusive equations of species and energy within the liquid droplet and compared with the spherically symmetric d^2-*law* model of Law and Law [17] in Fig 14 and model of Tong and Sirignano and infinite diffusivity model in Fig 15. For the present model and that of Tong and Sirignano, $Re_g = 100$. The relevant properties of heptane-dodecane mixture for the present work were evaluated as function of temperature and composition using relations provided elsewhere [37].

It was observed that for model of Law and Law, there was a steady decrease of heptane surface mass fraction due to preferential vaporisation of the more volatile species. Eventually, later in the droplet lifetime, the droplet is completely depleted of the heptane component. Another reason for the steadily decreasing heptane mass fraction may be the lower value of Le_l chosen. Because of this lower value of 10, there is a smaller resistance offered to heptane, hence its faster evaporation from the liquid surface. The droplet lifetime is about 221 ms.

For the present model, droplet lifetime is evaluated as 65 ms. It is observed that there is a sudden initial decrease in the heptane liquid surface concentration and heptane is depleted completely from the droplet when the droplet lifetime is about 10%. The trend of the present model is quite similar when compared with the model of Tong and Sirignano [12] which is also known as vortex diffusion limit model. For both, the decrease in the liquid surface concentration of heptane is sudden, occurring earlier in the lifetime but the result of Tong and Sirignano indicates that all of the heptane is not depleted from the droplet, rather it levels off at a constant value of about 0.03.

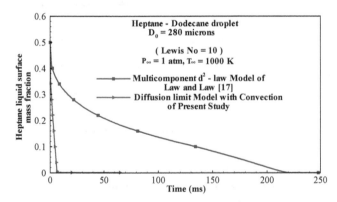

Figure 14. *Vaporisation behavior of a heptane-hexane-decane droplet (comparison between models)*

It is important to note that both the d^2-law model as well as infinite diffusivity model are quite approximate when compared to the present model and that of Tong and Sirignano. However, the present model offers less complexity, is quite accurate and can serve successfully in MC spray calculations.

The sudden decrease in heptane surface mass fraction earlier in the lifetime for both the studies may be due to the initially high fractional vaporisation rate of heptane which later becomes smaller as the droplet surface temperature increases but at the same time convection becomes more important leading to more efficient vaporisation which could have reduced the diffusional resistance. The difference in the results may be due to the value of Lewis number, which the authors [12] have not mentioned. The droplet lifetime for [12] is also a little higher (about 88 ms). This may be due to the bigger droplet size (360 μm).

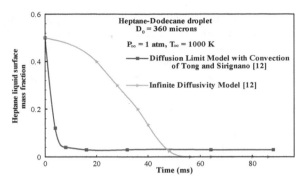

Figure 15. *Comparison of multicomponentdroplet vaporisation models*

5. Conclusions

A comprehensive, single component, diffusion controlled droplet combustion model was developed by solving numerically the time dependent energy and species conservation equations in the gas phase. It was observed that first the flame moves away from the droplet surface and then towards it, further, the flame to droplet diameter ratio was found to increase throughout the droplet burning period. These results are in conformity with the experimental observations and do not agree with the quasi-steady theory where the flame to droplet diameter ratio takes a large constant value.

The present droplet combustion model was further used to obtain the effect of fuels on important combustion parameters like dimensionless flame diameter, square of dimensionless droplet diameter and mass burning rate and a qualitative trend of emission characteristics such as CO_2, CO, NO and H_2O. It was noted that dimensionless flame diameter was influenced primarily by the fuel's boiling point. The droplet mass burning rate was the lowest for ethanol as compared to methyl linoleate (biodiesel) and n-heptane. Whereas, gasohol and diesohol proved out to be better blends than gasoline and diesel oil respectively with respect to NO and CO emissions.

In case of multicomponent droplet combustion, it was observed that for spherically symmetric combustion of a heptane-dodecane droplet, mass fraction of heptane remained nearly constant from droplet center till the value of dimensionless radius was approximately 0.76, then as droplet surface was approached, heptane began to vaporise and its mass fraction decreased rapidly to a lower value at a particular time of droplet burning.

For a 200 μm hexane-decane droplet (at its boiling point), vaporising in conditions of 1 atm and 1000 K with $Le_l = 10$, it was observed that when mixing of air and fuel vapour was ignored, a small amount of hexane mass fraction was left at the droplet surface at the end of droplet lifetime, whereas when mixing was considered, a higher concentration of hexane was present at the surface thereby altering the vaporisation behaviour. Also, an increase in Lewis number resulted in a higher mass fraction of hexane being present at the droplet surface.

A detailed multicomponent (MC) droplet vaporisation model (diffusion limit model with convection and no internal liquid circulation) was evolved by solving numerically (using finite difference technique) transient-diffusive equations of species and energy with appropriate boundary conditions for a 280 μm (heptane-dodecane) droplet vaporising at 1 atm and 1000 K with Re_g =100, and Le_l = 10. Roult's law was used for expressing phase equilibrium.Thermodynamic and transport properties were evaluated as a function of temperature and composition.

It was noted that there was a sudden initial decrease in the heptane liquid surface concentration from the initial assumed value of 0.5 and was completely depleted from the droplet surface very early in its lifetime. However, a 280 μm spherically symmetric heptane-dodecane droplet vaporising at its boiling point showed a steady decrease in heptane liquid surface mass fraction till the end of lifetime when it was vaporised completely. The present MC model was compared with other existing models and was found to be simpler but quite robust.

The predictions of the present work agreed well when compared with the experimental and theoretical results of other authors who have used more complicated models. The submodels developed in the present work were found to be accurate and yet simpler (requiring less CPU time) for their incorporation in spray codes where CPU economy plays a vital role.

Table 1. *Thermophysical Properties of Fuels at 298 K*

Fuel	Chemical formula	Molecular weight $kg/kmol$	Liquiddensit y kg/m^3	Latent heat kJ/kg	Liquid sp.ht kJ/kgK	Boilingpoint K	Heat of comb kJ/kg	Enthalpy of formation kJ/mol
n-Heptane	C_7H_{16}	100.205	684	316.54	2.245	371.6	44557	-187.82
n-Octane	C_8H_{18}	114.232	703	301.92	2.23	399	44425	-250.105
n-Decane	$C_{10}H_{22}$	142.284	730	272	2.21	447.1	44239	-249.65
n-Hexane	C_6H_{14}	86.177	659	335	2.27	342	44733	-167.19
n-Dodecane	$C_{12}H_{26}$	170.34	749	256.36	2.21	489.5	44109	-292.16
Methanol	CH_4O	32.042	787	1101	2.55	337.8	19910	-239.22
Ethanol	C_2H_6O	46.069	783	841.56	2.46	351.5	26811	-273.77
Methyl Linoleate	$C_{19}H_{34}O_2$	294.476	880	242.43	2.08	700.66	37830	-446.94

Table 2. *Comparison of Properties with Reported Values*

Fuel	Methyl Linoleate		Methyl Oleate		Methyl Stearate	
Property	Calculated[*]	Reported[**]	Calculated[*]	Reported[**]	Calculated[*]	Reported[**]
Critical Pressure (bar) P_c	11.91	11.91	11.68	11.68	11.46	11.46
Critical Temperature (K) T_c	871.80	795.3	864.57	772.3	857.41	774.2
Critical Volume(cm^3/mol) V_c	1085.5	1085.5	1105.5	1105.5	1125.5	1125.5

[*]Present study, [**]Yuan et al. {31]

Table 3. *Computed Values of Burning Parameters*

(T_∞=298 K ,P_∞=1atmosphere,$Y_{o,\infty}$=0.232) Fuel	$(A/F)_{stoich}$	B_T	k_b (mm^2/sec)	α_g (cm^2/sec)
n-Heptane	15.1008	8.765	1.036	1.403
n-Octane	15.053	8.947	0.832	1.23
n-Dodecane	14.94	9.66	1.0	1.66
Methanol	6.435	2.72	0.392	1.087
Ethanol	8.953	3.384	0.5582	1.096
Methyl Linoleate	11.16	9.78	0.823	1.973

Table 4. *Variation of CO, CO₂, H₂O and NO Concentrations for Different Fuels*

Fuels	Chemical formula	Adiabatic flame temperature(K)	NO conc(%)	CO conc(%)	CO₂ conc(%)	H₂O conc (%)
n-Heptane	C_7H_{16}	2396.17	3.446	3.522	9.768	14.112
Ethanol	C_2H_6O	2289.14	2.895	0.268	9.635	18.124
Methyl linoleate	$C_{19}H_{34}O_2$	2454.0	4.028	1.475	11.47	12.55
Gasoline	C_7H_{17}	2721.63	5.866	4.66	7.666	13.021
Diesel oil (DF2)	$C_{14}H_{30}$	2626.14	5.175	1.979	8.394	13.142
ATF (JP5)	$C_{12}H_{24}$	2707.31	5.758	2.704	7.979	12.406
Gasohol	70% gasoline + 30% ethanol	2387.25	3.399	0.75	9.977	14.527
Diesohol	80 % DF2 +20 % ethanol	2409.0	3.522	1.196	10.757	13.586

References

[1] S. Okajima and S. Kumagai, "Further Investigation of Combustion of Free Droplets in a Freely Falling Chamber Including Moving Droplets", Proceedings of the Fifteenth Symposium (International) on Combustion, The Combustion Institute, pp. 401-407, 1974.

[2] S. Kumagai and H. Isoda, "Combustion of Fuel Droplets in a Falling Chamber", Proceedings of the Sixth Symposium (International) on Combustion, Reinhold, N.Y., pp. 726-731, 1957.

[3] H. Isoda and S. Kumagai, "New Aspects of Droplet Combustion", Proceedings of the Seventh Symposium (International) on Combustion, The Combustion Institute, pp. 523-531, 1959.

[4] C. H. Waldman, "Theory of Non-Steady State Droplet Combustion", Fifteenth Symposium (International) on Combustion, The Combustion Institute, pp. 429-442, 1974.

[5] S. Ulzama and E. Specht, "An Analytical Study of Droplet Combustion Under Microgravity: Quasi-Steady Transient Approach", Proceedings of the Thirty First Symposium (International) on Combustion, The Combustion Institute, pp. 2301-2308, 2007.

[6] I. K. Puri and P. A. Libby, "The Influence of Transport Properties on Droplet Burning", Combustion Science and Technology, Vol. 76, pp. 67-80, 1991.

[7] M. K. King and "An Unsteady-State Analysis of Porous Sphere and Droplet Fuel Combustion Under Microgravity Conditions", Proceedings of the Twenty Sixth Symposium (International) on Combustion, The Combustion Institute, pp.1227-1234, 1996.

[8] J. R. Yang and S. C. Wong, "An Experimental and Theoretical Study of the Effects of Heat Conduction through the Support Fiber on the Evaporation of a Droplet in a Weakly Convective Flow", Int. J. Heat Mass Transfer, Vol. 45, pp. 4589-4598, 2002.

[9] C. K. Law, "Multicomponent Droplet Combustion with Rapid Internal Mixing", Combustion and Flame, Vol. 26, pp. 219-233, 1976.

[10] R. B. Landis and A. F. Mills, "Effects of Internal Diffusional Resistance on the Vaporization of Multicomponent Droplets", Fifth International Heat Transfer Conference, Paper B7.9, Tokyo, Japan, 1974.

[11] C. K. Law, "Internal Boiling and Superheating in Vaporizing Multicomponent Droplets", AIChE Journal, Vol.24, pp. 626-632, 1978.

[12] A. Y. Tong and W. A. Sirignano, "Multicomponent Droplet Vaporization in a High Temperature Gas", Combustion and Flame, Vol. 66, pp. 221-235, 1986.

[13] P. Lara-Urbaneja and W. A. Sirignano, "Theory of Transient Multicomponent Droplet Vaporization in a Convective Field", Proceedings of the Eighteenth Symposium (International) on Combustion, The Combustion Institute, pp.1365-1374, 1980.

[14] J. W. Aldred, J. C. Patel, and A. Williams,

[15] Combust. Flame. 17 (2) (1971) 139-148.

[16] A. J. Marchese and F. L. Dryer, "The Effect of Liquid Mass Transport on the Combustion and Extinction of Bicomponent Droplets of Methanol and Water", Combustion and Flame, Vol. 105, pp. 104-122, 1996.

[17] C. K. Law and H. K. Law, "On the Gasification Mechanisms of Multicomponent Droplets", pp. 29-48, Modern Developments in Energy, Combustion and Spectroscopy, Pergamon Press, 1993.

[18] C. K. Law and H. K. Law, "A d^2 – law for Multicomponent Droplet Vaporization and Combustion", AIAA Journal, Vol. 20, No. 4, pp. 522-527, 1982.

[19] S. K. Aggarwal and H. C. Mongia, "Multicomponent and High-Pressure Effects on Droplet Vaporization", Transactions of ASME, Vol. 124, pp. 248-255, 2002.

[20] W. A. Sirignano "Fluid Dynamics and Transport of Droplets and Sprays", Cambridge University Press, 1999.

[21] S. L. Lerner, H. S. Homan, and W. A. Sirignano, "Multicomponent Droplet Vaporization at High Reynolds Numbers: Size, Composition, and Trajectory Histories", Seventy Third Annual AIChE Meeting, 1980.

[22] S. K. Aggarwal, Y. A. Tong, and W. A. Sirignano, "A Comparison of Vaporization Models in Spray Combustion", AIAA Journal, Vol. 22, No.10, pp. 1448-1457, 1984.

[23] B. D. Shaw, "Studies of Influences of Liquid-Phase Species Diffusion on Spherically Symmetric Combustion of Miscible Binary Droplets", Combustion and Flame, Vol. 81, pp. 277-288, 1990.

[24] M. Mawid and S. K. Aggarwal, "Analysis of Transient Combustion of a Multicomponent Liquid Fuel Droplet", Combustion and Flame, Vol. 84, pp.197-209, 1991.

[25] N. W. Sorbo, C. K. Law, D. P. Y. Chang, and R. R. Steeper, "An Experimental Investigation of the Incineration and Incinerability of Chlorinated Alkane Droplets", Proceedings of the Twenty Second Symposium (International) on Combustion, The Combustion Institute, p. 2019, 1989.

[26] C. K. Law, T. Y. Xiong, and C. H. Wang, "Alcohol Droplet Vaporization in Humid Air", Int. J. Heat Mass Transfer, Vol. 30, p. 1435, 1987.

[27] Y. Zeng and Chia-fon F. Lee, "Modelling Droplet Breakup Processes Under Micro-explosion Conditions", Proceedings of the Thirty First Symposium (International) on Combustion, The Combustion Institute, pp. 2185-2193, 2007.

[28] R. E. Sonntag, C. Borgnakke, and G. J. Wylen, "Fundamentals of Thermodynamics", Sixth Edition, John Wiley and Sons, 2003.

[29] Choi et al, "Ethanol Droplet Combustion at Elevated Pressures and Enhanced Oxygen Concentrations", AIAA, 41st Aerospace Sciences Meeting and Exhibit, Reno, Nevada, pp. 1-7, 2003.

[30] A. K. Agarwal, "Biofuels (alcohols and biodiesel) Applications as Fuels for Internal Combustion Engines", Progress in Energy and Combustion Science, Vol. 33, pp. 233-271, 2007.

[31] A. Srivastava and R. Prasad, "Triglycerides-Based Diesel fuels", Renewable and Sustainable Energy Reviews, Vol. 4, pp. 111-113, 2000.

[32] W. Yuan, A. C. Hansen, and Q. Zhang, "Predicting the Physical Properties of Biodiesel for Combustion Modeling", ASAE Transactions, Vol. 46(6), pp.1487-1493, 2003.

[33] L. D. Clements, "Blending Rules for Formulating Biodiesel Fuel", Biological System Engineering, University of Nebraska - Lincoln, NE 68583-07266, pp. 44-53, 1996.

[34] W. Yuan, A. C. Hansen, and Q. Zhang, "Vapour Pressure and Normal Boiling Point Predictions for Pure Methyl Esters and Biodiesel Fuels", Fuel, Vol. 84, pp. 943-950, 2005.

[35] C. Olikara and G. L. Borman, "A Computer Program for Calculating Properties of Equilibrium Combustion Products with some Application to I.C. Engines", SAE Paper 750468, SAE Trans, 1975.

[36] M. K. Jain, "Numerical Solution of Differential Equations", Second Edition, Wiley-Eastern Limited, 1984.

[37] C. K. Law and W. A. Sirignano, "Unsteady Droplet Combustion with Droplet Heating – II: Conduction Limit", Combustion and Flame, Vol. 28, pp.175-186, 1977.

[38] R. C. Reid, J. M. Prausnitz, and B. E. Poling, "The Properties of Gases and Liquids", Fourth Edition, McGraw Hill Book Company, 1989.

12

Computational and Artificial Intelligence Study of the Parameters Affecting the Performance of Heat Recovery Wheels

Ahmed F. Abdel Gawad, Muhammad N. Radhwi, Asim M. Wafiah, Ghassan J. Softah

Mech. Eng. Dept., College of Eng. & Islamic Archit., Umm Al-Qura Univ., Makkah, Saudi Arabia

Email address:
afaroukg@yahoo.com (A. F. A. Gawad), mnradhwi@uqu.edu.sa (M. N. Radhwi), am907@hotmail.com (A. M. Wafiah), gjs_1982@hotmail.com (G. J. Softah)

Abstract: Heat recovery wheels represent key components in air handling units (*AHU*) that can be used in commercial and industrial building air-conditioning-systems for energy saving. For example, in health facilities, heat transfer process is to be applied in air-conditioning systems for heat recovery of the exhaust (return) air from the patient's room without contamination. Thus, heat recovery wheels are much suitable for such applications. Heat recovery wheels are also known as heat conservation wheels. A conservation wheel consists of a rotor with permeable storage mass fitted in a casing, which operates intermittently between two sections of hot and cold fluids. The rotor is driven by a low-speed electric motor. Thus, the two streams of exhaust and fresh air are alternately passed through the wheel. The present investigation considers computationally the different parameters that affect the operation of heat recovery wheels. These parameters signify actual operating conditions such as flow velocity, shape of cross-section of flow path, and wall material. Moreover, both the artificial neural network (*ANN*) and adaptive neuro-fuzzy inference system (*ANFIS*) techniques were utilized to predict the critical characteristics of the heat exchange system. These artificial intelligence techniques use the present computational results as training and verification data.

Keywords: Heat Recovery Wheels, Air Handling Units, Computational Study, Artificial Intelligence

1. System Description

1.1. Heat Recovery System

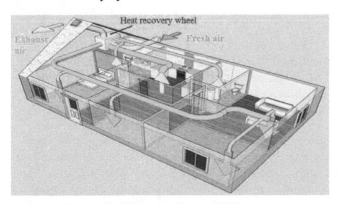

Fig. 1. Heat recovery system [1].

Figure1 shows a typical heat recovery system installed in a ventilation system. In the core, the fresh air stream is automatically preheated or pre-cooled (depending on the season) by the exhausted air and redirected to the cooling, heating or any demanding alternatives. The outgoing and incoming air streams pass next to each other but do not mix in the heat wheel; this providing guaranteed contaminant-free pre-heated or pre-cooled fresh air.

1.2. Heat Recovery Wheel

A thermal wheel consists of a matrix of heat-absorbing material, Fig. 2, which is slowly rotated within the supply and exhaust air streams of an air handling system. When cooling the interior of a building, as the thermal wheel rotates, heat is picked up from the fresh air stream in one-half of the rotation, and given up to the exhaust air stream in the other half of the rotation. Thus, heat energy from the hot fresh air stream is transferred to the matrix material and then from the matrix material to the cool exhaust air stream lowering the temperature of the supply air stream by an amount proportional to the temperature differential between air streams, or 'thermal gradient', and depending upon the efficiency of the device. Heat exchange is most efficient when the streams flow in opposite directions (counter flow), since this causes a favorable temperature gradient across the thickness of the wheel.

Because of the nature of thermal wheels in the way that heat is transferred from one stream to another without having to pass directly through or via an exchange medium, the gross efficiencies are usually much higher than that of any other air-side heat recovery system [3, 4].

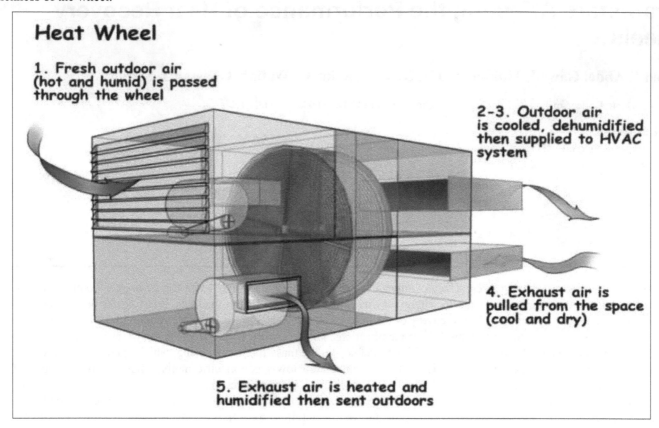

Fig. 2. Heat recovery wheel [2].

The heat recovery wheel consists of air channels and exchanger walls, which contain not only the adsorbent but also the base materials or the binders. The base material of the exchanger wall may be metal or fiberglass. As an example, the matrix pictures of exchanger wheels and the close-up of the front surface are shown in Fig. 3. As can be seen, they may have different values of width (20, 60, 100, 200, 400 *mm*) for the diameter of 350 *mm* [5].

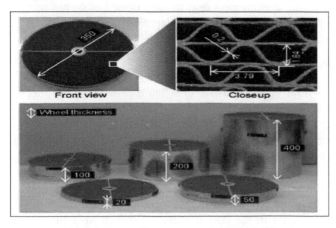

Fig. 3. Pictures of heat recovery wheels (mm) [5].

The major advantages of heat recovery wheel are [6]:
1. The system significant electrical power savings.
2. Chlorofluorocarbons (*CFCs*) free; thus, the system is environmentally friendly.
3. Construction and maintenance are simple.

2. Background

2.1. Previous Investigations

The design of *HVAC* (Heating, ventilation and air conditioning) systems for thermal comfort requires increased attention, especially matters arising from recent regulations and standards on ventilation [7]. The optimum temperature and humidity range for human comfort is presented by *ASHRAE* [8]. Building energy consumption and savings were highlighted by some investigators [9-11]. Much effort was devoted to the research and application of heat conservation wheel components, especially desiccant wheels [2, 12-24]. Also, many studies concentrated on the potential use of heat conservation wheel systems in various locations in the USA and Europe [5, 6, 9, 25-28]. Other researchers considered the mathematical models of rotary wheels [20, 28-30]. Moreover, artificial neural network (*ANN*) and adaptive neuro-fuzzy inference system (*ANFIS*) techniques were used by researchers in air conditioned systems [31-35] and other mechanical applications [36-42].

2.2. Present Work

In the present investigation, a computational study of the channels of heat recovery wheels was carried out for different materials and shapes. The commercial software-package (*ANSYS-F*luent 13) was exploited for parametric study of the influence of different parameters on heat transfer performance.

The investigated channels of the heat conservation wheel had five cross-sectional shapes, namely: circular, hexagonal, quadrangular, lozenge and sinusoidal. In addition, four different materials; steel, aluminium, nickel and copper were considered at different values of Reynolds number.

On the other hand, both the artificial neural network (*ANN*) and adaptive neuro-fuzzy inference system (*ANFIS*) methods were adapted to predict the important thermal characteristics. The present computational results were used as training and verification data for the artificial intelligence models.

3. Mathematical Model and Computational Features

3.1. Mathematical Model of Flow and Heat

It is assumed that the flow is incompressible, one-dimensional and turbulent through a straight channel, Fig.4. Thermal properties of the fluid (such as specific heat, thermal conductivity, density) are assumed to have bulk average values and be uniform at any cross-section. Mass flow rate is constant. Heat transfer is in the flow direction.

Based on these assumptions, differential equations relating the flow and thermal fields can be derived from the energy balance applied to a unit element through the channel of the heat wheel as a control volume. The details of the mathematical model can be found in [43-45].

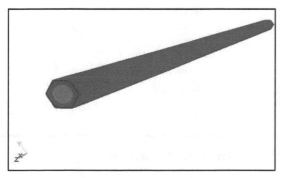

Fig. 4. Channel model.

3.2. Turbulence Modeling

The realizable k-ε turbulence model was used in the present study. The realizable k-ε model differs from the standard k-ε model in two important aspects: *(i)* the realizable k-ε model contains an alternative formulation for the turbulent viscosity. *(ii)* A modified transport equation for the dissipation rate, ε, has been derived from an exact equation for the transport of the mean-square vorticity fluctuation. Further details about the realizable k-ε turbulence model are found in [46].

3.3. Computational Domain and Boundary Conditions

The conservation wheel contains flow passages that form a matrix of similar channels. The present study concerns the flow and thermal fields through these channels. As these channels are similar, only one complete channel is considered. The flow was treated as steady without the effect of wheel rotation. The flow in the channel was kept hotter than the surroundings.

The computational domain, for all test cases, is a three dimensional (*3-D*) domain, Fig. 5. The ratio between the length of the channel and its hydraulic diameter (*L/d*) is (100/3).

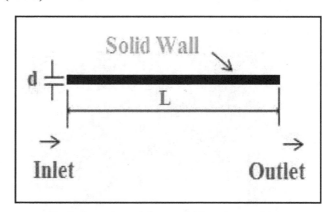

Fig. 5. Computational domain and boundary conditions.

Typically, there are three types of boundary conditions that can be listed as:

- Inlet boundary condition: inlet uniform velocity is specified at the entrance of the channel. According to *ASHRAE* handbook 2005 [47], the inlet temperature was taken as 319 *K*. This temperature value represents the annual average ambient temperature in Makkah.
- Outlet boundary condition: Pressure outlet flow condition and temperature of 300 *K* as outlet temperature.
- Wall boundary condition: the walls of the channel were treated as smooth. The no-penetration and no-sliding conditions were applied at the channel walls. The temperature at the inner surface of the walls was kept at 309 *K* for all test cases. This makes a temperature difference of 10 *K* between the inlet flow and the inner surface. This difference value was recommended in the literature [43]. As will be seen in the following sections, the wall material changes depending on the test case.

3.4. Computational Grid (Mesh)

Due to the complexity of the shapes of the channels, the

computational domain was discretized using unstructured grids. This type of grids usually guarantees the flexibility to generate enough computational points in locations of severe gradients. Unstructured grids adopt themselves easily to irregular geometries with minimum programmer's effort.

The computational domain was covered by tetrahedral-shaped elements, (Fig. 6). The grid is very fine next to the solid boundary. The dimensionless distance between the wall and first computational point is y+ ≈ 2. y+ is calculated as $y^+ = \frac{u_\tau y}{v}$. Where, y is the distance to the first point off the wall, v is the kinematic viscosity, u_τ is the friction velocity $(u_\tau = \sqrt{\frac{\tau}{\rho}})$, τ is the wall shear stress, and ρ is the flow density. The value of y+ = 2 ensures the resolution of the complex turbulent flow [44, 45].

Careful consideration was paid to ensure the grid-independency of the computational solution. Thus, three grid sizes were used to test the grid-independency, namely: 100,000,150,000 and 200,000 elements (cells). The results of both the flow and thermal fields show that the difference between the results of the second and third grid is in the range of 1-2 %. Thus, the second grid size (150,000) was used for all the test cases.

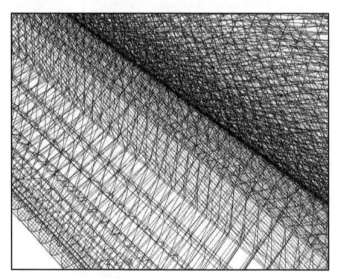

Fig. 6. Enlarged grid structure (sample).

3.5. Numerical Treatment

SIMPLE (semi-implicit method for pressure-linked equations) algorithm was employed to solve both the velocity and pressure fields. Each equation of the momentum and energy was solved by "first-order upwind" scheme. The "standard wall function" was used as the near-wall treatment technique in the turbulence model. The solution continues until the numerical residual (error) of all quantities gets below 10^{-5}.

3.6. Test Cases

A parametric study was carried out to obtain the best case for optimum operation. Different cases were computationally tested and evaluated. The parametric study covered the shape of the channel cross-section, the wall material, and the flow

Reynolds number.

Five cross-sectional shapes were considered; circular, hexagonal, quadrangular, lozenge, and sinusoidal. The sinusoidal cross-section is widely used in conservation wheels. It is used here mainly for comparison. The other four shapes are proposed by the authors. The circular shape has good characteristics when considering flow and thermal fields. On the other hand, the hexagonal, quadrangular, and lozenge shapes have good structural properties.

Four materials were tested as wall of the channels, namely: copper, aluminium, nickel and steel. The calculations were performed at six values of Reynolds numbers; 1200, 1540, 1830, 1950, 2200, and 2300. Reynolds number is calculated based on inlet velocity and the hydraulic diameter of the channel, *i.e.*, $Re = \frac{u_{in} d}{v}$. Where, u_{in} is the inlet velocity , d is the hydraulic diameter (3 *mm*), v is the kinematic viscosity. This combination of parameters leads to twenty four (24) test cases. Table.1 shows a summary of all the studied test cases.

***Table 1.** Summary of the test cases.*

Shape	Material	Reynolds Number
Circular	Copper	1540
	Aluminium	1950
	Nickel	
	Steel	2300
Hexagonal	Copper	1540
	Aluminium	1950
	Nickel	
	Steel	2300
Quadrangular	Copper	1200
	Aluminium	1830
	Nickel	
	Steel	2200
Lozenge	Copper	1200
	Aluminium	1830
	Nickel	
	Steel	2200
Sinusoidal	Copper	1200
		1540
	Aluminium	1830
	Nickel	1950
	Steel	2200
		2300

3.7. Validation of the Numerical Results

The present numerical results compare very well to the data obtained by others [5] as shown in Fig. 7 for the temperature contours. Generally, the same trend of temperature distribution and gradient is obtained in both Figs. 7a and 7b. The red color means the highest temperature and the blue color means the lowest temperature. More gradient contours appear in Fig. 7b. This may be attributed to the secondary flow that is not considered in the present computational investigation.

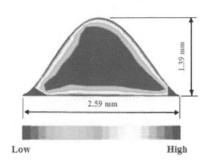

Fig. 7a. *Present computational predictions.*

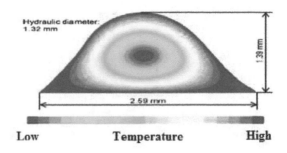

Fig. 7b. *Results of others [5].*

Fig. 7. *Cross-sectional view of the temperature field.*

4. Computational Results and Discussions

This section demonstrates the results of the computations of the test cases that were mentioned in Sec. 3.6 and Table 1. The carried-out parametric study aims to find the combination of parameters that gives the optimum performance of the wheel. Optimum performance, here, means best heat transfer exchange rate. However, the results must cover both the flow and thermal fields. The flow filed concerns the working fluid (air) that transfers heat. Thus, the flow field of the different cases is presented firstly. Then, the thermal filed results are presented. Naturally, concentration is paid to the thermal filed as heat transfer optimization is the principal goal of the study.

4.1. Flow Field

Figure 8 shows the velocity contours along the channels as well as at the entrance and exit of the channel for different cross-sections (shapes) and Reynolds number (*Re*) with aluminium wall. It is clear from the velocity contours at exit that the flow becomes fully-developed before the channel exit. For non-circular cross-sections, the highest velocity is seen in the middle of the channel cross-section. The same pattern is repeated typically for all test cases in Fig. 8. This ensures the fully-developed nature of the flow inside the channel. However, the entrance length may vary from one case to another. This variation of entrance length affect greatly the thermal filed as will be explained in the coming sections.

Figure 9 shows the pressure distribution along the channels as well as at entrance and exit for different cross-sections (shapes) and Reynolds number (*Re*) with aluminium wall. It is clear from Fig. 9 that the flow in the channel causes a considerable pressure drop from the entrance to the exit. The same pattern is repeated typically for all test cases in Fig. 9.

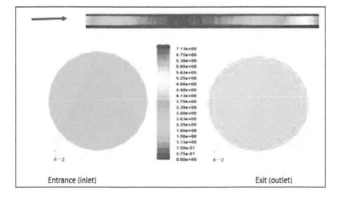

Fig. 8a. *Circular, Re = 1950.*

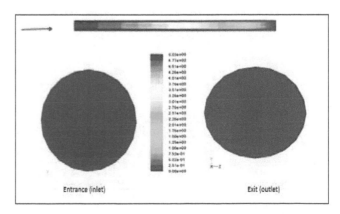

Fig. 9a. *Circular, Re = 1950.*

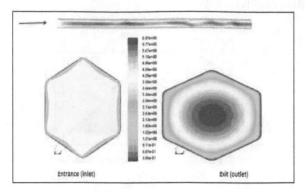

Fig. 8b. *Hexagonal, Re = 1950.*

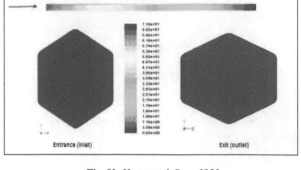

Fig. 9b. *Hexagonal, Re = 1950.*

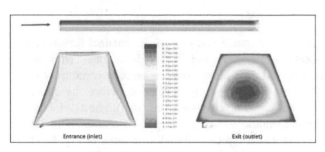

Fig. 8c. *Quadrangular, Re = 1830.*

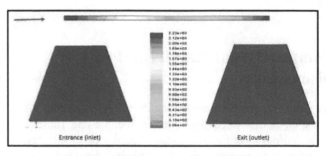

Fig. 9c. *Quadrangular, Re = 1830.*

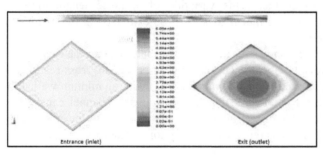

Fig. 8d. *Lozenge, Re = 1830.*

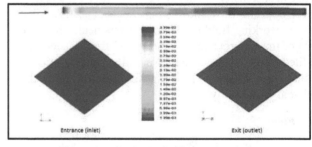

Fig. 9d. *Lozenge, Re = 1830.*

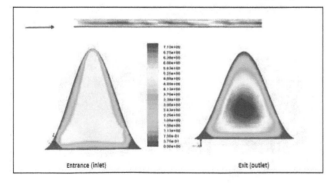

Fig. 8e. *Sinusoidal, Re = 1830, 1950,.*

Fig. 8. *Velocity contours, aluminium wall.*

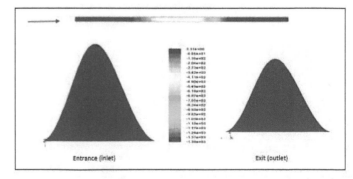

Fig. 9e. *Sinusoidal, Re = 1830, 1950.*

Fig. 9. *Pressure contours, aluminium wall.*

4.2. Thermal Field

To estimate the effect of different parameters on the heat transfer along the channel a certain criterion should be established. This criterion concerns the different thermal zones along the channel in the streamwise direction. Thus, the length of the channel was divided into three consequence zones; hot, middle and cold. For generality the lengths of the three zones are normalized by the total length of the channel. Hence, these lengths can be defined as, Fig. 10:

- Hot zone percentage relative length:
$$L_h = \frac{\text{Length of hot zone}}{\text{Channel total length}} \times 100$$
- Middle (Transitional) zone percentage relative length:
$$L_m = \frac{\text{Length of middle zone}}{\text{Channel total length}} \times 100$$
- Cold zone percentage relative length:
$$L_c = \frac{\text{Length of cold zone}}{\text{Channel total length}} \times 100$$

Apparently, the greater the value of L_c, the better the heat transfer process is. This criterion was implemented in the present study to determine the combination of parameters that gives the optimum heat transfer process.

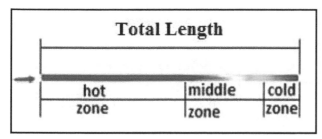

Fig. 10. Thermal zones along the channel length.

Figure 11 shows the temperature results for the different cross-sections (shapes) at corresponding values of Reynolds number with various wall materials. The results cover temperature contours as well as the relative lengths of the three thermal zones along the channels.

The first column of Fig. 11 shows the temperature contours along the channel as well as at entrance and exit sections at the corresponding Reynolds number and aluminium wall. Generally, the temperature inside the channel drops from the inlet temperature to the wall temperature well-before exiting the channel. The pattern of temperature contours at entrance and exit of the channel is repeated for all cases. However, the temperature contours along the channel length changes from one case to another depending on the case parameters. The second column of Fig. 11 shows the temperature contours along the channel length for all cases. Based on the results of the second column of Fig. 11, the percentage relative lengths of the thermal zones

along the channel for different cases are shown in the third column of Fig. 11.

Figure 11a shows the results of the temperature contours along the channel as well as at entrance and exit sections for the circular cross-section. It is clear that L_c represents most of the length of the channel for all cases. The maximum value of L_c is 0.91 at $Re = 1540$ with aluminium wall. Hence, the circular shape gives a noticeable good thermal performance.

Figure 11b shows the results for the hexagonal cross-section. Figure 11b(*iii*) shows that, at the same Reynolds number of 1950, the longest cold zone has a percentage relative lengths of $L_c = 25\%$ and $L_h = 59\%$ for the copper material. When considering aluminium as the wall material at different values of Reynolds number, the longest cold zone has a percentage relative lengths of $L_c = 31\%$ and $L_h = 58\%$ at $Re = 1540$.

Figure 11c shows the results for the quadrangular cross-section. It is clear from Fig. 11c(*iii*) that, at the same Reynolds number of 1830, the longest cold zone has a percentage relative length of $L_c = 94\%$ with corresponding length of hot zone $L_h = 3\%$ for the copper material. When considering aluminium as the wall material at different values of Reynolds number, the longest cold zone has a percentage relative length of $L_c = 95\%$ with corresponding length of hot zone $L_h = 3\%$ at $Re = 1200$.

Figure 11d shows the results for the Lozenge cross-section. It is obvious from Fig. 11d(*iii*) that, at the same Reynolds number of 1830, the longest cold zone has a percentage relative length of $L_c = 21\%$ with corresponding length of hot zone $L_h = 71\%$ for copper. When considering aluminium as the wall material at different values of Reynolds number, the longest cold zone has a percentage relative length $L_c = 75\%$ with corresponding length of hot zone $L_h = 15\%$ at $Re = 1200$.

Figure 11e shows the results for the sinusoidal cross-section at $Re = 1830$ and 1950. Figure 11e(*iii*) illustrates that, at the same Reynolds numbers of 1830 and 1950, the longest cold zone has a percentage relative length of $L_c = 65\%$ with corresponding length of hot zone $L_h = 18\%$ for copper. When considering aluminium as the wall material at different values of Reynolds number, the longest cold zone has a percentage relative length of $L_c = 88\%$ with corresponding length of hot zone $L_h = 7\%$ at $Re = 1200$ and 1540, while for $Re = 2200$, $L_c = 35\%$ and $L_h = 45\%$.

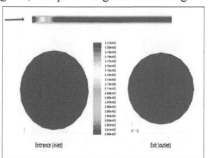

Fig. 11a(i). Temperature contours, Re=1950, aluminium wall.

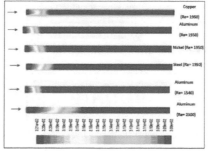

Fig. 11a(ii). Temperature contours for all materials.

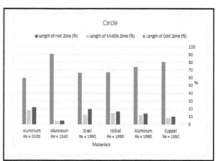

Fig. 11a(iii). Bar chart for identifying the three zones.

Fig. 11a. Circular cross-section.

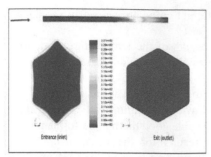

Fig. 11b(i). *Temperature contours, Re=1950, aluminium wall.*

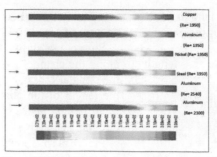

Fig. 11b(ii). *Temperature contours for all materials.*

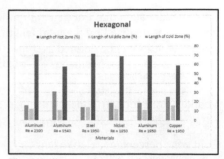

Fig. 11b(iii). *Bar chart for identifying the three zones.*

Fig. 11b. *Hexagonal cross-section.*

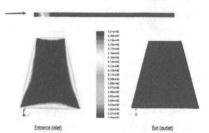

Fig. 11c(i).*Temperature contours, Re=1830, aluminium wall.*

Fig. 11c(ii). *Temperature contours for all materials.*

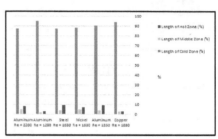

Fig. 11c(iii). *Bar chart for identifying the three zones.*

Fig. 11c. *Quadrangular cross-section.*

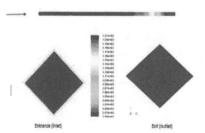

Fig. 11d(i). *Temperature contours, Re=1830, aluminium wall.*

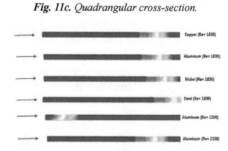

Fig. 11d(ii). *Temperature contours for all materials.*

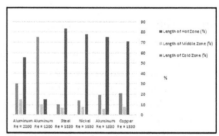

Fig. 11d(iii). *Bar chart for identifying the three zones.*

Fig. 11d. *Lozenge cross-section.*

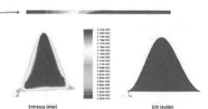

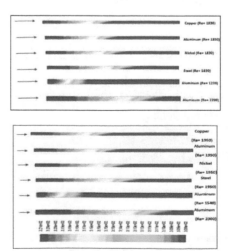

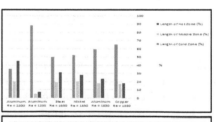

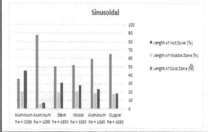

Fig. 11e(i). *Temperature contours, Re=1830, 1950, aluminium wall.*

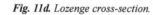
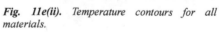

Fig. 11e(ii). *Temperature contours for all materials.*

Fig. 11e(iii). *Bar chart for identifying the three zones.*

Fig. 11e. *Sinusoidal cross-section.*

Fig. 11. *Temperature results.*

4.3. Overall View of All Cases

Based on the results of the previous section an overall view of all cases can be demonstrated. Table 2 illustrates overall results of the percentage relative lengths of all cases.

Table 2. Overall study results.

Shape	Material	Length of Cold Zone (%)	Length of Middle Zone (%)	Length of Hot Zone (%)
Circular	Copper Re = 1950	81	9	10
	Aluminium Re = 1950	74	12	14
	Nickel Re = 1950	68	15	17
	Steel Re = 1950	67	13	20
	Aluminium Re = 1540	91	4	5
	Aluminium Re = 2300	60	18	22
Hexagonal	Copper Re = 1950	25	16	59
	Aluminium Re = 1950	19	11	70
	Nickel Re = 1950	19	12	69
	Steel Re = 1950	14	14	72
	Aluminium Re = 1540	31	11	58
	Aluminium Re = 2300	16	12	71
Quadrangular	Copper Re = 1830	94	3	3
	Aluminium Re = 1830	90	4	9
	Nickel Re = 1830	88	5	7
	Steel Re = 1830	87	4	9
	Aluminium Re = 1200	95	2	3
	Aluminium Re = 2200	87	5	8
Lozenge	Copper Re = 1830	21	8	71
	Aluminium Re = 1830	19	6	75
	Nickel Re = 1830	14	8	78
	Steel Re = 1830	10	7	83
	Aluminium Re = 1200	75	10	15
	Aluminium Re = 2200	30	15	55
Sinusoidal	Copper Re = 1830, 1950	65	17	18
	Aluminium Re = 1830, 1950	59	18	23
	Nickel Re = 1830	52	20	28
	Steel Re = 1830, 1950	50	19	31
	Aluminium Re = 1200, 1540	88	5	7
	Aluminium Re = 2200, 2300	35	20	45

From Table 2, the value of L_c decreases with Reynolds number. Thus, decreasing the value of Reynolds number improves the heat transfer process. However, the value of Reynolds number is controlled by the actual operating conditions of the heat recovery wheel. In addition, at the same Reynolds number, copper shows the best heat transfer performance. Nevertheless, copper is more expensive than aluminium.

Generally, the "Circular" and "Quadrangular" are the two shapes that give the best heat transfer response in comparison to the "Sinusoidal" shape, which is commonly used. However, they may be not considered as the optimum choice from construction point of view.

It is concluded that the optimum thermal cooling cases can be stated as:

1. When considering the same range of Reynolds number (1830-1950) with different materials: Re = 1830, copper wall, quadrangular shape, L_c = 94%.
2. When considering the same material (aluminium) with different Reynolds numbers: Re = 1200, quadrangular shape, L_c = 95%.

5. Artificial Intelligence (*AI*)

Both the two artificial intelligence techniques (*AI*); artificial neural network (*ANN*) and adaptive neuro-fuzzy inference system (*ANFIS*) were used to predict the heat transfer performance of the present cases. This strategy was intended to demonstrate and judge the ability of both *ANN* and *ANFIS* to predict the heat transfer performance. Moreover, the advantages/disadvantages of the two *AI* techniques can be clearly noticed. Thus, the predictions of the investigated cases were divided between the two *AI* techniques. The results of the present computations were used as input data to the two *AI* techniques in their training phase.

5.1. Artificial Neural Network (ANN)

An *ANN* was used to predict the percentage relative length L_c for different materials (copper, aluminium, nickel, steel), different Reynolds numbers (1540, 1950, 2300), and different shapes (circular, hexagonal, sinusoidal). A robust learning heuristic for multi-layered feed-forward neural network, called the Generalized Delta Rule (*GDR*) or "Back Propagation Learning Rule", was implemented in the present

work. The back propagation learning rules are used to adjust the weights and biases of *ANN*s to minimize the sum-squared error of the network. This is done by continually changing the values of the network weights and biases in the direction of steepest descent with respect to error. The used back propagation of a multilayer *ANN* is shown in Fig. 12.

The neural network toolbox of the Matlab 7.6 package was chosen to train the *ANN* [48]. The input data comprises three groups (vectors) of data, which are wall materials, Reynolds numbers, and shapes. The output data comprises one group (vector) of data, which is L_c. The computational input and output data were used to train the *ANN* till it can approximate a function that associates input vectors with specific output. The initial weights as well as the initial biases employed random values between 1 and -1. The nodes in the hidden layer were varied from 2 to 18 for every input pattern and the evaluation of the performance of the network in determining the optimum hidden nodes was carried out. Also, many training cases were operated to get the optimum transfer functions arrangement.

ANN was trained for eighteen different cases. After the *ANN* was trained, a separate set of unseen test data was supplied as input to the *ANN* and its performance was evaluated. Table 3 shows the input and the output for each training case. Results for many training cases that were carried out are presented in Table 4. These results are obtained after 1000 epochs. The epoch is a complete iteration of *ANN*.

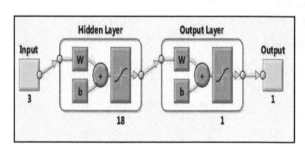

Fig. 12. *Back propagation of two-layer network with one output layer and one hidden layer [4].*

Table 3. *ANN training cases.*

Case	1	2	3	4	5	6	7	8	9	10	11	12	13	14	15	16	17	18
Shape	Circular						Hexagonal						Sinusoidal					
Input-1	0.1	0.1	0.1	0.1	0.1	0.1	0.2	0.2	0.2	0.2	0.2	0.2	0.3	0.3	0.3	0.3	0.3	0.3
Input-2	1	2	3	4	2	2	1	2	3	4	2	2	1	2	3	4	2	2
Input-3	1950	1950	1950	1950	1540	2300	1950	1950	1950	1950	1540	2300	1950	1950	1950	1950	1540	2300
Output	81	74	68	67	91	60	25	19	19	14	31	16	65	59	52	50	88	35

Key: Input-1 (Shape): (0.1) Circular, (0.2) Hexagonal, (0.3) Sinusoidal. Input-2 (Material): (1) Copper, (2) Aluminium, (3) Nickel, (4) Steel. Input-3 (Reynolds number): 1540, 1950, 2300. Output: L_c.

Table 4. *Results of ANN test cases.*

CASE	F1	F2	S1	SSE	SSET
Circular	Logsig	Purelin	6	0.0573	0.0734
Hexagonal	Logsig	Purelin	6	0.0607	0.0764
Sinusoidal	Logsig	Purelin	6	0.0418	0.0697
All Shapes	Logsig	Purelin	18	0.0217	0.0415

Key: F1: Transfer function of the first layer. F2: Transfer function of the second layer. S1: Number of neurons in the hidden layer. SSE: Sum-squared error for training vector. SSET: Sum-squared error for testing vector.

It is obvious from table 4 that the increase of number of neurons in the hidden layer (S1) enhances the performance of *ANN*. Thus, when training cases of "*All Shapes*" are considered, the lowest values of errors (SSE, SSET) are obtained. The predictions of L_c based on new set of input data that has not been used in the training of the neural networks are shown in Table 5.

Figure 13 shows a comparison between computational results and *ANN* predictions. Based on the results (outputs) of Table 5, Fig. 13, and the previous Tables 3 and 4, it can be said with confidence that the predictions of Table 5 are very good and acceptable. Hence, *ANN*s are attractive tools for quick and reliable prediction.

Table 5. *New input predictions.*

Inputs			Output
Material	Reynolds number	Channel shape	L_c (%)
2	1600	0.3	84
2	2000	0.3	56
1	2000	0.3	61
2	1600	0.2	27
2	1850	0.2	22
1	1850	0.2	28
2	1440	0.1	93
2	1850	0.1	77
1	2000	0.1	78

Key: Material: (1) Copper, (2) Aluminium, (3) Nickel, (4) Steel. Reynolds number: 1540, 1950, 2300. Channel shape: (0.1) Circle, (0.2) Hexagonal, (0.3) Sinusoidal.

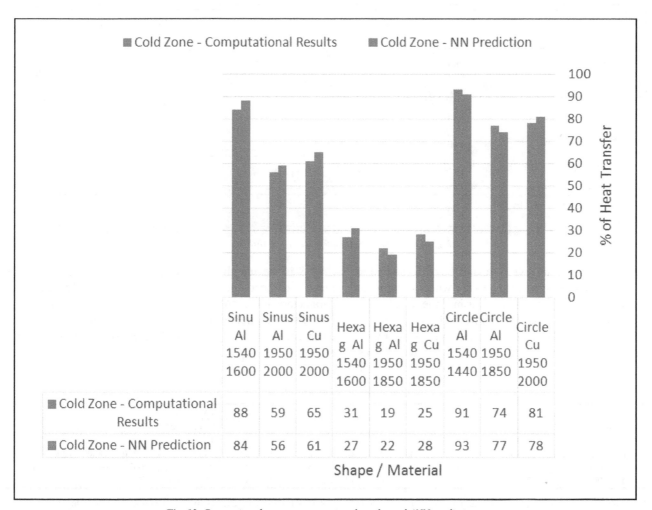

Fig. 13. *Comparison between computational results and ANN predictions.*

5.2. Adaptive Neuro-Fuzzy Inference System (ANFIS)

The basic structure of the fuzzy inference system (*FIS*), Fig. 14, is a model that maps input characteristics to input membership functions (*MF*), input membership function to rules, rules to a set of output characteristics, output characteristics to output membership functions (*MF*), and the output membership functions to a single-valued output or a decision associated with the output [48].

The parameters (*e.g.*, material, Reynolds number, and channel shape) associated with a given membership function could be chosen so as to tailor the membership functions to a collection of input/output data in order to account for the variations in the data values. This is where the so-called neuro-adaptive learning techniques can help. These techniques provide a method for the fuzzy modeling procedure to learn information about a data set in order to compute the best membership function parameters.

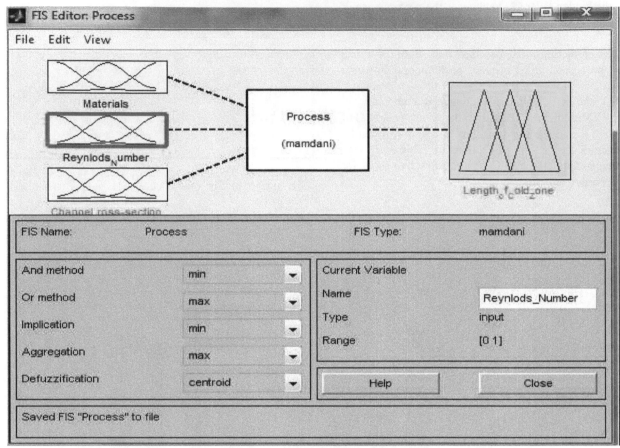

Fig. 14. *Structure of fuzzy inference system (FIS).*

In the present study, four *ANFIS* models were developed to predict the performance of a heat recovery wheel in terms of its heat transfer efficiency (output) using the *toolbox of Matlab 7.6* [48]. The predictions were based on the present computational results. There are three input sets to the *ANFIS* predictors, namely: material (*Input 1*), Reynolds number (*Input 2*), and channel shape (*Input 3*). The output of the *ANFIS* is the length of cold zone (L_c). Table 6 shows the input and the output for each training case. The different parameters of the *ANFIS* models are given in Table 7. All the four *ANFIS* models were trained until the sum of the squared error reaches 0.006. Thirty epochs were sufficient for the four models.

Table 6. *ANFIS training cases.*

Case	1	2	3	4	5	6	7	8	9	10	11	12	13	14	15	16	17	18
Shape	Quadrangular						Lozenge						Sinusoidal					
Input-1	1	2	3	4	2	2	1	2	3	4	2	2	1	2	3	4	2	2
Input-2	1830	1830	1830	1830	1200	2200	1830	1830	1830	1830	1200	2200	1830	1830	1830	1830	1200	2200
Input-3	0.4	0.4	0.4	0.4	0.4	0.4	0.5	0.5	0.5	0.5	0.5	0.5	0.6	0.6	0.6	0.6	0.6	0.6
Output	94	90	88	87	95	87	21	19	14	10	75	30	65	59	52	50	88	35

Key: Input-1 (Materials): (1) Copper, (2) Aluminium, (3) Nickel, (4) Steel. Input-2 (Reynolds number): 1200, 1830, 2200. Input-3 (Channel shape): (0.4) Quadrangular, (0.5) Lozenge, (0.6) Sinusoidal. Output: L_c

Table 7. *Different parameters for the four ANFIS models.*

ANFIS Model	Number of *MF* (Inputs)	Input *MF* type	Output *MF* type	Rules of Fuzzy	Optimization Method	No. of Epochs
Quadrangular	2×6	Gaussmf	Linear Grid	Partition	Hybrid	30
Lozenge	2×6	Gaussmf	Linear Grid	Partition	Hybrid	30
Sinusoidal	2×6	Gaussmf	Linear Grid	Partition	Hybrid	30
All Shapes	3×18	Gaussmf	Linear Grid	Partition	Hybrid	30

The structure of the *ANFIS* model for quadrangular shape (as an example of the four *ANFIS* models) is shown in Fig. 15. The output surface of performance, Fig. 16, explains the variation of quadrangular predictions with the three inputs (material, channel shape, Reynolds number).

After *ANFIS* had been trained, it was tested by checking its predictions by a set of data that had not been seen by *ANFIS* before, Table 8. The comparison between present computational results and *ANFIS* predictions for values of L_c is shown in Fig. 17.

When comparing the results (output) of Table 8 and Fig.

17, with the training Table 6, it is clear that the results (output) of Table 8 was very reasonable and well-accepted. Thus, *ANFIS* proves to be an effective and easy-to-use prediction technique.

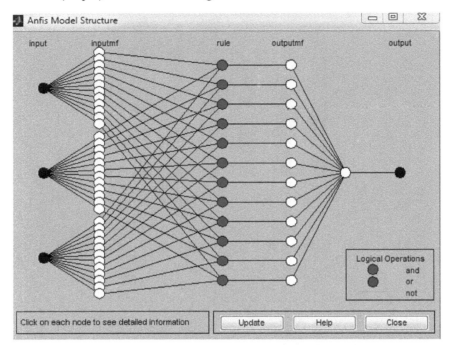

Fig. 15. *Structure of the ANFIS model (quadrangular prediction).*

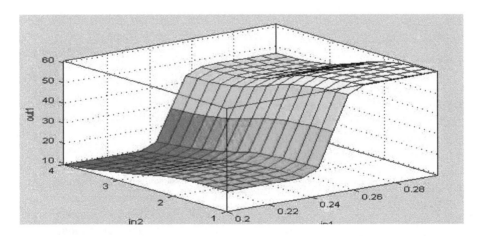

Fig. 16. *Output surface of performance of the ANFIS model (quadrangular prediction).*

Table 8. *New input predictions.*

Inputs			Output
Material	**Reynolds number**	**Channel shape**	L_c (%)
2	1300	0.6	82
2	1750	0.6	61
1	1900	0.6	62
2	1100	0.4	96
2	1750	0.4	92
1	1750	0.4	96
2	1300	0.5	70
2	1900	0.5	15
1	1900	0.5	19

Key: Materials: (1) Copper, (2) Aluminium, (3) Nickel, (4) Steel. Reynolds number: 1100, 1300, 1750, 1900. Channel shape: (0.4) Quadrangular, (0.5) Lozenge, (0.6) Sinusoidal.

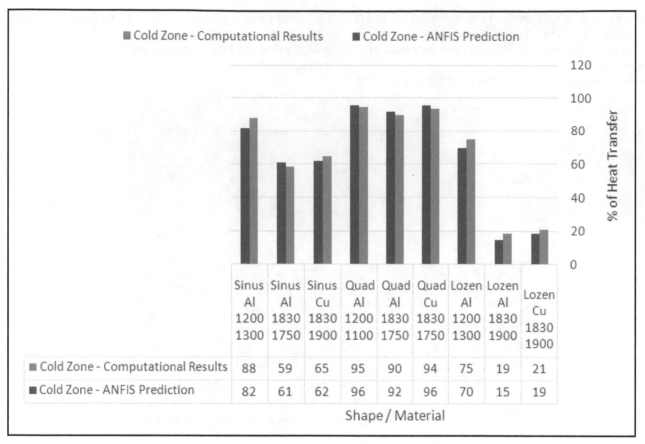

Fig. 17. Comparison of heat transfer between present computational results and ANFIS predictions.

6. Conclusions

In the present investigation, a computational study of heat recovery wheel was carried out for different materials and cross-sectional shapes. Also, the Reynolds number of flow was considered as important parameter. The main objective of the study is to find the combination of parameters that give the optimum performance (best heat transfer). The performance was evaluated by modeling the channel flow of the wheels. Also, artificial intelligence techniques (*ANN, ANFIS*) were proposed to predict the thermal behavior of the investigated cases.

Based on the above discussions, the following concluding points can be stated:

1. The present computational model and solution scheme were effectively validated. Hence, the outputs of present study are to be considered with confidence.

2. Generally, the circular and quadrangular shapes are the two shapes that give the finest heat transfer response in comparison to the "Sinusoidal" shape, which is commonly used. However, they may be not considered as the optimum choice from construction point of view.

3. It can be stated that the optimum thermal cooling cases are as follows:
   - When considering the same range of Reynolds number (1830-1950) with different materials: $Re = 1830$, copper material, quadrangular shape, $L_c = 94\%$.
   - When considering the same material (aluminium)

with different Reynolds numbers: $Re = 1200$, quadrangular shape, $L_c = 95\%$.

4. Although the hexagonal shape has good structural advantages, it is not the best choice when concerning the thermal performance.

5. The circular cross-section (shape), which is simple, gives much better thermal performance than the sinusoidal cross-section that is commonly used in heat exchange wheels. However, when using circular channels to construct the recovery wheel, there will be channels between them that are not circular. Thus, the overall performance may deteriorate.

6. For all cases, the wall materials can be arranged from best to worst as: copper, aluminium, nickel, and steel when considering the thermal performance.

7. It is very important to state that the above concluding points are based on the stated flow and thermal fields. Other concluding remarks may be obtained for other operating conditions.

8. The artificial intelligence techniques (*ANN, ANFIS*) proved to be very effective in predicting the thermal characteristics. They are much easier to program and operate than other computational techniques. However, they need computational and/or experimental data for training.

9. Energy saving approach adapted in this study guarantees contaminant-free pre-heated or pre-cooled fresh air, which is the major concern in health care

facilities and large crowded commercial structures.

Nomenclature

d	: Channel hydraulic diameter
F1	: Transfer function of the first layer
F2	: Transfer function of the second layer
L	: Channel length
L_c	: Cold zone percentage relative length: $= \frac{\text{Length of cold zone}}{\text{Channel total length}} \times 100$
L_h	: Hot zone percentage relative length: $= \frac{\text{Length of hot zone}}{\text{Channel total length}} \times 100$
L_m	: Middle zone percentage relative length: $= \frac{\text{Length of middle zone}}{\text{Channel total length}} \times 100$
Re	: Reynolds number $= \frac{u_{in}\,d}{v}$
S1	: Number of neurons in the hidden layer
u_τ	: friction velocity $= \sqrt{\frac{\tau}{\rho}}$
u_{in}	: Inlet velocity
y	: Distance to the first computational point off the wall
y^+	: Dimensionless distance between the wall and first computational point

Greek

v	: Kinematic viscosity
ρ	: Flow density
τ	: Wall shear stress

Abbreviations

AHU	: Air Handling Unit
AI	: Artificial intelligence
ANFIS	: Adaptive neuro-fuzzy inference system
ANN	: Artificial neural network
ASHRAE	: American Society of Heating, Refrigerating and Air Conditioning Engineers
CFC	: Chlorofluorocarbons
FIS	: Fuzzy inference system
GDR	: Generalized delta rule
HVAC	: Heating, Ventilation and Air Conditioning
MF	: Membership functions
SIMPLE	: Semi-implicit method for pressure-linked equations
SSE	: Sum-squared error for training vector
SSET	: Sum-squared error for testing vector

References

[1] S. B. Riffat, and G., Gan, "Determination of Effectiveness of Heat-pipe Heat Recovery for Naturally-ventilated Buildings", Applied Thermal Engineering, Vol. 18, No. 3-4, pp. 121-130, 1998.

[2] L. A. Sphaier, and W. M. Worek, "Analysis of Heat and Mass Transfer in Porous Sorbents Used in Rotary Regenerators", International Journal of Heat and Mass Transfer, Vol. 47, pp. 3415-3430, 2004.

[3] G. J. Softah, Numerical Analysis and Neuro-Fuzzy Investigation of the Performance of Heat Recovery Wheels in AHU Systems, M.Sc. Graduation Project, Mechanical Engineering Department, Umm Al-Qura University, Saudi Arabia, Spring 2014.

[4] A. M. Wafiah, Computational and Neural Investigation of the Operation of Heat Exchange Wheels in AHU Systems, M.Sc. Graduation Project, Mechanical Engineering Department, Umm Al-Qura University, Saudi Arabia, Spring 2014.

[5] S. Yamaguchi, and K. Saito, "Numerical and Experimental Performance Analysis of Rotary Desiccant Wheels", International Journal of Heat and Mass transfer, Vol. 60, pp. 51-60, 2013.

[6] W. Shurcliff, "Air-to-air Heat Exchangers for Houses", Annual Review Energy, Vol. 13, pp. 1-22, 1988.

[7] P. Mazzei, F. Minichiello, and D. Palma, "Desiccant HVAC Systems for Commercial Buildings", Applied Thermal Engineering, Vol. 22, No. 5, pp. 545-560, 2002.

[8] 2005 ASHRAE Handbook Fundamentals SI Edition, ASHRAE, USA, 2005.

[9] L. Pérez-Lombard, J. Ortiz, and C. Pout, "A Review on Buildings Energy Consumption Information", Energy and Buildings, Vol. 40, No. 3, pp. 394-398, 2008.

[10] K. K. W. Wan, D. H. W. Li, D. Liu, J. C. Lam, "Future Trends of Building Heating and Cooling Loads and Energy Consumption in Different Climates," Building and Environment, Vol. 46, No. 1, pp. 223-234, 2011.

[11] L. Shao, S. B. Riffat, and G. Gan, "Heat Recovery with Low Pressure Loss for Natural Ventilation", Energy and Buildings, Vol. 28, No. 2, pp. 179-184, 1998.

[12] A. Pesaran, and A. F. Mills, "Moisture Transport in Silica Gel Packed Bed I: Theoretical Study", International Journal of Heat and Mass Transfer, Vol. 30, pp. 1037-1049, 1987.

[13] A. Pesaran, and A. F. Mills, "Moisture Transport in Silica Gel Packed Bed II: Experimental Study", International Journal of Heat and Mass Transfer, Vol. 30, pp. 1051-1060, 1987.

[14] W. Zheng, and W. M. Worek, "Numerical Simulation of Combined Heat and Mass Transfer Process in a Rotary Dehumidifier", Numerical Heat Transfer Part A, Vol. 23, pp. 211-232, 1993.

[15] Y. J. Dai, R. Z. Wang, and H. F. Zhang, "Parameter Analysis to Improve Rotary Desiccant Dehumidification Using a Mathematical Model", International Journal of Thermal Sciences, Vol. 40 pp. 400-408, 2001.

[16] J. L. Niu, and L. Z. Zhang, "Performance Comparisons of Desiccant Wheels for Air Dehumidification and Enthalpy Recovery", Applied Thermal Engineering, Vol. 22, No. 12, pp. 1347-1367, 2002.

[17] F. E. Nia, D. V. Paassen, and M. H. Saidi, "Modeling and Simulation of Desiccant wheel for Air Conditioning", Energy and Buildings, Vol. 38, pp.1230-1239, 2006.

[18] G. Heidarinejad, H. P. Shahri, and S. Delfani, "The Effect of Geometrical Characteristics of Desiccant Wheel on Its Performance", International Journal of Engineering Transactions B: Application, Vol. 22, No. 1, pp. 63-75, 2009.

[19] A. Kodama, T. Hirayama, M. Goto, T. Hirose, and R. E. Critoph, "The Use of Psychometric Charts for the Optimization of a Thermal Swing Desiccant Wheel", Applied Thermal Engineering, Vol. 21, No. 16, pp. 1657-1674, 2001.

[20] T. S. Ge, Y. Li, R. Z. Wang, and Y. J. Dai, "A Review of the Mathematical Models for Predicting Rotary Desiccant Wheel, Renew. Sustain. Energy Revs., Vol. 12, No. 6, pp. 1485-1528, 2008.

[21] E. M. Sterling, A. Arundel, and T. D. Sterling, "Criteria for Human Exposure to Humidity in Occupied Buildings", ASHRAE Trans., Vol. 91, No. 1, pp. 611-622, 1985.

[22] M. H. Ahmed, N. M. Kattab, and M. Fouad, "Evaluation and Optimization of Solar Desiccant Wheel Performance", Renewable Energy, Vol. 30, No. 3, pp. 305-325, 2005.

[23] X. J. Zhang, Y. J. Dai, and R. Z. Wang, "A Simulation Study of Heat and Mass Transfer in a Honeycombed Rotary Desiccant Dehumidifier", Applied Thermal Engineering, Vol. 23, No. 8, pp. 989-1003, 2003.

[24] L. G. Harriman III, "The Basics of Commercial Desiccant Systems", Heating/Piping/Air Conditioning, Vol. 66, No. 7, pp. 77-85. 1994.

[25] H. M. Henning, T. Erpenbeck, C. Hindenberg, and I. S. Santamiria, "The Potential of Solar Energy Use in Desiccant Cooling Cycles", International Journal of Refrigeration, Vol. 24, pp. 220-229, 2001.

[26] P. Mavroudaki, C. B. Beggs, P. A. Sleigh, and S. P. Haliiday, "The Potential for Solar Powered Single-stage Desiccant Cooling in Southern Europe", Applied Thermal Engineering, Vol. 22, pp. 1129-1140, 2002.

[27] S. P. Halliday, C. B. Beggs, and P. A. Sleigh, "The Use of Solar Desiccant Cooling in the UK: a Feasibility Study", Applied Thermal Engineering, Vol. 22, pp. 1327-1338, 2002.

[28] J. Y. San, and S. C. Hsiau, "Effect of Axial Solid Heat Conduction and Mass Diffusion in a Rotary Heat and Mass Regenerator", Int. J. Heat Mass Transfer, Vol. 36, No. 8, pp. 2051-2059, 1993.

[29] R. B. Holmberg, "Combined Heat and Mass Transfer in Regenerators with Hygroscopic Materials", ASME J. Heat Transfer, Vol. 101, pp. 205-210, 1979.

[30] D. Charoensupaya, and W. M. Worek, "Effect of Adsorbent Heat and Mass Transfer Resistances on Performance of an Open Cycle Adiabatic Desiccant Cooling System", Heat Recov. Sys. CHP, Vol. 8, No. 6, pp. 537-548, 1988.

[31] N. Nassif, "Modeling and Optimization of HVAC Systems Using Artificial Neural Network and Genetic Algorithm", Building Simulation, Vol. 7, No. 3, pp. 237-245, June 2014.

[32] Z. Li, X. Xu, S. Deng, and D. Pan, "A Novel Neural Network Aided Fuzzy Logic Controller for a Variable Speed (VS) Direct Expansion (DX) Air Conditioning (A/C) System", Applied Thermal Engineering, Vol. 78, pp. 9-23, 2015.

[33] B. C. Ng, I. Z. M. Darus, H. Jamaluddin, and H. M. Kamar, "Application of Adaptive Neural Predictive Control for an Automotive Air Conditioning System", Applied Thermal Engineering, Vol. 73, No. 1, pp. 1244-1254, 2014.

[34] H. M. Factor, and G. Grossman, "A Packed Bed Dehumidifier/Regenerator for Solar Air Conditioning with Liquid Desiccants", Solar Energy, Vol. 24, No. 6, pp. 541-550, 1980.

[35] A. Al-Alili, Y. Hwang, and R. Radermacher, "Review of Solar Thermal Air Conditioning Technologies", International Journal of Refrigeration, Vol. 39, pp. 4-22, 2014.

[36] A. F. Abdel Gawad, "Numerical and Neural Study of the Turbulent Flow around Sharp-Edged Bodies", 2002 Joint US ASME-European Fluids Engineering Division Summer Meeting, Montreal, Canada, July 14-18, 2002.

[37] O. E. Abdellatif, and A. F. Abdel Gawad, "Experimental, Numerical and Neural Investigation of the Aerodynamic Characteristics for Two-Dimensional Wings in Ground Effect", Al-Azhar Engineering 7th International Conference, Cairo, Egypt, 7-10 April, 2003.

[38] A. F. Abdel Gawad, "Study of Both Airflow and Thermal Fields in a Room Using K-ε Modelling and Neural Networks", Al-Azhar Engineering 7th International Conference, Cairo, Egypt, 7-10 April, 2003.

[39] A. F. Abdel Gawad, M. M. Nassief, and N. M. Gurguis, "Numerical and Neural Study of Flow and Heat Transfer Across an Array of Integrated Circuit Components", Journal of Engineering and Applied Science (JEAS), Vol. 52, No. 5, pp. 981-1000, October 2005.

[40] A. F. Abdel Gawad, "Computational and Neuro-Fuzzy Study of the Effect of Small Objects on the Flow and Thermal Fields of Bluff Bodies", 8th Biennial ASME conference on Engineering Systems Design and Analysis (ESDA2006), Torino, Italy July 4-7, 2006.

[41] A. F. Abdel Gawad, "Investigation of The Dilution of Outfall Discharges Using Computational and Neuro-Fuzzy Techniques", 2007 ASME International Mechanical Engineering Congress & Exposition (IMECE2007), Seattle, Washington, USA, 5-11 November 2007.

[42] A. Hosseinzadeh, and M. Karimi, "Prediction of J-Integral Dependence to Residual Stress and Crack Depth on NACA 0012-34 Using FE and ANN", Engineering Solid Mechanics, Vol. 3, No. 2, pp. 103-110, 2015.

[43] Z. Wu, R. V. N. Melnik, and F. Borup, "Model-based Analysis and Simulation of Regenerative Heat Wheel", Energy and Buildings, Vol. 38, pp. 502-514, 2006.

[44] A. M. Wafiah, A. F. Abdel Gawad, and M. N. Radhwi, "Computational Investigation of the Operation of Heat Conservation Wheels in AHU-Systems", Umm Al-Qura University Journal of Engineering and Islamic Architecture (UQU-UJEA), Vol. 5, No. 2, Ragab 1435 H - May 2014.

[45] G. J. Softah, M. N. Radhwi, and A. F. Abdel Gawad, "A Parametric Study of the Performance of Heat Recovery Wheels in HVAC System", Umm Al-Qura University Journal of Engineering and Islamic Architecture (UQU-UJEA), Vol. 5, No. 2, Ragab 1435 H - May 2014.

[46] Fluent guide manual, 2011.

[47] R. Parsons, ASHRAE Handbook 2005: Fundamentals, American Society of Heating, Refrigeration and Air-Conditioning Engineers.

[48] Matlab guide manual, 2011.

13

Application of Phase-Change Materials in Buildings

Nader A. Nader, Bandar Bulshlaibi, Mohammad Jamil, Mohammad Suwaiyah, Mohammad Uzair

Department of Mechanical Engineering, Prince Mohammad Bin Fahd University, Al Khobar, Kingdome of Saudi Arabia

Email address:

nnader@pmu.edu.sa (N. A. Nade), ban_bus2011@hotmail.com (B. Bulshlaibi), uzairmehta69@gmail.com (M. Uzair),
mohammad.aljamil@gmail.com (M. Jamil), mks.90@hotmail.com (M. Suwaiyah)

Abstract: Phase-Change Materials (PCMs) are substances with a high heat of fusion that melt and solidifyat a certain temperature range. Theyare capable of storing and releasing large amounts of energy and have a high capacity of storing heat. PCMs preventenergy loss during material changes from solid to liquid or liquid to solid. They have several advantages such as its self-nucleating properties, and disadvantages such as having low thermal conductivity [4].There are different types of PCM with a wide range of applications. This paperstudies the potential application of PCMs in building as energy conservation materials.The analysis shows the result shows that the use of BioPCM materialas insulation layer in building can decrease the cooling load by 20% in comparison to standard one.

Keywords: Phase-Change Materials, Construction, Cooling Load, Insulation

1. Introduction

Researchers have investigated Phase-Change Materials (PCMs)heavily over the last two decades as energy conservation materials in buildings.The tremendous increase in energy prices has motivated researcher to search for a new materials that has the capacity to reduce energy demand.PCMs thermophysical properties that include melting point range, heat of fusion, thermal conductivity, and density; have promoted it as a valid option for energy conservation.Additionally, PCMs can freeze with little supercoiling; compatible with construction material; chemically stable; recyclable; and can reduce HVAC load.However PCMS have a few drawbacks include the requirements of a freezing cycle in order to transfer high heat; and the capacity of its volumetric latent heat storage (LHS) is low making PCMs flammable. Therefore, PCMs need to be stored in a proper container to avoid any of these disadvantages.

PCM can be classified under three categories: organic, inorganic and eutectic [15]. Organic PCMs have technical grade paraffin's or paraffin mixtures made of oil that help PCMs obtain reliable phase change points [7].Paraffin is also available in large temperature ranges, making them accessible, especially because they have a long freeze-melt cycle. PCMscan also be made out of non-paraffin compounds. The benefits for using this organic material in heat storage are that it has the advantage of not being corrosive or undercooled, but still do cause a lower phase change enthalpy, low thermal conductivity and inflammability.In using inorganic PCM materials for heat storage such as hydrated salts and metallic, have a greater phase change enthalpy despite its disadvantages of undercooling, corrosion, phase separation and lack of thermal stability.The third classification of PCM is eutectic which can be organic or inorganic [8]. They have a sharp melting point and their volumetric storage capacity is higher than organic paraffin compounds. However, its thermophysical properties are limited and are still rather new to thermal storage [9].To increase heat conductivity in PCM without affecting the energy storage, Mehling et al. [5] and Py et al. [6] designed a compound-material made of a graphite matrix to embed the PCM in it. Graphite decreases the sub cooling of PCMs hydrated salts and decreases the volume change of paraffin's. This produces high thermal conductivity where there is 8% of the latent heat of fusion per unit mass of the paraffin [3].

When heat is applied to a substance, the energy transfers in one of two ways. The first is that the substance gains heat [2, 3]. For example, if heat is applied to water, it will rise in temperature to a maximum of 100°C, its boiling point. Likewise, if heat is removed, the temperature of the water will fall, to a minimum of 0°C, or its freezing point. This type of heat transfer, or storage, is called sensible heat as seen in Figure 1.The temperature of a substance however, doesn't always rise when heat is added.Boiling water for example, remains at 100°C no matter how much heat is added which is why the water turns into vapour [2, 3]. Therefore, any substance that absorbs heat reaches either its melting or evaporation point without getting any hotter.Latent heat is this type of heat storage. Without latent heat, PCMs would not be able to act alone in controlling room temperature because when used in construction they

change from solid to liquid at 23-26°C. When PCM melts, they absorb heat from the room to keep the room temperature stable. PCMs only return to its original solid state during ventilation at night.These valuable properties of PCMs shall dramatically reduce cooling and heating energy demand if properly managed and implemented.This paper studies the potential application of PCMs in building as energy conservation materials.

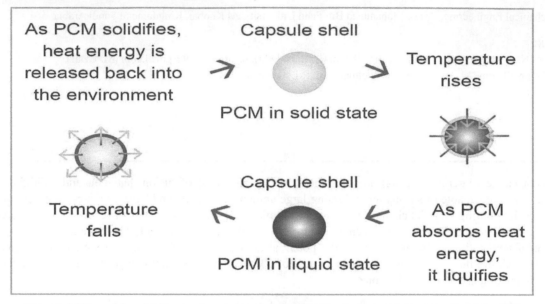

Figure 1. *PCMs work by solidifying to release heat energy and absorbing heat energy before being liquidities. [20].*

2. Comparison of Pcms

2.1. Water and Gel Packs

Ice and gel packs have become extremely popular for keeping materials cold around 0°C. These devices have the advantages of good performance, low cost, nontoxic, not flammable, environmentally friendly and easy to use. The only disadvantage to these ice and gel packs is that they are only useful at maintaining their surroundings or thermal load at 0°C. If one wishes to obtain a water-based PCM lower than 0°C, then a salt can be added to the water. This will depress the freezing point. However, this significantly decreases the latent heat and broadens the melt/freeze temperature. This technique is familiar by many in applications of adding salt to ice for making homemade ice cream or adding salt to roads during icy conditions in order to lower the freezing point of water.

Similarly, gel packs are types of salts. The predominate compound in many gel packs is sodium polyacrylate. Sodium polyacrylate is the predominant chemical used in diapers. The addition of sodium polyacrylate does change the crystal structure of ice and therefore diminishes the effectiveness of ice as a PCM.

2.2. Salt Hydrates

Salt hydrates as PCMs have been among the most researched latent heat storage materials. Salt hydrates are

often times the lowest cost PCM behind water and gel packs. Numerous trials and sub-scale tests have been carried out on these compounds. The material comprises $M \cdot nH_2O$, where M is an inorganic compound.

Salt hydrates have several problems related to practical applications as PCMs. Foremost, is the limited temperature ranges available to salts in meeting specific temperature needs at the desired temperature. There are few salts that melt between 1 and 150°C. In the absence of salts melting at temperatures between 1 and 150°C, eutectic mixtures and salt hydrates are pursued for these missing temperature ranges. The main problems experienced in the past with salt hydrate-based thermal storage applications were due to the fact that most salt hydrates melt incongruently (e.g., they melt to a saturated aqueous phase and a solid phase, which is generally of a lower hydrate of the same salt)[7]. Abhat[8] specifies the following problem with salt hydrates, "The major problem in using salt hydrates as PCMs is that most of them melt incongruently, i.e. they melt to a saturated aqueous phase and a solid phase which is generally a lower hydrate of the same salt. Due to density differences, the salt phase settles out and collects at the bottom of the container, a phenomenon called decomposition. Unless specific measures are taken, this phenomenon is irreversible, i.e. during freezing, the solid phase does not combine with the saturated solution to form the original salt hydrate." This observation was repeated by Verner[9] and by Lane[10]. Another common problem found in salt hydrates is with the salt hydrates' poor nucleating properties that result in what is termed supercooling of the liquid salt hydrate prior to freezing. This must be overcome by the addition of a suitable nucleating agent that has a crystal structure similar to that of the parent substance.

Other issues of using salt hydrates involve the volume change, the corrosive nature of the salt hydrate and the toxicity of some of these materials. Often times the volume change in the solid/liquid phase change of a salt hydrate is up to 10%. While this is significant, it can be accommodated in special packaging. The packaging needs to be specific to the type of salt hydrate used. Many salt hydrates are corrosive to metals. The toxicity of salt hydrates also varies widely. Therefore, the safety data sheets should be carefully checked for human toxicity and environmental damage when disposed.

2.3. Paraffins

Paraffins are high-molecular-mass hydrocarbons with a waxy consistency at room temperature. Paraffins are made up of straight chain hydrocarbons. The paraffins are separated into two main sub-groups, evenchained (n-Paraffin) and odd-chained (iso- Paraffin). Typically, iso-paraffins do not make good PCMs because of stearic hindrances in their molecular packing. Therefore, we will only focus on the n-paraffins. The melting point of paraffins is directly related to the number of carbon atoms within the material structure with alkanes containing 12-40 C-atoms possessing melting points between 6 and 80 degrees centigrade. These are termed 'pure paraffins' and should not be confused with paraffin waxes. Paraffin waxes contain a mixture of hydrocarbon molecules with various carbon numbers with lower melting points and poorer latent heats than pure paraffins. Paraffin waxes are often considered a low-grade PCM.

Paraffins form a good PCM candidate for certain applications and certain select temperature ranges. Paraffins have good thermal storage capacity plus the materials are proven to freeze without supercooling. Paraffins also have the advantages of chemical stability over many heating and freezing cycles[11], high heat of fusion, they are non-corrosive, compatible with most all materials and non-reactive to most materials of encapsulation.

Paraffins do have some disadvantages associated with them. Literature reports on how pure paraffin products are often reported to have very high latent heats. However, the details are most important in this assessment. Commercially cost-effective paraffins are mixtures of alkanes and therefore do not have sharp, well-defined melting points. For example, 98% hexadecane has a latent heat of about 230 J/g at a melting point of 18°C, but due to the very small amount of hexadecane in petroleum crude oil (small fraction of 1%) and the enormous separation costs (repeated high-vacuum, multi-stage distillation) to achieve a 98% pure product, hexadecane is not a viable commercial product as exemplified by the price of over $10 per pound. Wax products are commercially viable for a variety of applications including uses in canning and candle making. A C20-C24 canning wax has a latent heat of 150 J/g and melts over a 7°C temperature range— considerably poorer performance than +98% n-paraffin products found in most literature. Pure paraffins are also

limited in their range of melting points that they can target. Table 1 lists 16 of the most common paraffins and their melting points. However, from a practical standpoint, only the evennumberedparaffins are in any abundance in crude petroleum. This takes the number of common paraffins for use as PCMs down to 8 commercially viable paraffins. Many researchers try to expand this temperature range above the 8 common paraffins by creating mixtures of 2 or more paraffins. This, however, has the detrimental effect of broadening the melt/freeze temperature and lowering the latent heats even more.

Other concerns with paraffins used as PCMs are social dynamics. Paraffins are made from petroleum products, which increases our reliance on crude oil. Paraffins prices have followed the unstable price of petroleum. Furthermore, petroleum-derived paraffins have geopolitical consequences and contribute to the increase in carbon emissions blamed for the global warming crisis.

2.4. Vegetable-based PCMs

In the past, the phase change material market has been dominated by paraffin products and salt hydrates. Paraffins have recently become more popular than the salt hydrates. However, the use of vegetable compounds has started to make strong inroads in this PCM market. The ability of vegetable compounds to compete in this market depends on their price:performance characteristics relative to paraffins.

The Department of Agriculture and the National Science Foundation has sponsored research to investigate the potential vegetable-derived compounds becoming significant in the PCM market[12-14]. During these investigations the researchers were able to produce around 300 different fat- and vegetable oil-based PCMs ranging from - 90°C to 150°C with latent heats between 150 and 220 J/g.

The safety, environmental and social benefits of using vegetable-based PCMs are significantly greater than those of paraffins. Paraffins can be toxic and have laxative effects when ingested. Many vegetable-based PCMs can be considered, "food grade" or have no effect when ingested. Researchers under the Department of Agriculture research program also discovered that many vegetable-based PCMs had lower flash points and 10-20% longer horizontal flame propagation rates than did their temperature-comparable, paraffin-based counterparts. Paraffins are long chain alkanes. This means that there is no "active" site on the paraffin molecule for microbes and bacteria to begin their breakdown. While this is attractive for the life of the PCM, it presents environment problems when the paraffin is disposed in a landfill. The paraffin would stay undegraded for decades. Many vegetablebased PCMs have shown stability after 30 years of accelerated aging tests[15]. However, these same vegetable-based PCMs will degrade in six months or less when discarded in a landfill. Many vegetable based PCMs can be derived locally using common agricultural crops. This leads to a more stable price and regionally based feedstocks. Vegetable-based PCMs are considered to be nearly carbon neutral, as recently ranked by the USDA Bio Preferred program[16]. Vegetable-based PCMs are able to provide price: Performance characteristics superior to paraffin PCM chemicals because 1) the natural vegetable-based feedstocks are available at relatively low prices and 2) multiple reaction-modification schemes provide many degrees of freedom to produce a variety of premium products. These vegetable-derived chemical building blocks are available at concentrations considerably greater than any particular n-paraffin is available in crude oil. Because of this, vegetable-based PCMs have a distinct competitive advantage over paraffins and could corner the PCM market. Vegetable-based PCMs are able to provide a price: Performance characteristic superior to salt hydrate PCM chemicals over most of the temperatures between -90°C and 150°C because viable salts or salt hydrates simply are not available over much of that temperature range. For the limited temperatures where viable salt PCM chemicals are available, vegetablebased PCMs have advantages of 1) being renewable and more environmentally friendly and; 2) being compatible with wall boards or similar materials that would absorb water from hydrates and render the salt hydrates ineffective as PCM chemicals; 3) vegetablebased PCMs are capable of being microencapsulated, whereas salt hydrates are not.

3. Pcms Applications in Buildings

More than 500 natural and synthetic PCMs are known in addition to water [11]. They differ from each other by their altering phase change temperature ranges and their heat storage capacities. Other properties of PCM for a high efficient cooling system with thermal energy system (TES) include [9]:

- A melting temperature in the desired operating range, in construction this would be 23°C or 26°C.

- A high latent heat of fusion per unit volume. In other words, they can store a large amount of heat per unit of volume, minimizing the area of PCM tiles that are needed.
- High thermal conductivity. The quicker the PCM reacts to changes in temperature, the more effective the phase changes will be.
- Minimal changes in volume. Substances expand or contract when they change state. Because PCMs in construction need to be contained within a cassette, large changes in volume could create problems.
- Congruent melting. This means that the composition of the liquid is the same as that of the solid, which is important to prevent separation and super cooling.
- A completely reversible freezing/melting cycle.
- Durability over a large number of cycles.
- Non-corrosiveness to construction materials.
- Non-flammable.

PCMS can be for heating and cooling in buildings as seen in Figure 2[12]. PCM can be placed in porous construction material such as plasterboard to increase heat [13]. As for cooling, air conditioners with PCM collect and store ambient air during the night.PCMs can also be placed in thermoelectric refrigeration to improve effectiveness of the heat sink such as using a window with a PCM curtain [13, 14].A PCM curtain fills up the double sheeted gap between the window and the air vent, and upon freezing the PCMs would prevent the temperature of the air from decreasing and this reduces overheating around the window [15].

By altering the state of aggregation of a PCMs in a specific temperature range, applications of PCMs can be placed in apparel, blankets, surgical tools, antibacterial and hygiene applications, insulation, clothing and many more [11].

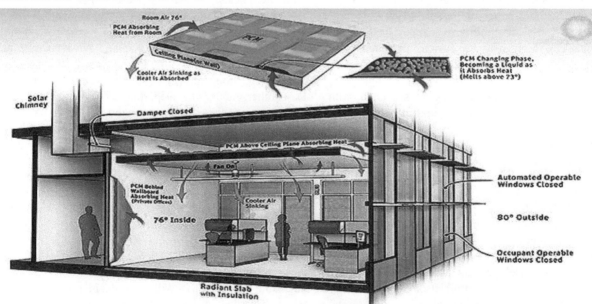

Figure 2. All types of PCM applications in a building [19].

Additionally, PCMs have multiple applications for solar energy storagedue to their hydrated salts, alkanes, waxes or paraffin's [10]. Many of which are organized in Table 1 below [17, 18].

Table 1. *PCM-TES applications [4].*

Applications
1. Thermal storage of solar energy
2. Passive storage in bioclimatic building/architecture (HDPE + paraffin)
3. Cooling use of off-peak rates and reduction of installed power, ice-bank
4. Heating and sanitary hot water: using off-peak rate and adapting unloading curves
5. Safety: temperature maintenance in rooms with computers or electrical appliances
6. Thermal protection of food, transport, hotel trade, ice-cream, etc.
7. Food agroindustry, wine, milk products (absorbing peaks in demand), greenhouse
8. Thermal protection of electronic devices (integrated in the appliance)
9. Medical applications: transport of blood, operating tables, hot – cold therapies
9. Cooling of engines (electric and combustion)
10. Thermal comfort in vehicles
11. Softening of exothermic temperature peaks in chemical reactions
12. Spacecraft thermal systems
13. Solar power plants

4. Case Study of Pcms Applications in Buildings

PCMs can smooth temperature variability's in a building especially when used in air conditioning applications. This can have substantial economic and environmental benefits. To evaluate the value of PCMs materials in air conditioning application the cooling load needed for the building shown in Figure 3 was estimated with the application of PCMs materials and without. The building is located in the Dammam City in the Eastern province of Saudi Arabia where the average outdoor temperature (To) be about 45 °C and indoor temperature (Ti) is about 24 °C. The cooling Load (CL) was estimated based on the following equation:

$$C.L = Q(t) + Q(s) + Q(l) + Q(o) + Q(e) + Q(v) \quad (1)$$

Where, the heat transmitted through structures is represented by Q (t), the heat emitted by the sun is Q(s), the light gained is Q (l), the heat of the occupants is Q (o), the heat of the equipment is Q (e), and the heat of ventilation is Q (v). Therefore:

$$Q(t) = sum\ U * A * (To-Ti) \quad (2)$$

Where, U is the overall heat transmission coefficient (w/m2*k), A is the area, to is the outdoor temperature, and Ti is the indoor temperature.

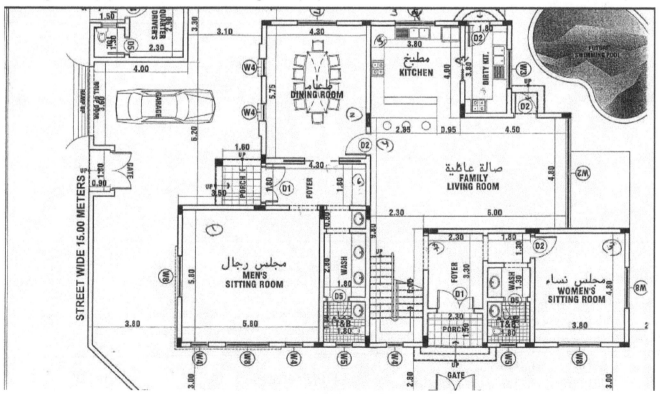

Figure 3.Comparing a normal wall with a PCM wall.

The standard materials for the walls are made 0f 20mm brick wall, 2mm insulation of polyurethane. Whereas the BioPCM properties are given in Table 2 below:

Table 2. BioPCM properties.

Description	BioPCM	GR27	Water
Melting Point (°C)	29	28	0
Density (kg/m3)	860	710	1000
Specific Heat (kJ kg-1 °C-1)	1.97	1.125	4.179
Latent Heat (kJ/kg)	219	72	334
Viscosity @ 30 °C (cp)	7		0.798
Boiling Point (°C)	418		100
Thermal Conductivity (W m-1 °C-1)	0.2	0.15	0.6

5. Results

The cooling load from a normal wall and the cooling load

from a PCM wall were calculated for each room in the building. The analysis shows a total reduction of needed cooling load for the whole building of about 20% when utilizing PCMs in the building. The total cooling load for the building with normal construction materials was estimated to be about 16 tonnes while with PCMs was estimated to be about 13 tonnes. More details are provided for the men sitting room in the building. Table 3 shows the cooling load calculation results for Men sitting room with normal construction materials is presented in Table 3. While the cooling loads with PCMs material for the same room is presented in Table 4.

Table 3. Cooling load for men sitting room estimate for a normal wall.

Location	Area	U	SC	CLF	Tin-To	Q (w)	
Wall	N(inner)	17.4	2.36			0	0
	S (inner)	11.46	2.36			0	0
	E (inner)	17.4	2.36			0	0
	W (inner)	14.7	2.36			0	0
	N(outer)	17.4	2.34			21	855.04
	S (outer)	11.46	2.34			21	563.14
	E (outer)	17.4	2.34			21	855.04
	W (outer)	14.7	2.34			21	722.36
Floor		33.64	1.5			0	0.00
Ceiling		33.64	1.5			21	1059.66
Window	S	5.94	2.07	0.57		21	147.18
window	W	2.7	2.07	0.57		21	66.90
Door	N	2	0.26			0	
					Sum of Q(w)	4269.32	

Table 4. Cooling load for men sitting room estimate for a PCMs wall.

Location	Area	U	SC	CLF	Tin-To	Q (w)	
Wall	N(inner)	17.4	2.36			0	0
	S (inner)	11.46	2.36			0	0
	E (inner)	17.4	2.36			0	0
	W (inner)	14.7	2.36			0	0
	N(outer)	17.4	1.1			21	401.94
	S (outer)	11.46	1.1			21	264.73
	E (outer)	17.4	1.1			21	401.94
	W (outer)	14.7	1.1			21	339.57
Floor		33.64	1.5			0	0.00
Ceiling		33.64	1.5			21	1059.66
Window	S	5.94	2.07	0.57		21	147.18
window	W	2.7	2.07	0.57		21	66.90
Door	N	2	0.26			0	
					Sum of Q(w)	2681.92	

6. Conclusion

The potential use of PCMs in construction material, heat transfer and other applications are promising given the magnificent it thermophysical properties. PCMs should be incorporated further in global energy management solutions due to the stress for innovations with a low impact on the environment. The results gained from analysis shows a 20% reduction of cooling load when utilizing PCMs materials in comparison to standard construction materials. Accordingly, PCM wall is a promising solution for the problem of depleting fuel resources in the form of latent heat storage

materials. PCM should be used in buildings, ceiling tiles, air conditioners, thermal heating and many other applications. Further research is recommended in PCMs application in the construction industry due to its magician properties and potential wide applications.

References

[1] "Manufacturing innovative thermal storage technologies for smart & sustainable buildings." PhaseChange energy solutions, 2013. Accessed on from: http://www.phasechange.com/index.php/en/

[2] Chair P. P., Lee T., Reddy A.. "Application of Phase Change Material in Buildings: Field Data vs. EnergyPlus Simulation. Arizona State University, 2010. Accessed on from: http://repository.asu.edu/attachments/56138/content/Murugana ntham_asu_0010N_10151.pdf

[3] "Phase-change materials." Building, 2013. Accessed on from: http://www.building.co.uk/business/cpd/cpd-1-2013-phase-change-materials/5050027.article

[4] "Phase-change material." Wikipedia, 2012. Retrieved from: http://en.wikipedia.org/wiki/Phase-change_material

[5] Marin J.M., Zalba B., Cabeza L.F., Mehling H., Determination of enthalpy-temperature curves of phase change materials with the T-history method—improvement to temperature dependent properties, Measurement Sci. Technol., in press.

[6] Py X., Olives R., Mauran S., Paraffin/porous-graphite-matrix composite as a high and constant power thermal storage material, Int. J. Heat Mass Transfer 44 (2001) 2727-2737.

[7] Shapiro M., Feldman D., Hawes D., BanuD.. 1987. PCM thermal storage in drywall using organic phase change material. Passive Solar J 4: 419-438

[8] Perez, A. D. P., "Situacion y future de los PCM (Phase Change Material)", Centro de DesarrolloTecnologico – Fundacion LEIA, (2010).

[9] Alkan, C., Sari, A., Karaipekli, A. and Uzun, O. 2009. "Preparation, characterization and thermal properties of microencapsulated phase change material for thermal energy storage." Solar Energy Materials and Solar Cells, 93: (1), 143-147.

[10] Dincer, M.A. Rosen. Thermal energy storage. Systems and Applications, 2002. England: John Wiley & Sons. Print.

[11] MondalS.. 2008. Phase changing materials for smart textiles – An overview. Applied Thermal Engineering, 28: 1536-1550. Retrieved from: http://ucheg.ru/docs/5/4074/conv_1/file1.pdf

[12] Streicher, W., Cabeza, L., Heinz, A. 2005. A Report of IEA Solar Heating and Cooling programme - Task 32 "Advanced storage concepts for solar and low energy buildings." Solar Heating & Cooling Programme, 1-33.

[13] Zalba, B., Marin, J.M., Cabeza, L.F., Mehling, H. Review on thermal energy storage with phase change: materials, heat transfer analysis and applications. Applied Thermal Engineering 23, (2003) 251-283.

[14] Sharma, A., Tyagi, V.V., Chen, C.R., Buddhi, D. Review on thermal energy storage with phase change materials and applications. Renewable and Sustainable Energy Reviews 13, (2009) 318-345.

[15] Juarez, D., Balart, R., Ferrandiz, S., Peydro, M.A. Classification of phase change materials and his behaviour in SEBS/PCM blends. Manufaturing Engineering Society International Conference, 2013.

[16] Rousse, D.R., Salah, N.B., Lassue, S. An overview of phase change materials and their implication on power demand. National Science and Engineering Research Council of Canada, 1-6. 2009.

[17] Kosny, J., Kossecka, E. Understanding a Potential for Application of Phase-Change Materials (PCMs) in Building Envelopes. ASHRAE, 2013.

[18] Rai, A.K., Kumar, A. A Review on Phase Change Materials & Their Applications. International Journal of Advanced Research in Engineering and Technology, 3 (2) 214-225. 2012.

[19] Infinite R Company, http://www.phasechangetechnologies.com/

[20] Accessed from Daikin, http://www.daikin.pl/vrv-iv/continuous_heating/

[21] Merriam-Webster Collegiate Dictionary. 11th Edition. "Sensible Heat." 7.

[22] Lane.G.A, "Phase Change Material for Energy Storage Nucleation to Prevent Supercooling", Solar Energy Materials and Solar Cells, 27,(1991), 4.

[23] Abhat, A. "Low Temperature Latent Heat Thermal Energy Storage: Heat Storage Materials," Solar Energy, Vol. 30, No. 4, 313-332, 1983.

[24] Verner, C. Phase Change Thermal Energy Storage. Thesis, Brighton University (Seehttp://freespace.virgin.net/m.eckert/carl_vener's_dissertation.htm), May, 1997.

[25] Lane.G.A. "Phase Change Material for Energy Storage Nucleation to Prevent Supercooling," Solar Energy Materials and Solar Cells 27 135-160, p4, 1991.

[26] S.D. Sharma, D. Buddhi, R. L. Sawhney, "Accelerated Thermal Cycle Test of Latent Heat Thermal Storage Material", Solar Energy, 66,(1999), 483.

[27] National Science Foundation, SBIR, Award Number - 0750470 1

[28] United States Department of Agriculture, SBIR, Project Number MOK-2003-05519

[29] United States Department of Agriculture, NRI, Project Number MOR-2005-02692

[30] Internal Studies Conducted by Phase Change Energy Solutions-2009, Entropy Solutions Inc-2009, University of Missouri-Columbia -2007.

[31] http://www.biopreferred.gov

14

Economic Analysis of Solid Waste Treatment Plants Using Pyrolysis

Huseyin M. Cekirge[1], Omar K. M. Ouda[2], Ammar Elhassan[3]

[1]Department of Mechanical Engineering, Prince Mohammad Bin Fahd University, Al Khobar, KSA
[2]Department of Civil Engineering, Prince Mohammad Bin Fahd University, Al Khobar, KSA
[3]Department of Information Technology, Prince Mohammad Bin Fahd University, Al Khobar, KSA

Email address:

hmcekirge@usa.net (H. M. Cekirge), oouda@pmu.edu.sa (O. K. M. Ouda), aelhassan@pmu.edu.sa (A.Elhassan)

Abstract: Municipal Solid Waste (MSW) management is a chronic environmental and economic problem in urban areas worldwide and more specifically in developing countries. Waste-to-Energy (WTE) technologies show a great potential to convert this problem to a revenue source. Pyrolysis is a promising technology and is currently utilized in many regions of the world for MSW disposal and energy generation. The economic value of pyrolysis has been insufficiently evaluated. This paper introduces and discusses the economic value of pyrolysis as MSW management disposal method and energy source. The return period of investments is considered for various pricing policies with respect to end product of process. Hypotheses and conclusions of the model works are briefly reported.

Keywords: Waste to Energy, Pyrolysis, Municipal Solid Waste Management

1. Introduction

Municipal Solid Waste (MSW) refers to domestic solid waste such as food scraps, paper, cardboard, plastics, clothing, glass, metals, wood, street sweepings, landscape and tree trimmings and general wastes from parks and other recreational areas. The world urban areas generated about 1.3 billion tons of solid waste in 2012. This volume is expected to increase to 2.2 billion tons by 2025. Waste generation rates will more than double over the next twenty years in developing countries. Globally, solid waste management costs will increase from today's annual US $ 205.4 billion to about US $ 375.5 billion in 2025. Cost increases will be most severe in developing countries such as Pakistan [1, 2]. In developing countries, urban MSW is usually a city's single largest budgetary item and it can be a valuable source of biomass, recycled materials, energy and revenue if properly and wisely managed. Several energy recovery or waste-to-energy (WTE) technologies such as pyrolysis, anaerobic digestion (AD), incineration and refused derived fuel (RDF) have been developed in order to generate energy and value-added products in the form of electricity, transportation fuels, heat, fertilizers and chemicals[3,4]. Studies show that WTE can contribute substantially to energy demand especially in heavily populated urban areas [5-13]. Additionally, the WTE environmental value is quite significant with several factors including, but not limited to, greenhouse gas emission reduction, energy saving, landfill area saving, and soil and groundwater protections [14-16].

Pyrolysis is a promising technology and is currently utilized in many regions of the world for MSW disposal and energy generation. The economic value of pyrolysis has been insufficiently evaluated. This paper introduces and discusses the economic value of pyrolysis as MSW management disposal method and energy source. Fast and slow pyrolyses are considered as thermal processes, essential final products are gases, liquid fuel and electricity. Models are proposed to cover and analyze all these products. In these models fast pyrolysis and its products pyrolysis oil is not considered. The paper has three sections:

1) Estimation of income from solid waste
2) Investment calculations
3) Maintenance costs

The estimations were made by considering realistic input values and the return periods for each element were calculated. The models can be used with multi product estimation such as electricity, gas and liquefied gas.

2. Estimation of Income from Solid Waste

The capacity of a plant can be determined by considering input solid waste. Typically a person produces MSW at a rate of 0.5 kg to 2.5 kg per day and the waste has carbon content at 20-35 percent[1,2].These numbers are dependent on area, culture and income levels.

The products, defined by Table 1, are electricity, gas and liquid fuel. The reactor may produce these products in pre-defined percentages. Because electricity, gas and liquid fuel can be produced in different percentages, and the system can be designed for one product otherwise the percentages of the products must be defined.

The thermodynamic constants are taken from Cengel and Boles [2014]. The composition and the caloric value of MSW, MSW per capita are taken from Hoornweg and Bhada-Tata [2012]. The part of MSW for pyrolysis is about 15 to 20

percent of the MSW, which varies according to the recycling rates of the MSW; these parameters are defined by Table 2. The population, MSW per capita and the percentage of pyrolysis material dictate the production capacity of the plant.

Table 1. Definition of products from the process.

ELECTRICITY_PRODUCTION? (YES - NO)	YES	1
PERCENTAGE_OF_ELECTRICITY (0 - 100)	100	%
GAS_GAS_PRODUCTION? (YES - NO)	NO	0
PERCENTAGE_OF_GAS (0 - 100)	0	%
LIQUID_FUEL PRODUCTION? (YES - NO)	NO	0
PERCENTAGE_OF_LIQUID_FUEL (0 - 100)	0	%

Table 2. Definition of system parameters.

1_KG_TRASH_CALORIC_VALUE	2,000.00	KCAL
1_KCAL	0.001163	KWH
1_KG_ORGANIC_TRASH_KWH_VALUE	2.33	KWH
POPULATION	200,000	
MSW_PER_CAPITA	1.60	KG
PYROLYSIS_MATERIAL_PERCENTAGE(0–25)	20	%
PLANT_MATERIAL_DAILY_FOR_PYROLYSIS	64.00	TON

If the plant produces electricity, the income and the power of the plant can be seen on Table 3.

Table 3. Estimation of the income from electricity production and the power of the plant.

EFFICIENCY_FOR_TRASH_PERCENT (80 - 100)	90	%
KWH_PRICE_$	0.08	$
DAILY_ENERGY_FROM_ELECTRICITY	133,977.60	KWH
INCOME_$_PER_YEAR_FROM_ELECTRICITY	3,912,145.92	$
POWER_MW	5.58	MW

The incomes from gas and liquid fuel are presented by Tables 3 and 4, respectively. It should be noted that the calculations are performed per gas or liquid fuel production only.

Table 4. Estimation income from only gas production during pyrolysis.

GAS_EFFICIENCY_PERCENT (80 - 100)	90	%
PRICE_OF_GAS_PER_M3	0.35	$
ENERGY_PER_M3_IN_GAS	45.00	MEGAJOULE
1_KWH	3.60	MEGAJOULE
ENERGY_PER_M3_IN_GAS	12.50	KWH
AMOUNT_GAS_PER_DAY	9,646.39	M3
YEARLY_INCOME_FROM_GAS	1,243,705.87	$

Table 5. Estimation income from only liquid fuel production during pyrolysis.

LIQUID_FUEL_PODUCTION_EFFICIENCY_PERCENT (80 - 100)	90	%
LIQUID_FUEL_PRICE_PER_LITER	0.70	$
OPERATING COST	0.06	$
LIQUID_FUEL_PER_TON(100 - 300)	100.00	LITER
DAILY_PRODUCTION_LIQUID_FUEL	5,760.00	LITER
YEARLY_INCOME_FROM_LIQUID_FUEL	1,345,536.00	$

The efficiency values presented by Tables 3, 4 and 5 depend on the system and the contents of the MSW system and these values vary 80-100 % of the whole pyrolysis

material. The prices of electricity, gas and liquid fuel are according to prevailing market values of these commodities. However, the price of electricity is affected by government regulations and subsidies and varies from 3 to 20 US cents. Liquid fuel is produced through Fischer-Tropsch process at a rate of 100 to 300 liters per ton [17-20].

Table 6. *The incomes from the sale of recyclables, biochar, carbon credit, gate (tipping) and brown water.*

RECYCLABLE_SALES_PER_TON_$	0.00	$
PERCENTAGE_OF_RECYCLABLE (0 - 20)	0	%
DAILY_AMOUNT_OF_RECYCLABLE	0.00	TON
YEARLY_RECYCLABLE_SALES_$	0.00	$
BIO_CHAR_$_PER_TON	0.00	$
PERCENTAGE_OF_CHARCOAL (0 - 2)	0	%
DAILY_AMOUNT_OF_CHARCOAL	0.00	TON
YEARLY_BIO_CHAR_SALES_$	0.00	$
CARBON_CREDIT_$_PER_TON	0.00	$
PERCENTAGE_FOR_CARBON_CREDIT (0 - 100)	0	%
DAILY_AMOUNT_OF_CARBON_CREDIT	0	TON
YEARLY_CARBON_CREDIT_$	0.00	$
GATE_(TIPPING)_FEE_$_PER_TON	0.00	$
PERCENTAGE_OF_GATE (TIPPING) (90 - 100)	0	%
DAILY_AMOUNT_OF_GATE (TIPPING)	0.00	TON
YEARLY_GATE_(TIPPING)_FEE_$_	0.00	$
BROWN_WATER_$_PER_TON	0.00	$
PERCENTAGE_OF_WATER (0- 80)	0	%
DAILY_AMOUNT_OF_BROWN_WATER	0	TON
YEARLY_INCOME_BROWN_WATER_$	0.00	$

The incomes from of recyclables, biochar, carbon credit, gate (tipping) and brown water are presented in Table 6. The sale values can be determined through the agreements with local municipalities. The percentage of recyclable(0-20); charcoal percentage (0 – 2), carbon credit percentage (0 - 100), tipping percentage (90 – 100) and percentage of water (0 – 80) are all dependent on the content of the MSW.

3. Total Income

Total income is estimated by adding these incomes which are possible if the sale of these products are present. The income from one tone of household waste (*trash*) can be estimated. In this calculation, only the sale of electricity is considered; if gas and liquid fuel are also to be produced and their production rates are as per Table 1.

Table 7. *Total income.*

TOTAL_INCOME_YEARLY_$	3,912,145.92	$
INCOME_OF_PROCESSED_ONE_TON_TRASH	33.49	$

4. Investment Calculations

The list of equipment is determined by considering the amount of the trash and the capacity necessary equipment, see Table 8. The system uses slow pyrolysis [22] where the obtained gas product runs electric generators; Table 8 is set for only electricity production. Equipment for fuel liquefiers which uses Fischer–Tropsch process [17, 19] and gas filters are not considered. The dryers are required to eliminate moisture in MSW [23], and finally the moisture is used for water production. The exhaust gases are used in dryers to increase the efficiency of the system. The other components are MSW sorting unit, waste handling unit, de-sulfurization unit, gas filters and granulation system. The other components of the capital investments are civil works, engineering design, installation and commissioning. This needs rewriting, very confusing.

Table 8. *List of the major items of the plant.*

CAPITAL INVESTMENT			
ITEMS	Pcs.	Unit Price USD	Total Price USD
SYSTEM EFFICIENCY	0.9		
DAILY CAPACITY OF PYROLYSIS REACTOR, TON	24		
PYROLYSIS SYSTEM	2	2,000,000.00	4,000,000.00
DAILY CAPACITY OF DRYING UNIT, TON	70		
DRYING SYSTEM	2	250,000.00	500,000.00
DAILY CAPACITY OF CONDENSER, TON	70		
CONDENSER	2	300,000.00	600,000.00
DAILY CAPACITY OF MSW PRESORTING, TON	100		
MSW PRESORTING	2	400,000.00	800,000.00
DAILY CAPACITY OF WASTE HANDLING, TON	200		
WASTE HANDLING	1	400,000.00	400,000.00
DAILY CAPACITY OF DE-SULFURIZATION UNIT, TON	200		
DE-SULFURIZATION	1	350,000.00	350,000.00
DAILY CAPACITY OF GAS GENSET, TON	24		
GAS GENSET	2	1,000,000.00	2,000,000.00
DAILY CAPACITY OF GAS FILTER UNIT, TON	24		
GAS_FILTER	0	350,000.00	0.00
DAILY CAPACITY OF LIQUIFIER UNIT, TON	24		
LIQUIFIER	0	3,000,000.00	0.00
DAILY CAPACITY OF GRANULATION UNIT, TON	24		
GRANULATION SYSTEM	1	250,000.00	250,000.00
CIVIL WORKS	1	1,000,000.00	1,000,000.00
PROJECT AND ENGINEERING	1	500,000.00	500,000.00
INSTALLATION AND COMISSIONING	1	1,000,000.00	1,000,000.00
TOTAL			11,400,000.00

5. Operating Expense

Operating costs can be seen in Table 9, where the yearly profit and payback periods are also presented. The costs of the operation are payments of electricity, miscellaneous maintenance, water treatment, salaries, lubrication, and cost on unseen expenses.

Table 9. Operating expenses.

OPERATING EXPENSES			
	Unit Cost USD	Daily Cost	Annual Cost
ELECTRICITY COST_@_365_DAYS	0.03	3,000.00	32,850.00
GENSET MAINTENANCE	2	18,000.00	36,000.00
GAS_FILTER_MAINTENANCE	0	20,000.00	0.00
LIQUIFIER_MAINTENANCE	0	20,000.00	0.00
DRYER MAINTENANCE			60,000.00
CONDENSER MAINTENANCE			30,000.00
WATER TREATMENT			30,000.00
PYROLYSIS UNIT MAINTENANCE			157,500.00
SALARIES			200,000.00
OTHERS			50,000.00
MSW PRESORTING MAINTENANCE			22,000.00
HANDLING MAINTENANCE			1,800.00
GAS GENSET LUBRICATION COST	2	24,000.00	48,000.00
TOTAL			668,150.00
GROSS PROFIT		2,299,784.72	USD/YEAR
PAYBACK PERIOD		4.96	YEAR
		59.48	MONTHS

The water may have some odors and these odors may be avoided by odor control technologies which are widely available on the market. Since only electricity is produced in this scenario, the cost of maintenance of gas filters and liquefiers is not withstanding.

6. Various Scenarios and Return Period

The important variables profit and return period for investment are the sale price of electricity and population; the various cases are presented by Table 10 and can be extended further. In these scenarios, the income from recycling, charcoal, carbon credit, tipping and produced water are not considered. If these incomes are to be taken into consideration, the profit and return period of the investment will be shortened considerably.

Table 10. Various scenarios.

POPULATION	PRICE OF ELECTRICITY, $/kWh	PROFIT $	RETURN_PERIOD, YEAR
200,000	0.08	3,189,245.92	3.57
300,000	0.08	5,074,118.88	2.95
200,000	0.10	4,145,382.40	2.75
300,000	0.10	6,511,973.60	2.30

7. Conclusion

MSW is a chronic problem in urban areas. WTE technologies such as pyrolysis can be utilized to convert this problem to a revenue source if properly managed and implemented. This paper presented an economic analysis of the Pyrolysis technology as an MSW management option. The analysis showed that the determining factor in WTE investment is the selling price of electricity. However, more comprehensive scenarios can be developed where electricity, gas and liquid fuel production are considered with their selected production percentages.

References

[1] D. Hoornweg and L. Thomas, 1999. What a Waste: Solid Waste Management in Asia. East Asia and Pacific Region. Urban and Local Government Working Paper. World Bank. 1818 H Street, NW, Washington, DC 20433 USA.

[2] Daniel Hoornweg and Perinaz Bhada-Tata, What A Waste a Global Review of Solid Waste Management, Urban Development & Local Government Unit, World Bank, 1818 H Street, NW, Washington, DC 20433 USA, www.worldbank.org/urban, No. 15, March 2012.

[3] P. Costi, R. Minciardi, M. Robba, M. Rovatti, M. and R. Sacile, (2004). An environmentally sustainable decision model for urban solid waste management, waste management, 24(3), 277-295.

[4] S. J. Burnley, (2007) "A review of municipal solid waste composition in the United kingdom" Vol. 27, No. 10. pp. 1274-1285.

[5] ASME, American Society of Mechanical Engineers (2008), Waste-to-Energy: A renewable Energy Source from Municipal Solid Wastes, White paper submitted to the Congress, 2008.

[6]   UNEP, United Nations Environmental Program (1996), International Source Book on Environmentally Sound Technologies for Municipal Solid Waste Management, Osaka/Shiga.

[7]   O. K. M. Ouda, S. A. Raza, R. Al-Waked, J. F. Al-Asad and Nizami, A-S.,(2015).Waste-to-Energy Potential in the Western Province of Saudi Arabia, Journal of King Saud University - Engineering Sciences, doi: http://dx.doi.org/10.1016/j.jksues.2015.02.002

[8]   O. K. M. Ouda, H. M. Cekirge and R. Syed, (2013), An assessment of the potential contribution from waste-to-energy facilities to electricity demand in Saudi Arabia, Energy Conversion and Management 75, pp. 402-406.

[9]   O. K. M. Ouda and H. M. Cekirge, (2014), Roadmap for Development of Waste-to Energy Facility in Saudi Arabia, American Journal of Environmental Engineering 3(6), pp. 267-272.

[10]   B. S. Tawabini, O. K. M. Ouda and S. A. Raza, (2014). Investigating of Waste to Energy Potential as a Renewable Energy Resource in Al-Hasa Region, Saudi Arabia, 5th International Symposium on Energy from Biomass and Waste(VENICE 2014). 17-20 November 2014, Venice, Italy.

[11]   O. Aga, O. K. M. Ouda and S. A. Raza, (2014), Investigating Waste to Energy Potential in the Eastern Region, Saudi Arabia, Renewable Energy for Developing Countries (REDEC 2014), November 26-27, 2014, Beirut, Lebanon,

[12]   R. Al-Waked, O. K. M. Ouda and S. A. Raza, (2014), Potential Value of Waste to Energy Facility in Riyadh City - Saudi Arabia, the Eighth Jordanian International Mechanical Engineering Conference (JIMEC 8), September 22-23, 2014, Amman, Jordan.

[13]   O. K. M. Ouda and S. A. Raza, (2014). Waste-to-Energy: Solution for Municipal Solid Waste Challenges- Global Annual Investment, 2014 IEEE International Symposium on Technology Management and Emerging Technology (ISTMET), May 27-29, 2014, Bandung, West Java, Indonesia.

[14]   M. M. Gilbert and P. E. Wendell, Introduction to Environmental Engineering and Science, Chapter 9: Solid Waste Management and Resource Recovery, Third Edition, Pearson Education Inc., 2008, ISBN-13:978-0-13-233934-6.

[15]   O. K. M. Ouda and H. M. Cekirge, (2014). Potential Environmental Values of Waste-to-Energy Facilities in Saudi Arabia, the Arabian Journal for Science and Engineering (AJSE) 39 (11), pp 7525-7533.

[16]   O. K. M. Ouda, (2013), Assessment of the Environmental Values of Waste-to-Energy in the Gaza Strip, Current World Environment 8(3), pp. 355-364.

[17]   L. Mahjoob and H. Nezihi Ogul, (private communication) American Combustion Technologies, Inc., Los Angeles, California, 2014.

[18]   Yunus A. Cengel and Michael A. Boles Thermodynamics: An Engineering Approach, 8th Edition, New York, McGraw-Hill, 2014.

[19]   Jin Hu Fei Yu and Yongwu Lu, Application of Fischer–Tropsch Synthesis in Biomass to Liquid Conversion, Catalysts, 2, 303-326, 2012.

[20]   Gary C. Young, (2010), Municipal Solid Waste to Energy Conversion Processes: Economic, Technical and Renewable Comparisons, John Wiley & Sons, Hoboken, New Jersey, 2010.

[21]   Maura Farver and Christopher Frantz (2013), Garbage to Gasoline: Converting Municipal Solid Waste to Liquid Fuels Technologies, Commercialization, and Policy Duke University Nicholas School of the Environment April, 2013.

[22]   Anh N. Phan, Changkook Ryu, Vida N. Sharifi and Jim Swithenbank (2008), Characterisation of Slow Pyrolysis Products from Segregated Wastes for Energy Production, Journal of Analytical and Applied Pyrolysis, Pages 65–71, Volume 81, Issue 1, January 2008.

[23]   Chen Shu, Ma Xiao-Qian and Liang Zeng-Ying, (2014), Moisture Transfer Models and Drying Characteristics of MSW Containing High Moisture, TELKOMNIKA Indonesian Journal of Electrical Engineering, Vol.12, No.3, 1741 – 1750, March 2014.

15

The Impact of the Environmental Condition on the Performance of the Photovoltaic Cell

Ehsan Fadhil Abbas Al-Showany

Refrigeration & Air Conditioning Engineering Department, Kirkuk Technical College, Northern Technical University, Kirkuk, Iraq

Email address:

ehsanfadhil@gmail.com

Abstract: The present study investigated the impact of weather conditions on the production performance of the photovoltaic (PV) module. The experiments have been conducted by using two identical PV module of 75 Watt each and they were placed in the same weather conditions in the summer season in Kirkuk city-Iraq. One of them was used as a conventional module as a reference panel and the other unit which has been used in all required tests. Water circulation has been used for cooling of the PV module to and very fine soil used to estimate the effect of each of hot weather and dust deposition on the performance of PV respectively. The results show that the fill factor (FF) and PV efficiency affected inversely with increasing in temperature, on the other hand cooling process contributed to increase the voltage generation across PV panel by 11.8%, while the reduction in voltage generation by unclean panel due to natural pollution deposition on the front of the panel for a period of three months was about 3.8% compared with clean panel and 13.8% if it has been compared with voltage production by panel when it has been cooled by water.

Keywords: Photovoltaic, Performance of PV, Efficiency of PV

1. Introduction

Renewable energy is generated from natural resources such as the sun, wind, and water, using technology which ensures that the energy stores are naturally replenished, and they are becoming more and more attractive especially after deterioration of the environment pollution and the constant fluctuation in oil price. Solar energy has been encouraged to substitute conventional energy. It has good potential and the direct conversion technology based on solar photovoltaic has several positive attributes especially in remote area [1, 2]. The physical process in which a photovoltaic (PV) cell converting sunlight directly into electricity through semiconductor diode tight into direct current (DC). One single PV cell produces up to 2 watts of power at approximately 0.5V DC, many PV cells are connected together to form modules (panels) to increase power output which are further assembled into large units called arrays [3]. The inherent material property of these semi-conductor limits of PV system efficiency of the photovoltaic system within 15-20% [4]. The performance of PV systems is affected by

several parameters including environment conditions. There are several studies in the field of environmental conditions such as (temperature, dirt accumulation and wind speed). Herein we review some of them: Kaldellis and Kokala [5] investigated an experimental analysis was conducted in the laboratory soft energy application and environmental protection (SELAB) to determine the effect of pollution on the energy performance of PV-panels quantitatively. Majid et al. [6] showed that the efficiency of PV cell decrease 0.485% per $1°C$, after the surface temperature increase than $25°C$. Elhab B. R. et al. [7] presented a mathematical approach to determine optimum tilt angles for solar collector. Kaldellis et al. [8] introduced an experimental study to determine the reduction in efficiency of photovoltaic due to deposition of natural air pollution on the PV panels, where they concluded that the efficiency has been decreased by 0.4% per month, when it is exposure to outdoor without cleaning. Dincer F. and Meral M. E. [9] studied several factors effect on solar cell efficiency such as changing of cell temperature, using maximum power point tracking (MPPT) with solar cell and energy conservation efficiency for solar cell. Bhattacharya et al. [10] presented a statistical analysis method to determine the effect of ambient temperature and wind speed on the performance of the photovoltaic model. Tianze et al. [11] presented a new technologies to improve the conversion efficiency of PV model and to reduce the cost of it. Hamrouni et al. [12] and Hosseini et al. [13] improved the operation PV systems by spraying water over front of the photovoltaic cells. Rajput D. S. and sudhakar [14] studied the effect of environment dust on power production by PV experimentally, and they obtain that the power production reduced by 92.11%. Sulaiman et al. [15] provided an experimental study by using different obstruction materials conducted under controlled conditions using spotlight substitute for sunlight. The result which was obtained from external resistance could reduce the efficiency by 85%.

The goal of this study is to determine the effect of environmental conditions on the electrical power production of PV cells experimentally in Kirkuk city, Iraq. The study was accomplished by using two identical PV modules, one then used as a reference panel and its performance has been compared with the second module, when water circulation used for cooling the PV cell and very fine soil used as natural dust pollution deposition on the frontal surface for the PV module.

2. Experimental Methodologies

This study is conducted by using two Photovoltaic cells of 75W and they were mounted on the angular movable stand and control board as shown in Fig. (1). These two PV modules has been connected to battery source by control board which contains to solar charge controller (CMTP02), 4 channel signal recorder (velleman) was used for instantaneous recording of DC voltage generation from PV cells, electric DC lamp 50W used as electrical load and current flow in electrical circuit measuring by Professional digital multimeter (ST-9915) as shown in Fig. (2). Weather station type (Wireless weather station HP2000) as shown in Fig. (3), has been used to record the required data about testing place (Al-Wasity, Kirkuk / Iraq) weather condition during all experiment and the latitude and longitude of the location are 35.24 N and 44.21 E respectively. Three sets of experiments to show the effects of environmental conditions on the performance of PV cells, as follows:-

Fig. 1. Installation of PV test modules on the base.

Fig. 2. Components of control board.

Fig. 3. Photographic of professional wireless weather station.

Fig. 4. Instruments used in test of PV characteristics.

2.1. The Influence of Temperature on the PV Characteristics

The effect of temperature was investigated by using two commercially PV cells of 75W. One of them is used as conventional reference panel and the other module has been used for executing experiments by adding a water circulation to it, as shown in Fig. (3), their purpose is to control surface temperature of PV cell through water flow on the upper layer of PV cell at constant temperature, to insure that the surface temperature has reached to steady state. The range of water temperatures were used at (5, 10, 20, 40, 50, 60, 70)°C. PV characteristic experiments in different surface temperatures have been conducted by connecting precision variable resistor, ammeter and voltmeter to the PV cell as shown in Fig. (4). The principle of the PV characteristic in each experiment is measuring each of voltage potential across variable resistance and current flow for each resistance value, where that the procedures of the test begins by changing the resistance from zero ohm (open circuit) up to maximum resistance value (closed circuit) gradually by variable resistor device. Will the obtained data has been used to sketch a relation between each of voltage vs electric current and voltage vs power production as shown in Fig. (5).

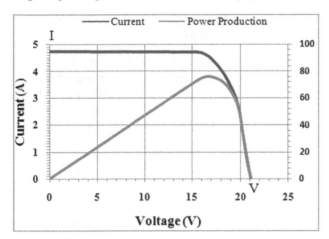

Fig. 5. *Typical I-V and P-V characteristics of solar cell at standard condition test.*

The maximum power (P_{max}) was calculated using measured maximum current I_{mp} and maximum voltage I_{mp} [16, 17].

$$P_{max} = I_{mp} \times I_{mp} \tag{1}$$

and maximum power can be calculated according to the short circuit current I_{sc} and open circuit voltage [9,18, 19].

$$P_{max} = FF(I_{sc} \times V_{oc}) \tag{2}$$

where *FF* Fill factor of solar cell. It is essentially measure of quality of the solar cell, can be interpreted graphically as the ratio of the rectangular areas depicted in Fig. (6).

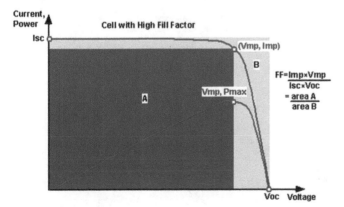

Fig. 6. *A typical current-voltage and maximum power curve [18].*

According to the above definition and Fig (6), the *FF* can be expressed as

$$FF = \frac{area\ A}{area\ B} = \frac{I_{mp} \times V_{mp}}{I_{sc} \times V_{oc}} \tag{3}$$

The solar cell efficiency η is calculated using the ratio of the maximum power produced divided by the input light irradiance (G, W/m²) and solar cell area (A_c, m²). Following

relation was used for the calculation of model efficiency.

$$\eta = \frac{P_{max}}{G \times A_c} \tag{4}$$

Substituting Eq. (3) into Eq. (4)

$$\eta = \frac{FF \times V_{oc} \times I_{sc}}{G \times A_c} \tag{5}$$

It turns out that both the open circuit voltage and the fill factor decrease substantially with temperature, while short circuit current increase slightly. Thus the net effect leads to a linear relation in the form [20, 21].

$$\eta = \eta_{Tref} \left[1 - \beta_{ref} (T_c - T_{ref}) + \gamma \log_{10} G \right] \tag{6}$$

where η_{Tref} is the module's electrical efficiency of 15.6% at a reference temperature T_{ref} of 25°C and solar radiation flux of 1000W/m², β_{ref} is the temperature coefficient, with respect to temperature 0.0045 K⁻¹ and γ is mainly material properties and equal to 0.12 for crystal silicon modules. Thus, the latter term from Eq. (6) is usually taken as zero [22]. Then Eq. (6) reduces to

$$\eta = \eta_{Tref} \left[1 - \beta_{ref} (T_c - T_{ref}) \right] \tag{7}$$

where the quantities of η_{Tref} and β_{ref} are given by PV manufactures. The actual value of the temperature coefficient, in particular, depends not only on the PV material but on T_{ref}, as well. It is given by the ratio

$$\beta_{ref} = \frac{1}{T_0 - T_{ref}} \tag{8}$$

T_c is the module operating temperature varies according to global solar irradiance, G and mean monthly ambient temperature, T_a are related as in Eq.(9), [23, 24]:

$$T_c - T_a = (219 + 823K_t) \frac{NOCT - 20}{800} \tag{9}$$

there is another method used to calculate cell temperature based on monthly mean of solar irradiance, G_b and ambient temperature, and *NOCT* is normal operating cell temperature [25].

$$T_c = T_a + \left(\frac{NOCT - 20}{800} \right) G_b \tag{10}$$

In which T_0 is the temperature at which the PV module's electrical efficiency drops to zero, for crystalline silicon cells this temperature is (270°C).

Also there is a new approach for estimating the operating temperature of photovoltaic model using statistical formula with an error of less than 3%, based on the steady state approach prediction as [26]:

$$T_c = 0.943 \times T_a + 0.195 \times G - 1.528 \times U + 0.3529 \tag{11}$$

2.2. Effect of Temperature on the Power Production of the PV Module

To investigate the performance of PV module in environment that is hotter than standard condition is done by using electric load which provided the power by power supply system which contains a battery (12V 36AH), PV module and solar charge controller (CMP02). For this purpose two tests on the PV module were done by starting the load of (100W) for period of 450 sec. and then shut down the power source, while the charging process to the battery was continuous until the battery becomes fully charge. To achieve this goal we used voltage data logger (Velleman PCLAB 2000SE) to obtain high accuracy data. The first test was done by using the PV module in normal condition and the other test was done by using the cooling process to the PV module. The performance of the module was estimated by reduction in charging time of the battery, and calculated as:

$$\% \ Reduction \ in \ time \ of \ battery \ charging = \frac{\Delta t_1 - \Delta t_2}{\Delta t_1} \times 100 \quad (12)$$

where Δt_1 and Δt_2 are time of battery charging for first and second cases respectively.

2.3. Estimation of Dust Impact on the Performance of the PV Module

The effect of dust deposition on the front surface of the PV was investigated by using two identical cells used in previous tests and it was estimated by comparing between performance of clean and dirty cells. The experiments were done based on the density of the dust on the front of the PV by using five different dust pollution mass depositions were selected as (0.05, 0.1 0.5, 1.0 and 2.0)g/m² with average diameter of (0.005mm). The dust deposition density (ΔM, g/m²) is expressed as [5]:

$$\Delta M = \frac{\Delta m}{A_c} \quad (13)$$

where Δm is the total mass of dust layer on the surface of the polluted PV cells in (g) and A_c is the cell area in (m^2). The output parameters of PV cells are recorded in each values of dust deposition density under four different environment conditions by expose PV cells directly to the atmospheric and when the temperature of the frontal surface of the PV cells range from 40 to 60°C. Start from 8am, and ends at 1pm for ten days in second-third of July, 2015. The average of the results of these experiments is used to estimate the influence of dust pollution on the PV performance. The reduction in performance of PV calculated by using following relations [14, 16]:

$$\% Reduction \ in \ power \ production = \frac{P_{clean} - P_{dirty}}{P_{dirty}} \times 100 \quad (14)$$

and

$$\% Reduction \ in \ PV \ Cell \ efficiency = \frac{\eta_{clean} - \eta_{dirty}}{\eta_{dirty}} \times 100 \quad (15)$$

3. Results and Discussion

A professional wireless weather station was used to record the input data of the environmental conditions during test times; also the DC data logger was used to record output

parameters from the PV cells with minimum error. From results which were obtained through three sets of experiments on the two identical PV cells to study environment effect from the following aspects:

3.1. The First Aspect: Study of the PV cell characteristics in Different Surface Temperature

Figs. (7 and 8) shows the I-V characteristics of PV cell tests at different temperature conditions, as we noted that the temperature greatly was effected by the power production from solar module, so any increases in ambient temperature than standard condition 25°C, it will have negative effect on the PV performance, and vice versa, also we noted that the PV cell gives the maximum power at 5°C, while the minimum power gives at 60°C, as shown in fig. (9), as shown in fig. (9), and from there, we find a mathematical expression for each of fill factor and actual efficiency of module (FF = -6E-06T² - 0.0007T + 0.7899) and (η = -3E-06T² - 0.001T + 0.2345) respectively.

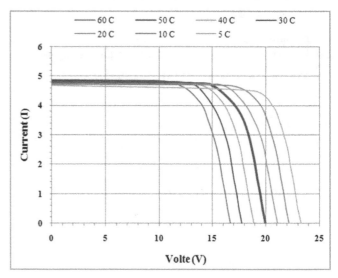

Fig. 7. I-V characteristic of different surface temperature.

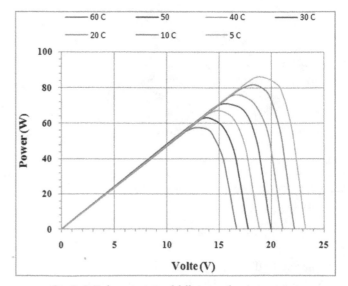

Fig. 8. P-V characteristic of different surface temperature.

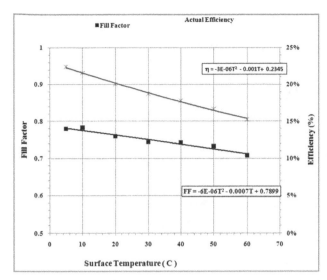

Fig. 9. Relation between fill factor and efficiency with surface temperature PV cell.

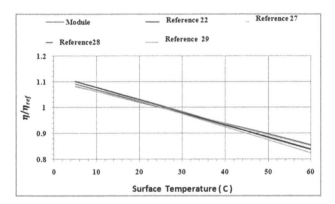

Fig. 10. The ratio (η/η_{ref}) for PV module compared with correlations predicted by previous References.

The actual efficiency of the present PV module has been compared with the empirical relations in Table (1) that put to connect the ratio of actual efficiency to the standard condition efficiency with surface temperature of the module. The comparison result appears they approximately conforming to the results of these correlations and with accuracy about ±2% as shown in Fig. (10).

Table 1. *Empirical Relations.*

Correlation	Ref.
$\eta = \eta_{Tref}\left[1 - \beta_{ref}\left(T_c - T_{ref}\right)\right]$	22
$\eta = \eta_{ref} - \mu\left(T_c - T_{ref}\right)$	27
$\eta = \eta_{25} - b(T_c - 25)$	28
$\eta_{(GT,T_C)} = \eta_{(GT,25)}[1 + c_3(T_c - 25)]$	29

where μ is overall cell temperature coefficient, $b = b(G_T)$ and $c_3 = -0.5\%\ loss\ per\ °C$

3.2. The Second Aspect: Study of Electrical Power Production at Different Surface Temperature

The PV unit has been operated in two cases in July 24, 2015. The first case at 9 am, PV module operated in conventional case when surrounding temperature was 48°C and solar radiation (746 W/m²) and battery charging watching by 4 channel signal recorder as shown in Fig. (11). At the beginning of the test, the battery was fully charged, and PV cell works as in open circuit and it generated a voltage of 18.36 V, when the power start-On to two lamps during 431 sec with voltage supply of 12.07V and then power source cut off from lamps and then we will wait until the battery charging reaches to initial level, and this needs to 1362 sec in this case, and these about 8.4 times of working time for lamps. In second case, time of experiment at 2pm, when ambient temperature is 51°C and solar radiation 930 W/m², where we used water circulating with controlled temperature of 25°C. This process contributed to increase the level of voltage generation from 17.8 to 19.9V, as shown in Fig. (12), and the ratio of improvement in voltage generation was about 11.8%. When power was supplied to the lamps for the same period in the first case, in this case, battery was fully charging faster than first case by 50%.

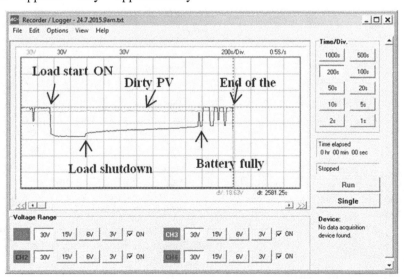

Fig. 11. System operation performance for normal use.

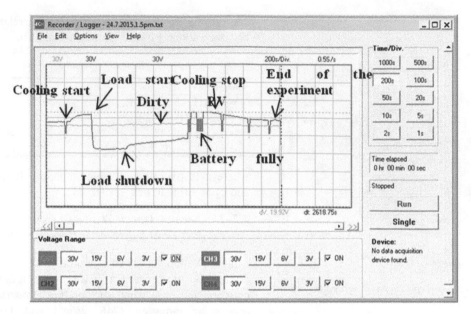

Fig. 12. System operation performance with cooling process.

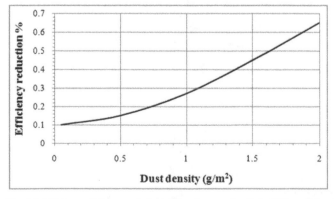

Fig. 13. Dust pollution disposition on the PV test module.

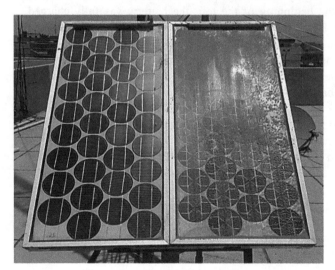

Fig. 14. Efficiency differences between the clean and polluted PV module at different of dust density on the frontal surface.

4. Conclusions

Experiments results refer to that the environmental conditions has great effect on the PV module performance, to improve its performance and should be done by periodic maintenance process on the PV module from dust deposition and add suitable size of PV panels to compensate for the shortfall, which occurs in power production of the PV unit due to hot weather and that cannot be controlled.

References

[1]   Bataineh K. and Dalalah D., Optimal configuration for design of stand-alone PV system, *Smart Grid and Renewable Energy*, V. 3, pp. 139-147 (2012).

[2]   Ma J., Man K. L., Ting T. O., Zhang E. G., Guan S., Wong P. W., Krilavicius T., Sauleviius D., and Lei C. U., Simple Computational method of predicting electrical characteristics in solar cell, *Elektronika IR Elektrotechnka*, V. 20, No. 1, pp.41-44, (2014).

[3]   Muzathik A. M., Photovoltaic modules operating temperature estimations using a simple correlation, *International Journal of Energy Engineering*, V. 4, Iss. 4, pp. 151-158, (2014).

[4]   Mani M. and Pillai R., Impact of dust on solar photovoltaic performance: Research status, challenges and recommendations, *Renewable and Sustainable Energy Reviews*, V. 14, pp. 3124-3131, (2010).

[5]   Kaldellis J. K. and Kokala A., Simulating the dust effect on the energy performance of photovoltaic generates based on experimental measurements, *Energy*, V. 36, pp.5154-5161, (2011).

[6]   Majid Z. A. A., Ruslan M. H., Sopian K., Othman M. Y. and Azmi M. S. M., Study on performance of 80 Watt floating photovoltaic panel, *Journal of Mechanical Engineering and Sciences*, V. 7, pp. 1150-1156, (2014).

[7]   Elhab B. R., Sopian K., Mat S., Lim C., Sulaiman M. Y., Ruslan M. H. and Saadation O., Optimizing tilt angles and orientation of solar panels for Kuala Lumpur Malaysia, *Scientific Research and Essays*, v. 7, No. 42, pp. 3758-3765, (2012).

[8] Kaldellis J. K. and Kokala A., Quantifying the decease of the photovoltaic panels energy yield due to phenomena of natural air pollution disposal, *Energy* V. 35, pp. 4862-4869, (2010).

[9] Dincer F. and Maral M. E., Critical factors that affecting efficiency of solar cells, *Smart Grid and Renewable Energy*, V.1, pp. 47-50, (2010).

[10] Bhattacharya T., Chakraborty A. and Pal K., Effects of ambient temperature and wind speed on performance of Monocrystalline solar photovoltaic module in Tripura, India, *Journal of Solar Energy*, Article ID 817078, (2014).

[11] Tianze L., Xia Z., Chuan J. and Luan H. Methods and analysis of factors impact on the efficiency of the photovoltaic generation, Journal *of Physics, Conference Series* 276, (2011).

[12] Hamroni N., Jradi M. and Cherif A., Solar radiation and ambient temperature effects on the performances of a PV pumping system, *Revue des Energies Renouvelables*, V. 11, No.1, pp. 95-16, (2008).

[13] Hosseini R., Hosseini N. and Khorasanizadeh H., An experimental study of combining system with a heating system, *Word Renewable Energy Congress 2011-Swiden*, pp. 2993-3000, (2011).

[14] Rajput D. S. and Sudhakar K., Effect of dust on the performance of solar panel, *International Journal of Chem Tech Research*, V. 5, No. 2, pp. 1083-1086,(2013).

[15] Sulaiman S. A., Singh A. K., Mokhtar M. M. and Bou-Rabee M. A., Influence of dirt accumulation on performance of PV panels, *Energy Procedia* V.50, pp. 50-56, (2014).

[16] Ali H. M., Zafar M. A., Bashir M. A., Nasir M. A., Ali M. and Siddiqui A. M., Effect of dust deposition on the performance of photovoltaic modules in Taxila, Pakistan, *Thermal Science, Online, First Issus* (00), pp-46-46, (2015).

[17] Mohamed A. O. and Hasan A., Effect of dust accumulation on performance of photovoltaic solar modules in Sahara environment, *Journal of Basic and Applied Science Research*, V. 2, No. 11, pp. 11030-11036, (2012).

[18] Raina G., Mandal S., Shinda S., Patil M. and Hedau R., A novel technique for PV panel performance prediction, *International Journal of Computer Application, International Conference and Workshop on Emerging Trends in Technology* 2013, pp. 19-24, (2013).

[19] Pandey A. K., Pant P. C., Sastry O. S., Kumar A. and Tyagi S.

K., Energy and exrgy performance evaluation of typical solar photovoltaic module, *Journal of Thermal Science, Online First Issus* (00), pp. 147-147, (2013).

[20] Skoplaki E. and Palyvos J. A., On the temperature dependence of photovoltaic module electrical performance: A review of efficiency/ power correlations", *Solar Energy*, V. 83, pp. 614-624, (2009).

[21] Dubey S., Sarvaiya J. N. and Seshadri B., Temperature dependent photovoltaic efficiency and its effect on PV production in the World- A review, *Energy Procedia*, V. 33, pp. 311-321, (2013).

[22] Evans D., Simplified method for predicting photovoltaic array output, *Solar Energy*, V. 27, pp. 555-560, (1981).

[23] Mehmood A., Waqas A. and Mahmood H. T., Stand-alone system assessment for major cities of Pakistan based on simulated results; A comparative Study, *Nust Journal Engineering Sciences*, V.1, No.1, pp. 33-37, (2013).

[24] Rahman H. A., Nor K. M. and Hassan M. Y., The impact of meteorological factors on energy yields for the building integrated photovoltaic system in Malaysia, *Solar09, the 47th ANZSES Annual Conference 2009 Townsville, Queensland, Australia, (2009).*

[25] Benatiallah A., Mostefaoui R., Boubekri M. and Boubekri N., A simulation model for sizing PV installations, *Desalination*, V.209, pp.97-101, (2007).

[26] Muzathik A. M., Photovoltaic modules operating temperature estimation using a simple correlation, *International Journal of Energy Engineering*, V. 4, Iss.4, pp. 151-158,(2014).

[27] Bazilian M., Prasad D. Modeling of a photovoltaic heat recovery system and its role in a design decision support tool for building professionals, *Renewable Energy* 27, pp. 57-68, (2002).

[28] Durisch W., Urbon J., Smestad G. Characterization of solar of solar cells and modules under actual operations, *Renewable Energy* 8, pp. 359-366, (1996).

[29] Mohring H. D., Stellbogen D., Schaffler R., Oelting S., Gegenwart R., Konttinen P., Carlsson T., Cendagorta M., Hermann W., Outdoor performance of polycrystalline thin fin film PV modules in different European climates, *Proceeding of 19th EC Photovoltaic Solar Energy Conference, June 7-11,2004, Paris, France.*

16

Identifying Defects in the Transmission Lines by Neural Networks

Kherchouche Younes, Savah Houari

Faculty of Electrical Engineering, University Djillali Liabes , ICEPS (Intelligent Control Electrical Power System) Laboratory, Sidi Bel Abbes, Algeria

Email address:

younes_elt@yahoo.fr (K. Younes), housayah@yahoo.fr (S. Houari)

Abstract: The paper presents a new technique for the identification of defects in transmission lines using artificial neural networks. The technique uses the amplitude at the fundamental frequency voltage and current signals to one end of the line. A study is conducted to evaluate the performance of the fault identifier.

Keywords: Defects, Identification, Transmission Line, Neural Networks

1. Introduction

A single-phase fault, a tripping and reclosing phase is necessary to maintain system stability.

The identification of defects is used to determine the phase of the faulted transmission line. Different approaches and techniques for identifying defects have been proposed in the literature. The most common technique used is based on the evaluation of signal amplitudes of currents and voltages at the fundamental frequency. The technique is based on the theory of symmetrical components which requires a calculation of Direct amplitudes, Negative and zero sequence components of This technique assumes that the transmission line is ideally transposed, She may have difficulty identifying a two-phase ground fault. Different techniques have been proposed to estimate the amplitudes of the signals, ie using the discrete Fourier transform, Walsh functions [1], the Kalman filter [2, 3], etc.

These techniques use a long computation time and do not have the capacity to adapt dynamically to the operating conditions of the system. The application of neural networks (RN) was too used for identification defects. In [4] a single neural network is used to identify the default after its occurrence. The inputs to the neural network 5 are samples of the signals of current and voltage at a frequency of 1 khz. The output layer consists of 11 neurons, 10 neurons for each type of defect and a neuron state without blemish. The use of a single neural network with multiple outputs demands large quantity of data in the step of learning and the results are not satisfactory if we consider the variation of network parameters. A single neuron can not identify all fault types correctly. In [5] the authors use 33 samples per cycle to 2 khz sampled signals as inputs to the neural network. A window of one cycle (or 3 cycles) is used to identify the fault and the fault is identified in 15 ms. In [6] the authors use 10 neural networks, one neuron for each type of defect. The inputs to the neural networks are the amplitudes of the voltages and currents after fault detection. Each neural network 10 is trained to identify one type of defect. The use of multiple neural networks demands learning each network regardless of where a longer learning time. In [7], [8] and [9] a single neural network is proposed to identify the fault. The inputs to the neural network are the amplitudes of currents and voltages after the detection of the fault. The neural network has four outputs, three outputs to indicate the faulty phase and an output to indicate if the fault is grounded or not. In this paper, we propose an identifier-based neural defects using the amplitudes of the signals at the fundamental frequency as inputs. Four neural networks are used to identify the defect. Each phase is provided with its specific detector fault based neurons and a neural network is used to identify the implication of the earth.

2. Electric Network Considered

Figure 1 shows the network considered in this study to evaluate the performance of the proposed default identifier. The transmission line of 400 [kv] and a length of 150 [km] from 2 extremities powered by sources GS and GR. The line is represented by parameters distributed by taking into account the dependence of the parameters as a function of frequency. The simulation of the fault line was carried out by the MTLAB software 7.7.0.

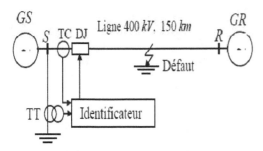

Fig. 1. Electrical network considered [1].

3. Identifier Defects Proposed

Designate by Ia, Ib, Ic the current and Va, Vb, Vc, the voltages of phases a, b and c. Current and zero sequence voltage are appointed by Io and Vo. The identifier of defects (IDEF) proposed consists of four independent networks of neurons: RNA, RNB, RNC and RNG. The first three networks are defect detectors in step (DDP) for detecting the phase (a, b or c) in default and the fourth network is a defect detector to ground (DDT) to indicate the involvement of the ground (G) on default. The outputs of the neural networks RNA, RNB, RNG and RNC are respectively A, B, C, and G (Figure 2).

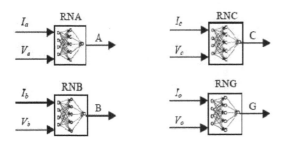

Fig. 2. Detectors component IDEF.

4. Input and Output of RNA

The signals calculated at a frequency of 100 khz are processed by a low-pass Butterworth filter of order 4 with a cut off frequency of 300 Hz A sample collection process of 1 khz (20 samples per cycle of 50 Hz) is applied. This sampling rate is consistent with rates currently used in numerical relays.

A window of data is used as input to the neural network defect detector. The identifier of the default IDEF amplitudes calculated using the fundamental frequency of the discrete Fourier transform DFT [1] (Ia, Ib, Ic, Io) current and voltage signals (Va, Vb, Vc, Vo) to the place relay as inputs to neural networks (Figure 3).

A series of four consecutive instantaneous amplitudes of the DFT [3] of each signal (sampled at 1 khz) is selected as

input to neural networks to identify the fault. This series is a mobile with a length of 3 ms window. The amplitudes of the DFT [3] are normalized to reach the input level of the neural networks.

The outputs A, B and C networks are indexed with a value of '1 '(presence of a fault in the corresponding phase) or '0' (situation without default). The output G is indexed with a value of '1 '(default connected to the earth) or '0' (default not related to the earth). The sigmoid transfer function is used in the hidden layer and the output layer. The structures of the neural networks of the IDEF for flaw detectors Phase (DDP) and ground faults (DDT) are summarized in Table 1.

Table 1. Structures flaw detectors phase and earth of the IDEF.

Detectors to phase faults (DDP) end earth (DDT)		Input variables	Number of neurons		
			Input layer	Hidden layer	Output Layer
DDP	RNA	$|Ia|,|Va|$	8	18	1
	RNB	$|Ib|,|Vb|$	8	18	1
	RNC	$|Ic|,|Vc|$	8	18	1
DDT	RNG	$|I0|,|V0|$	8	18	1

Architectures of neural networks for DDP (NAS) and DDT (RNG) are given in Figures 3 and 4 respectively.

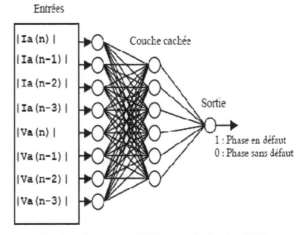

Fig. 3. Architectures of RNA network identifier IDEF.

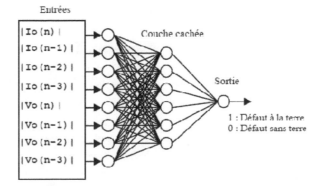

Fig. 4. Architectures RNG network identifier IDEF.

5. Tests and Results

After the final learning using Table 2, the neural network identifier IDEF defects are tested using Table 3 with patterns not seen during training using weights and biases the final learning.

Table 2. *Parameters for the learning of neural networks.*

Parameter	Learning
Fault location Lf (km)	3, 40, 80, 120, 147
Angle fault occurred θf (deg)	0, 45, 90
Fault resistance R (Ω)	0, 20, 40 ,100
Power sources (GVA)	9, 20
Voltage source (pu)	0,9, 1,1
Angles sources (load angles in deg.)	-20, 0, +20

Table 3. *Parameters for testing networks neurons.*

Parameter	Learning
Fault location Lf (km)	3,4 ,5 ,... 40, 80, 120, 147
Angle fault occurred θf (deg)	0, ,30,60, 90
Fault resistance R (Ω)	0, 5,10,20, 40 ,100
Power sources (GVA)	9, 10,20
Voltage source (pu)	0,9, 1.0,1,1
Angles sources (load angles in deg.)	-20, -10,0, 10,+20

We consider the network parameters following: SS = 10 GVA, US = $1\angle20°$ *pu*, τ_S = 20 *ms* et SR= 9 GVA, UR = $0.9\angle0°$ *pu*, τ_R = 20 *ms*.

The results in Figure 7a are obtained for a single-phase fault (a-g) 20 km with a fault resistance Rf = 40Ω and t = 30 ms ($\theta = 60^0$). Figure 7b shows the output of IDEF for phase fault no land (a-b) located 20 km with Rf = 40 Ω and t = 32 ms ($\theta=90^0$). Figure 7c are obtained for a three-phase fault no land (a-b-c) 100 km t = 28ms ($\theta=30^0$).

▬▬▬ Phase A,   ▬▬▬ Phase B
▬▬▬ Phase C,   ▬▬▬ La terre(g)

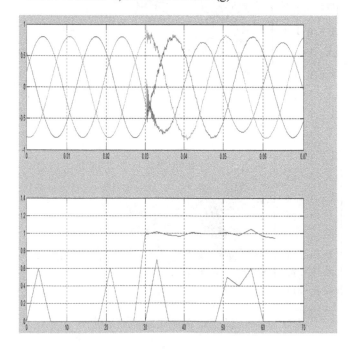

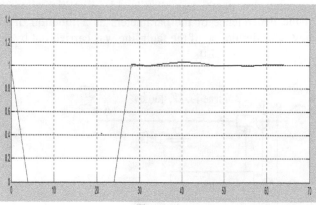

(a)Default (a-g), Lf = 20 km and t = 30 ms

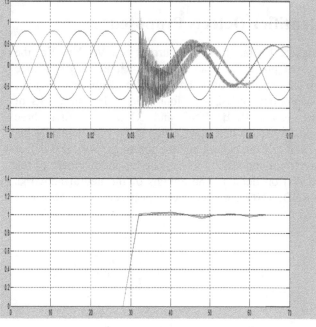

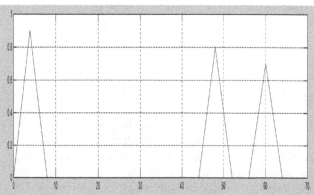

Time (ms)
(b)Default (a-b), Lf = 20 km and t = 32ms

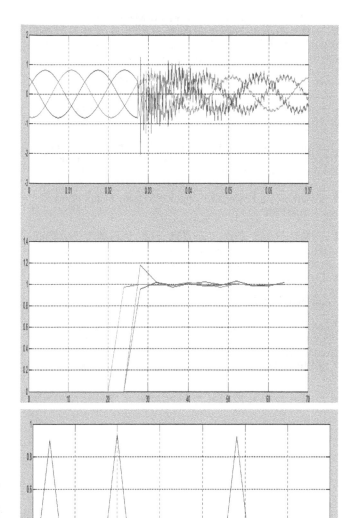

Time (ms)
(c)Default (a-b-c), Lf = 100 km and t = 28ms

Fig. 5. *Outputs the identifier IDEF.*

The results demonstrate the capacity of the fault identifier to pinpoint the type of fault for each test considered .

The response time t_r^{ident} of the identifier that is the difference between time t_o^{ident} obtained by the identifier and the actual time of the fault t_a^{def} is used as a criterion for evaluating the performance of the identification of the defect. The response time t_r^{ident} is defined by:

$$t_r^{\text{ident}} = t_o^{\text{ident}} - t_a^{\text{def}} \qquad (1)$$

Table 4 gives the values of t_r^{ident} for IDEF all fault types and for four values of θf (0 °, 30 °, 60 ° and 90 °) to a 40 km no fault resistance.

Table 4. *Results IDEF (tr ^{ident}) for a default 40 km no resistance.*

Type of fault	Response time tr ident IDEF in *ms*			
	T=0ms	T=28ms	T=30ms	T=32ms
	Tr^{ident}	Tr^{ident}	Tr^{ident}	Tr^{ident}
A-g	0	0	2	8
B-g	0	4	2	1
C-g	0	2	3	1
A-b-g	0	4	2	1
A-c-g	0	4	2	1
B-c-g	0	4	2	2
A-b	0	1	4	2
B-c	0	4	2	8
A-c	0	4	2	8
A-b-c	0	4	2	8

Table IV we can see that θ = 0 has its I'DEF gives good results and that the fault occurred angle θf has hardly a rule influence the detection time obtained to ^{ident} and the maximum value response time tr^{ident} is equal to 8 ms obtained for many types of default and angles θf.

Table 5. *Results IDEF (tr ^{ident}) for a fault at 100 km with resistance.*

Type of fault	Response time tr ident IDEF in *ms*					
	Θf = 0°			Θf = 90°		
	Rf = 0 Ω	Rf = 40Ω	Rf = 100 Ω	Rf = 0 Ω	Rf = 40 Ω	Rf = 100 Ω
A-g	1	0	2	4	2	2
B-g	1	2	8	4	2	8
C-g	1	4	4	4	4	4
A-b-g	1	4	1	4	2	2
A-c-g	1	1	1	6	8	2
C-b-g	1	4	4	6	4	4
A-b	1	1	1	8	2	2
A-c	1	1	1	4	2	2
C-b	4	4	4	4	2	2
A-b-c	3	3	3	2	4	4

Table V gives the values t_r^{ident} of IDEF for all fault types and θf two values (0 ° and 90 °) for a defect 100 km with three values Rf: 0 Ω, 40 Ω and 100 Ω.

Table 5 shows that the fault resistance has no specific rule of influence on the response time t_r^{ident}. The maximum value of tr ^{ident} is 8ms obtained for phase-phase faults (a-c-g) for θf = 90 ° and (b-g) for θf = 90 ° and (a-b) for θf = 90 °.

6. Conclusion

A new approach based neurons is proposed to identify faults in transmission lines in real time. This approach returned in the formation of fast numerical relay. The fault identifier consists of four independent neural networks, each network detects the fault on the phase concerned. The fault identifier uses the amplitudes of the fundamental frequency of the current signals in the place of the relay as the inputs to neural networks. The outputs are used to indicate the type of fault. The results show that the proposed identifier may identify the fault in less than 08 ms matter what the network parameters and fault conditions (fault resistance, fault inception angle, voltage sources, load angles, etc.. .).

Every the test results show that the defect identifier can be used to support a new generation of very fast numerical relays.

References

[1] Héctor J.A.F, Ismael D.V. and Ernesto V.M. "Fourier and Walsh digital filtering algorithms forDistance protection", IEEE Trans. On PAS, Vol. 11, No. 1, February, 1996, pp. 457-462.

[2] Barros J. And Drake J.M. "Real time fault detection and classification in power systems using Microprocessors", IEE Proc. Gener. Transm. Distrib., Vol. 141, No. 4, July 1994, pp. 315-322.

[3] Girgis A.A. and Brown, R.G.: Adaptive Kalman Filtering in Computer Relaying: "Fault Classification Using Voltage Models", IEEE Trans. On Power Apparatus and Systems PAS-104 (1985) no. 5, pp. 1168-1177.

[4] Dalstein T. And Kulicke B. "Neural network Approach to fault classification for high speed Protective relaying", IEEE Trans. On Power Delivery, Vol.10 No.2, April 1995 pp. 1002-1011.

[5] Kezunovic M., Rikalo I. And Sobajic D. J. "High speed fault detection and classification with neural nets", Electric Power Systems Research, Vol. 34, 1995 pp. 109-116.

[6] Poeltl A. And Fröhlich K.: "Two New Methods for Very Fast Type Detection by Means of Parameter Fitting and Artificial Neural Networks", IEEE Transactions on Power Delivery PWRD vol.14 (1999) no. 4, pp. 1269-1275.

[7] Pasand M.S and Zadeh H.K. "Transmission line fault detection & phase selection using ANN", International Conference on Power Systems Transients, New Orleans, USA, IPST 2003, cdrom.

[8] Aggarwal R. K., Xuan Q. Y., Dunn R.W., Johns A.T. and A. Bennett A.: "A Novel Fault Classification Technique for Double-circuit Lines Based on a Combined Unsupervised/Supervised Neural Network", IEEE Transactions on Power Delivery PWRD vol.14 (1999) no. 4, pp. 1250-1255.

[9] Lin W.M, Yang C.D., Lin J.H. and Tsa M.T. "A fault classification method by RBF neural network with OLS learning procedure", IEEE Transactions on Power Delivery vol.16, no. 4, October 2001, pp. 473-477.

[10] Logiciel matlab 7.7.0

[11] Bouthiba Tahar " Application des réseaux de neurones pour la détection des défauts dans les Lignes de transport '', Conférence Internationale Francophone d'Automatique, ENSEIRB, Bordeaux (France), 30 mai - 1er juin 2006

17

Evaluation of the energy-saving performance of heat-resistant paint

Takashi Oda[1, 2, *], **Kimihiro Yamanaka**[1], **Mitsuyuki Kawakami**[3]

[1]System Design of Tokyo Metropolitan University, Tokyo, Japan
[2]Nissin Sangyo Co., Ltd., Tokyo, Japan
[3]Human Sciences of Kanagawa University, Kanagawa, Japan

Email address:

oda-takashi1@ed.tmu.ac.jp (T. Oda), kiyamana@tmu.ac.jp (K. Yamanaka), kawakamim@kanagawa-u.ac.jp (M. Kawakami)

Abstract: We developed an empirical quantitative evaluation of the energy-saving performance of heat-resistant paint in winter. Specifically, heat-resistant paint and conventional wall paint were applied to steel boxes placed in a room kept at constant temperature and humidity, after which heat sources were placed in the boxes and the energy-saving performance of each type of paint was evaluated from the change in temperature at the box walls and inside the boxes. The experimental results show that the heat-resistant paint reduced the amount of heat escaping through the walls, and it can thus be expected to reduce heat loss. Furthermore, in the case of the heat-resistant paint, the amount of heat passing through the walls was 16% less than that in the case of the conventional paint. The conduction heat flux for the box with ceramic insulating paint was less than that for other boxes. It is thus thought that the thermal resistance of the ceramic insulation paint is higher than that of the other paints. We estimated the thermal resistance of each paint and found that thermal resistance of ceramic insulating paint was 12.4 times that of conventional energy-saving paint and twice that of high-reflectance paint.

Keywords: Energy-Saving, Heat-Resistant Paint, Comfortable Work Area

1. Introduction

In recent years, the importance of various approaches to save energy in residential and work spaces has increased as a result of apprehension about global warming and the electric power supply after the Great East Japan Earthquake[1]. Measures related to cooling and heating loads have become particularly important, and there is a trend toward improving building insulation to reduce the effect of the outside environment and to allow cooling and heating equipment to be used in an efficient manner[2]. With regard to the performance of building insulation, in the past, insulation materials were often used to direct the transfer of heat toward the inside of a structure. Recently, there has been a growing expectation that the application of heat-resistant or heat-insulating paint, which can improve energy-saving performance, can serve as a new energy-saving technique that differs from the conventional technique of installing insulation materials.[3] However, the different mechanism and properties of this technique mean that conventional methods of evaluating insulation performance cannot be used to evaluate the energy-saving performance. As a consequence, each manufacturer is presently using its own proprietary standards to evaluate energy savings. For end users selecting a product, the lack of common indicators of energy-saving performance obstructs the promotion of effective energy saving, and therefore, a common method of quantitative evaluation is desirable. Furthermore, although there have been numerous reports on the energy-saving performance of heat-resistant paint in summer, few such evaluations are available for winter.

Among energy-saving paints, many are so-called high-reflectance paints, which when applied to the roof and outer walls of a building can help regulate heat on the outside of the building by reflecting sunlight with high efficiency [4]. The standards JIS K 5602 "Determination of reflectance of solar radiation by paint film" and JIS K 5675 "High solar reflectance paint for roof" detail a quantitative method of evaluating high-reflectance paints [5,6]. The method described in these standards determines solar reflectance from the measurements of spectral reflectivity using the spectral radiation distribution of standard sunlight. In contrast to the situation for outdoor paints, there are energy-saving paints that are used on interior surfaces not exposed to solar radiation as a means of increasing cooling and heating efficiency, but there has been no agreement on a common standard with which to evaluate the energy-saving effect of an interior paint [7]. It is confirmed that the interior paint has insulation characteristics in simulation by a characteristic of penetration depth of far-infrared radiation [8].

Against this background, the present study directly measures the conduction heat fluxes of ceramic insulating paints with the greatest market share among energy-saving paints for indoor surfaces and two varieties of conventional energy-saving paint. A study of the physiological influence on human body by the ceramic insulation paint interior decoration painting is conducted [9]. From the results, a more unified verification of effectiveness is proposed. The verification estimates the heat-transfer inhibition effect of energy-saving paints intended for indoor surfaces, with the

ultimate aim of quantitatively evaluating the effectiveness of paints. Additionally, the effects of high-reflectance paints are compared.

2. Experimental Method

Iron boxes (sides: 500 mm; thickness: 1.6 mm) were used in the present study. Iron was chosen as the base material because it is a thermally stable raw material with low thermal resistance and the effects of individual differences are thus minimized. A ceramic insulating paint, conventional energy-saving paint, and high-reflectance paint were each applied to the inside of a separate box to create three different samples. The specifications of the three paints are given in table 1.

Table 1. Specifications of the three paints

	Conventional energy-saving paint	Ceramic insulative paint	High-reflectance paint
Diluent	Water	Water	Water
Coating method	Spray	Spray	Spray
Number of times recoating	2	2	2
Standard application amount	0.2~0.4 kg/m²/times	0.2~0.23 kg/m²/times	0.22 kg/m²/times
Specific gravity	1.24	0.78	1.24

In the experiment, the conduction heat flux from the inside to the outside of a box and the structural temperature on the inside and outside of a point on the box ceiling were measured while heat was generated inside the iron box by a small fan heater (FH-120, FUKADAC); the setup is shown in figure 1. The heat flux was measured by a heat flux sensor (MF-180, Eko Instruments), and the temperature inside and outside the box was measured by a thermocouple. The experiment was performed in a room with constant temperature and humidity. The ambient temperature set at 22 °C. Measurements were taken starting 30 min after the start of operation of the heater and the thermal behavior was studied at 10-min intervals from then. Additionally, it was verified before the start of each trial that the box and its interior were at the same temperature as the soundings (22 °C).

Figure 1 shows the installation point of the heat flux sensor for the measurement of the conduction heat flux (inside the box). The heat flux sensor was affixed with double-sided tape to the ceiling of the box, which was not directly exposed to the warm air from the small heater. The conduction heat flux to the outside was measured by a logger (3635-04, Hioki E. E. Corporation) at a sampling rate of 1 sample per minute while the heater was operating inside the box.

Figure 3 shows an iron box specimen and the installation points of the thermojunction temperature sensors (inside the box). The thermojunction temperature sensors were affixed to the inner and outer surfaces of the box ceiling with aluminum tape. The changes in the inside and outside surface temperatures of the box structure were measured by a logger (309, Centre) at a sampling rate of 1 sample per minute while the heater was operating inside the box.

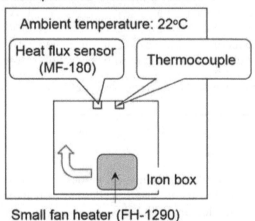

Figure 1. Experimental setup

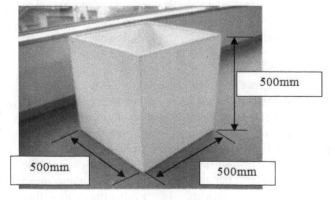

Figure 2. Iron box

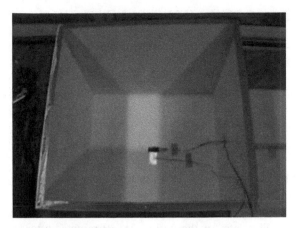

Figure 3. Installation points of the thermojunction temperature sensors (inside the box)

Figure 4. The setting situation

3. Results for Conduction Heat Flux

Figure 5 shows the conduction heat flux over time. The horizontal axis in the figure shows the elapsed time and the vertical axis shows the conduction heat flux from inside to outside the box. The three curves indicate the conduction heat fluxes of the three boxes. According to the figure, the trend in conduction heat flux appears to differ by paint between the first half and second half of the measurements. The measurement time was split into three periods (1–10 min, 11–20 min, and 21–30 min) to characterize these differences. Figure 6 shows the results of consolidating the conduction heat flux for each paint in each time period. The horizontal axis shows the time period, and the vertical axis shows the average conduction heat flux from inside to outside the box for each paint in each time period. The three bars show the results for each box. As can be seen in the figure, the conduction heat flux of the box to which ceramic insulating paint was applied tended to be smaller than that of other boxes in each time period. Table 2 gives the results of taking the values of the figure as characteristic values and conducting a two-way analysis of variance (ANOVA) with time period (A) and variety of paint (B) as factors. As seen in the table, both the measurement time period and the variety of paint were significant ($p < 0.05$ and $p < 0.01$, respectively), but a pairwise comparison employing the Holm method

showed that there were no significant differences in conduction heat flux among either the measurement time periods or the varieties of paint.

Table 2. ANOVA table for Conduction heat flux

Source of variation	SS	df	MS	F-value
Time period	2066.88	2.00	1033.44	4.23*
Variety of paint	8267.54	2.00	4133.77	16.92**
Interaction	2545.68	4.00	636.42	2.61
Within	19784.25	81.00	244.25	
Total	32664.35	89.00		

(**:$p<0.01$, *:$p<0.05$)

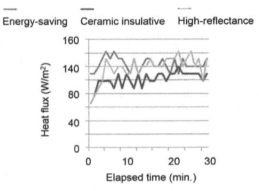

Figure 5. Results of measurement the Conduction heat flux

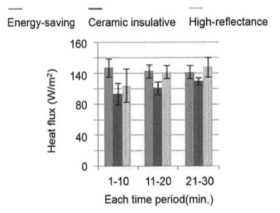

Figure 6. Results of the conduction heat in each time period

4. Results for Structural Temperatures of the Inner and Outer Walls

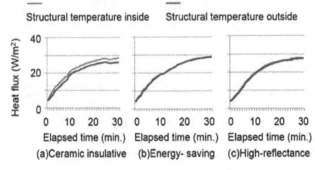

Figure 7. Changes in inner wall and outer wall temperatures

Figure 7 shows the changes in temperatures of the inner and outer walls of the boxes. The horizontal axes in the figures show the elapsed time and the vertical axes show the wall surface temperature. The two curves in each figure show the temperature changes for the inner and outer walls.

The figure reveals that the temperature of the inner wall tended to be higher, from immediately after the heater was turned on, in the box to which ceramic insulating paint was applied, and there was a temperature gradient between the inside and outside of that box. In contrast, the temperatures of the inner and outer walls of the other boxes changed almost identically.

Figure 8 shows the inner–outer-wall temperature difference for each box in each time period. The horizontal axis in the figure shows the paint and the vertical axis shows the inner–outer-wall temperature difference. The three curves show results for different measurement time periods. Table 3 gives the results of taking the values of the figure as characteristic values and conducting a two-way ANOVA with time period (A) and variety of paint (B) as factors. As noted in the table, both the measurement time period and the variety of paint were significant ($p < 0.05$ and $p < 0.01$, respectively). Furthermore, pairwise comparison employing the Holm method showed that the inner–outer-wall temperature difference was significantly larger for the ceramic paint.

These results indicate that heat loss from the box with ceramic insulating paint was the lowest, which is consistent with the variety of paint being an influencing factor in the results of the conduction heat flux, as shown in table 2.

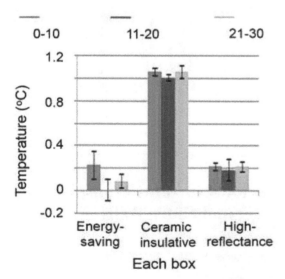

Figure 8. *The inner–outer wall temperature difference*

Table 3. *ANOVA table for inner–outer wall*

Source of variation	SS	df	MS	F-value
Time period	2066.88	2.00	1033.44	4.23*
Variety of paint	8267.54	2.00	4133.77	16.92**
Interaction	2545.68	4.00	636.42	2.61
within	19784.25	81.00	244.25	
total	32664.35	89.00		

5. Estimate of the Thermal Resistance

The conduction heat flux for the box with ceramic insulating paint was less than that for other boxes. It is thus thought that the thermal resistance level of the ceramic insulating paint is higher than that of other paint. We therefore estimate the thermal resistance of each paint in the following.

The thermal resistance R is expressed as [10]

$$R = \Delta T / Q \qquad (1)$$

where ΔT is the difference in temperature inside and outside the box and Q is the conduction heat flux.

We used expression (1) to estimate the thermal resistance of each paint. The data for the period 21–30 minutes after the beginning of heating are the amounts of heat flux and are relatively stable; we thus selected this time period for the calculation.

In general, the difference in structural temperature inside and outside the box becomes large as the insulation performance improves.

We estimated the thermal resistance of each box from the experimental data. The thermal resistance of the coating of each box was estimated by excluding the thermal resistance of the iron moiety. The thermal resistance levels were estimated as 0.0008 m2·K/W for conventional energy-saving paint, 0.0100 m2·K/W for ceramic insulating paint, and 0.0016 m2·K/W for high-reflectance paint. The thermal resistance level of the ceramic insulating paint was 12.4 times that of the conventional energy-saving paint, and twice that of the high-reflectance paint. An explanation for the ceramic insulating paint having the highest thermal resistance requires an evaluation of the physical properties of the coating and remains a problem for future research.

6. Conclusion

The present study of the heat transfer inhibition effect of energy-saving paints experimentally verified the effects of three varieties of paint—an ordinary paint, a ceramic insulating paint, and a high-reflectance paint—applied to iron boxes. The results obtained are summarized as follows.

1) The variety of paint was shown to be an influencing factor of the conduction heat flux, which is an index that represents the loss of heat to the outside of the box.
2) The inside–outside temperature gradient of the box was steeper for the box with ceramic paint than for the other boxes, indicating that heat loss was smallest for the box with ceramic paint.
3) We estimated that the thermal resistance of the ceramic paint was the highest among the paints.

References

[1] Agency for Natural Resources and Energy " Summary of the law concerning the rational use of energy" http://www.enecho.meti.go.jp/category/saving_and_new/savin g/summary/pdf/2014_gaiyo.pdf (24/8/2014 access).

[2] Ministry of Land, Infrastructure, Transport and Tourism "Energy-saving standard of a house and the building"http://www.mlit.go.jp/common/000996591.pdf (24/8/2014 access).

[3] SK KAKEN Co.,LTD "About the latest trend of the high reflectance paint"http://www.kenzai.or.jp/kouryu/image/42-04.pdf (24/9/2014 access).

[4] Japan Paint Manufacturers Association, "Guideline of manufacture and environment 2011", http://www. toryo.or.jp/ jp/anzen/reflect/reflect-info2.pdf, (30/6/2014 access).

[5] Japanese Industrial Standards, "Determination of reflectance of solar radiation by paint film", JIS K5602, 2008.

[6] Japanese Industrial Standards, "High solar reflectance paint for roof" detail a quantitative method of evaluating high-reflectance paints", JIS K 5675, 2011.

[7] Ministry of Land, Infrastructure, Transport and Tourism, "Energy-saving standard of architectural structure", ttp://www.mlit.go.jp/common/000996591.pdf, (01/7/2014 access).

[8] Takashi Oda, Masato Tazawa, Takeshi Kunishima, Kimihiro Yamanaka "Effect of penetration depth of far-infrared radiation into architectural walls" Grand renewable energy 2014

[9] Kazumi Tagami, Takashi Oda, Tastujiro Ishiko, Yuan Shenghua," Effects of inner wall thermal emissivity on human body temperature, and their metabolic hcat production" Japanese Journal of Biometeorology 50 (3): S64, 2013

[10] Katsunori Hanamura"Practice-Heat transfer engineering", pp.11, 2007

18

Evaluation on external economies of renewable energy resource utilization

Yanlin Qu[1, *], Yilan Su[2]

[1]Business College of Honghe University, Mengzi County, Yunnan Province, 661100, China
[2]College of Life Science and Technology, Honghe University, Mengzi County, Yunnan Province, 661100, China

Email address:
yanlinqu@gmail.com (Yanlin Qu)

Abstract: Different wind power projects have different external economics level, the practical evaluation on which can provide reference for wind power projects selection and qualify the role of completed projects. In this paper, the external economies of wind power engineering projects were evaluated by employing VIKOR method. The basic theory of VIKOR method was firstly introduced, and then the empirical evaluation was performed. After the evaluation index system was built from environmental, social and economic aspects, the external economies of three selected wind power engineering projects in Hebei province were studied. The evaluation result shows the external economies of wind power project #3 are the best one. The application of VIKOR method in external economies evaluation of wind power project is effective and feasible.

Keywords: Wind Power Engineering Project, External Economies Evaluation, VIKOR Method

1. Introduction

Renewable resources are a part of Earth's natural environment and the largest components of its ecosphere, which is an organic natural resource that can replenish in due time compared to the usage, either through biological reproduction or other naturally recurring processes [1]. Renewable energy resources are a kind of energy resources, which have the characteristics of cyclic regeneration and sustainable utilization. Renewable energy resources include wind energy, solar energy, biomass energy, and so on. With the development of human society, the important role of energy in people's daily lives is becoming increasingly prominent. Nowadays, the energy supply shortage and environmental pollution issues make exploiting and utilizing renewable energy as the focus of worldwide concerns [2, 3]. Energy solutions for the future will increasingly depend on the use of renewable energy resources.

In the past many years, China's wind power has developed rapidly, the cumulative installed capacity of which has increased from 0.3 GW in 2000 to 77 GW in 2013. In 2010, China surpassed the United States and ranked the first in terms of cumulative installed capacity of wind power [4]. As

a kind of renewable energy, wind energy has the advantages of huge reserves, widely distribution, renewable and pollution-free [5]. External economies are benefits that are created when an activity is conducted by a company or other type of entity, with those benefits enjoyed by others who are not connected with that entity [3]. The entity that is actually managing the activity does not receive the external economies, although the creation of these benefits for outsiders usually has no negative impact on that entity [6]. The benefits produced by wind power, such as environmental emission reduction and economic growth, can enjoys by outsiders, such as the residents, nation, and ecology. Therefore, wind power has external economics.

Currently, the study on external economies mainly focuses on the industry and trade. Ref. [7] estimated the indexes internal returns to scale and external economies for two-digit manufacturing industries in West Germany, France, the U.K., and Belgium countries; Ref. [8] reviewed the international trade literature on intra-industrial and inter-industrial production externalities; Ref. [9] proposed a two-country model of monopolistic competition in which differentiated products are produced subject to external economies of scale and two countries differ only in size measured by the factor endowment; Ref. [10] studied the impact on productivity growth of technological externalities, both inter- and intra-industry, national or international, at the industry level for the EU countries and the period 1995–2002.

This paper focuses on the external economies of wind power project, and employs the VIKOR method to evaluate the external economies of wind power project. The VIKOR method is a multi-criteria decision making (MCDM) method, which was originally developed by Serafim Opricovic [11]. As a MCDM method, VIKOR has been applied in many issues, such as supplier selection [12], water resources planning [13], service quality of airports [14], and so on.

This paper comprises the following: Section 2 introduces the basic theory regarding VIKOR method; taking three wind power engineering projects in Hubei province as example, the external economies evaluation of wind power engineering projects based on VIKOR model is performed in Section 3; Section 4 concludes this paper.

2. Basic Theory of VIKOR Method

VIKOR method is a MCDM method to solve decision problems with conflicting and non-commensurable criteria, assuming that compromise is acceptable for conflict resolution, the decision maker wants a solution that is the closest to the ideal, and the alternatives are evaluated according to all established criteria. VIKOR ranks alternatives and determines the solution named compromise that is the closest to the ideal [15].

Suppose the set of m feasible alternatives be $X = \{X_1, X_2, \cdots, X_m\}$, the set of n criterion functions be $O = \{O_1, O_2, \cdots, O_n\}$. The input data are the elements a_{ij} of the performance matrix $A = \{a_{ij}\}_{m \times n}$, where a_{ij} is the value of the j-th criterion function for the alternative X_i.

The basic step of VIKOR method is as follows.

2.1. Determine the Criterion Weight

The criterion weight determination is very important, which has great impact on the evaluation result. There are many criterion weight determination methods, such as expert consultation method, AHP method, entropy weight method, and so on.

2.2. Normalize the Performance Matrix

Normalize the initial performance matrix $A = \{a_{ij}\}_{m \times n}$ to the standardized decision matrix $R = \{r_{ij}\}_{m \times n}$ according to Eq. (1).

$$r_{ij} = \begin{cases} \dfrac{a_{ij} - \min\limits_i(a_{ij})}{\max\limits_i(a_{ij}) - \min\limits_i(a_{ij})}, i \in I_1, j \in N \\ \dfrac{\max\limits_i(a_{ij}) - a_{ij}}{\max\limits_i(a_{ij}) - \min\limits_i(a_{ij})}, i \in I_2, j \in N \end{cases} \quad (1)$$

where I_1 and I_2 represent the subscript set of benefit-type criterion and cost-type criterion, respectively; $N = \{1, 2, \cdots, n\}$.

2.3. Determine the Positive and Negative Ideal Solution

The positive ideal solution is

$$r^* = \{r^*_1, r^*_2, \cdots, r^*_n\} = \max_i \{r_{ij}, j = 1, 2, \cdots, n\} \quad (2)$$

The negative ideal solution is

$$r^- = \{r^-_1, r^-_2, \cdots, r^-_n\} = \min_i \{r_{ij}, j = 1, 2, \cdots, n\} \quad (3)$$

2.4. Calculate the Group Utility Value and Individual Regret Value

The group utility value can be calculated according to Eq. (4), and the individual regret value can be calculated according to Eq. (5).

$$S_i = \sum_j w_j \frac{r^*_j - r_{ij}}{r^*_j - r^-_j} \quad (4)$$

$$R_i = \max \left\{ w_j \frac{r^*_j - r_{ij}}{r^*_j - r^-_j}, j \in N \right\} \quad (5)$$

2.5. Calculate the Comprehensive Ranking Index

The comprehensive ranking index can be calculated according to Eq. (6).

$$Q_i = v \frac{S_i - S^*}{S^- - S^*} + (1-v) \frac{R_i - R^*}{R^- - R^*} \quad (6)$$

where

$$S^* = \min_i S_i \quad (7)$$

$$S^- = \max_i S_i \quad (8)$$

$$R^* = \min_i R_i \quad (9)$$

$$R^- = \max_i R_i \quad (10)$$

v is introduced as a weight for the strategy of maximum group utility, $v \in [0,1]$; $1-v$ is the weight of the individual regret. In VIKOR method, the value of v is usually set as 0.5, which indicates the decision making is performed based on the balanced compromising way.

Therefore, the decision making can be made according to the value of Q_i. The alternative with the minimum Q value is the best one. The calculation process of VIKOR method for evaluation is shown in Fig.1.

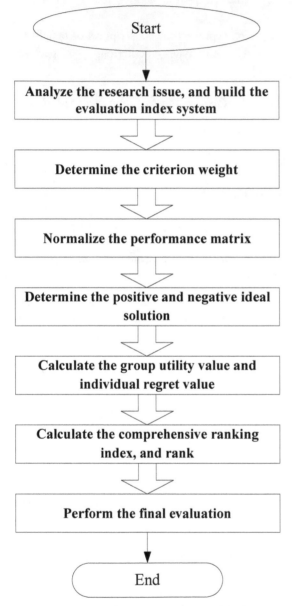

Figure 1. The calculation process of VIKOR method.

3. Empirical Evaluation

The wind resource is abundant in Hebei province. In 2013, the cumulative on-grid installed capacity of wind power amounted to 7.75 GW in Hebei province, which ranks the second in China. In the past few years, there are several approved wind power engineering projects, such as GUOHUA KUYUANBAYUAN wind farm. Therefore, three wind power engineering projects in Hebei province were selected in this paper.

Firstly, the evaluation index system of external economies of wind power engineering project was built, and then an evaluation on the external economies of wind power engineering projects was carried out by employing VIKOR model.

3.1. Build the Evaluation Index System

When building the evaluation index system, two methods are adopted, namely Questionnaire and Literature review. After reviewing the relative literatures, the Questionnaire was designed. Questionnaires about the external economies of wind power engineering project were dispatched to experts and engineers in the field of wind power. The external economies evaluation index system was obtained by analyzing the results of questionnaires, the result of which is shown in Fig.2. The external economies of wind power engineering project are divided into environmental benefit, social benefit, and economic benefit. All of the indices are the greatest-type index.

3.2. Evaluation Process

(1) Determine the criterion weight

In this paper, the expert consultation method was employed to determine the criterion weight. 50 criterion weight scoring tables are distributed to university professors, enterprise expertise and engineers. According to the statistic results, the criterions weights of external economies of wind power engineering project are determined, which are listed in Table 1.

(2) Normalize the performance matrix

The external economies criterions scoring of three wind power wind power engineering projects are listed in Table 2.

So, we can obtained the performance matrix $A = \{a_{ij}\}_{m \times n}$:

$$A = \begin{bmatrix} 6.7\% & 23.3\% & 11\% & 7\% & 3.8\% & 2\% & 43\% & 34\% \\ 8.9\% & 19.8\% & 10.5\% & 6\% & 4.4\% & 1.3\% & 46\% & 52\% \\ 7.7\% & 20.9\% & 14\% & 5\% & 4.3\% & 1.7\% & 47\% & 46\% \end{bmatrix}$$

According to Eq. (1), the standardized decision matrix $R = \{r_{ij}\}_{m \times n}$ can be obtained:

$$R = \begin{bmatrix} 0 & 1 & 0.14 & 1 & 0 & 1 & 0 & 0 \\ 1 & 0 & 0 & 0.5 & 1 & 0 & 0.75 & 1 \\ 0.45 & 0.31 & 1 & 0 & 0.83 & 0.57 & 1 & 0.67 \end{bmatrix}$$

(3) Determine the positive and negative ideal solution

According to Eq. (2), the positive ideal solution can be obtained, which is $r^* = \{1,1,1,1,1,1,1,1\}$, and the negative ideal solution is $r^- = \{0,0,0,0,0,0,0,0\}$.

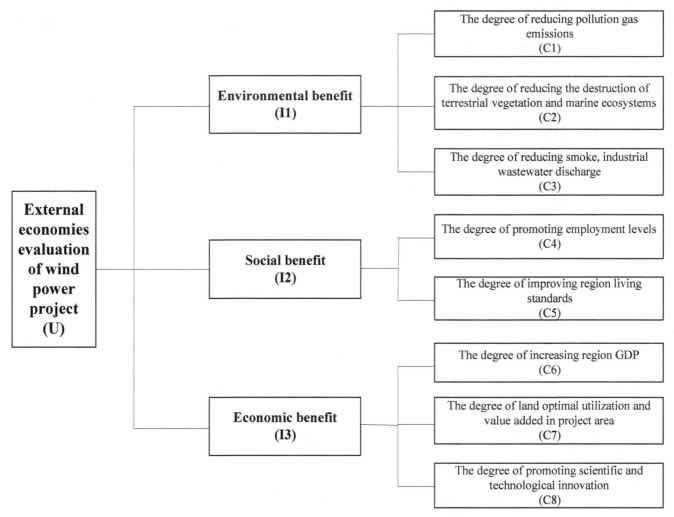

Figure 2. *External economies evaluation index system of wind power engineering project.*

Table 1. *Weights of external economies evaluation criterions*

Criterion	C1	C2	C3	C4	C5	C6	C7	C8
Weight	0.21	0.11	0.13	0.14	0.08	0.17	0.07	0.09

Table 2. *Criterion values of three wind power projects*

Criterion / Scoring	C1	C2	C3	C4	C5	C6	C7	C8
Wind power project #1	6.7%	23.3%	11%	7%	3.8%	2%	43%	34%
Wind power project #2	8.9%	19.8%	10.5	6%	4.4%	1.3%	46%	52%
Wind power project #3	7.7%	20.9%	14%	5%	4.3%	1.7%	47%	46%

(4) Calculate the group utility value and individual regret value

According to Eq. (4), the group utility value of three wind power projects can be calculated, which are

$$S_1 = 0.5614; S_2 = 0.4975; S_3 = 0.4462$$

According to Eq. (5), the individual regret values of three wind power projects can be calculated, which are

$$R_1 = 0.21; R_2 = 0.17; R_3 = 0.14$$

(5) Calculate the comprehensive ranking index

According to Eq. (6), set v=0.5, the comprehensive ranking index of three wind power projects can be calculated, which are

$$Q_1 = 1; Q_2 = 0.437; Q_3 = 0$$

Therefore,

$$Q_3 < Q_2 < Q_1 \Rightarrow Q_3 \succ Q_2 \succ Q_1$$

According to the calculation result, we can safely draw the conclusion that the external economies of wind power project #3 are the best one.

4. Conclusions

As a kind of renewable energy resource utilization, wind power engineering projects have developed rapidly in China in the past years. Different wind power projects have different external economics level, the practical evaluation on which can provide reference for wind power projects selection. In this paper, the external economies of three wind power engineering projects are evaluated based on the VIKOR method. After the evaluation index system was built, the weights firstly were determined, and then the evaluation was performed. The calculation result shows the external economies of wind power project #3 are the best one, which indicates the construction of wind power project #3 plays greater role in the promotion of environmental, social and economic development.

Acknowledgements

The author thanks editor and reviewers for their comments and suggestions.

References

[1]   http://en.wikipedia.org/wiki/Renewable_resource

[2]   Tükenmez M, Demireli E. Renewable energy policy in Turkey with the new legal regulations [J]. Renewable Energy, 2012, 39(1): 1-9.

[3]   Li H, Guo S. External economies evaluation of wind power engineering project based on analytic hierarchy process and matter-element extension model [J]. Mathematical Problems in Engineering, 2013, 2013.

[4]   Zhang S, Andrews-Speed P, Zhao X. Political and institutional analysis of the successes and failures of China's wind power policy [J]. Energy Policy, 2013, 56: 331-340.

[5]   Chingulpitak S, Wongwises S. Critical review of the current status of wind energy in Thailand [J]. Renewable and Sustainable Energy Reviews, 2014, 31: 312-318.

[6]   J. Markusen. Micro-foundation of external economies [J]. Canadian Journal of Economics, 1990, 23(1): 495–508.

[7]   Caballero R J, Lyons R K. Internal versus external economies in European industry[J]. European Economic Review, 1990, 34(4): 805-826.

[8]   Choi J Y, Yu E S H. External economies in the international trade theory: a survey[J]. Review of International Economics, 2002, 10(4): 708-728.

[9]   Suga N. A monopolistic-competition model of international trade with external economies of scale[J]. The North American Journal of Economics and Finance, 2007, 18(1): 77-91.

[10]   Claver N D, Castejón C F, Gracia F S. External economies as a mechanism of agglomeration in EU manufacturing[J]. Applied Economics, 2012, 44(34): 4421-4438.

[11]   Opricovic S, Tzeng G H. Extended VIKOR method in comparison with outranking methods [J]. European Journal of Operational Research, 2007, 178(2): 514-529.

[12]   Sanayei A, Farid Mousavi S, Yazdankhah A. Group decision making process for supplier selection with VIKOR under fuzzy environment[J]. Expert Systems with Applications, 2010, 37(1): 24-30.

[13]   Opricovic S. Fuzzy VIKOR with an application to water resources planning[J]. Expert Systems with Applications, 2011, 38(10): 12983-12990.

[14]   Kuo M S, Liang G S. Combining VIKOR with GRA techniques to evaluate service quality of airports under fuzzy environment[J]. Expert Systems with Applications, 2011, 38(3): 1304-1312.

[15]   http://en.wikipedia.org/wiki/VIKOR_method

19

Integrated Briquetting Plant: Study for Maichew Particle Board Factory

Temesgen Gebrekidan, Yonas Zeslase

Department of Chemical Engineering, Adigrat University, Adigrat, Ethiopia

Email address:

temekidan@gmail.com (T. Gebrekidan), yonbelete7@gmail.com (Y. Zeslase)

Abstract: Biomass briquetting is a process which converts agro residue and saw dust materials with low bulk density into a uniform size and more convenient households or industrial fuel product. The major advantages offered by biomass densification are related to handling improvement and increasing calorific value per unit volume. In general this process offers energy generation opportunity from any biomass residue. Maichew Particle Board Factory generates a huge amount of dust particles from its different lines which recovered using different cyclones. But this recovered waste doesn't use as a fuel for boiler or dryer and also not yet convert to any valuable product. It is simply disposed to the environment. Therefore, the objective of the project was to study an integrated briquetting plant for Maichew Particle Board Factory which converts the waste dust particle in to valuable product. This case study involved the overall energy review of Ethiopia and briquetting process in detail. According to the selection of appropriate technology which converts the eucalyptus dust to high quality briquette fuel screw press extruder with a pre heater and heated die coils has been chosen. In addition the amount of energy required to convert the dust particle to briquettes is also evaluated. Based on the results of this study, the economic evaluation of the integrated briquetting plant for Maichew Particle Board is feasible besides to its socio environmental benefits. It is an investment which can recover its capital with almost two years.

Keywords: Biomass Energy, Biomass Briquetting, Waste Dust Particle

1. Introduction

Historically, biomass has been a major source of household's energy in Ethiopian. Biomass meets the cooking energy needs of most rural households and half of the urban household's demands. Despite significant penetration of commercial energy in Ethiopia during last few decades, biomass continues to dominate energy supply in rural and traditional sectors. Biomass energy constitutes wood fuels (including charcoal, and wood wastes), crop residues (such as coffee husk, bagasse, rice husk and crop stalks) and animal dung (including biogas) [8, 10]. Ethiopia's energy system is characterized mainly by biomass fuel supply, with households being the greatest energy consumers. The total percentage total energy consumption in 1998 is estimated accordingly the following table.

From this total, biomass fuels accounted for 94% (77% from wood fuels, 17% from agricultural residues, dung and charcoal), petroleum fuels for 5% and electricity for less than 1% of supplies. The household sector takes up nearly 90% of the total energy supplies.

Table 1. Total Energy Consumption in Ethiopia in 1998.

Energy Type	Percentage (%)
Biomass	94
Petroleum	5
Electricity	1
Total	100

This dependence on biomass resources has been detrimental to their sustainability and adverse to the environment. The overall dependency of majority of the population on the charcoal and fuels wood as energy source has brought in its wake the threat of deforestation and this leads to decrease the soil fertility. This impasse can be sustainably resolved through the introduction and wide dissemination of new and renewable energy technologies. Such as:

- Use of densified agricultural residues for fuel;
- Demonstration projects on biogas, solar energy and thermal;

- Improved stoves;
- Improved charcoal production technologies; and
- Trials on fast growing tree species for fuel.

Processing of agricultural residue to produce good quality house hold cooking fuels had started in the early 1990 after the investigation of the opportunities to use biomass residue at state farms and agro industries as the substitute for the fuel used in households and industries. And a pilot plant had erected in different governmental farms such as Diksis (wheat straw), Amibra (cotton stalk) and Shoa (bagasse) and so on. But these plants were not successful while operation due to the instability of the country in the civil war in 1980's and some technical failure such wear of the pressing machine and lack of skilled man power to operate the plants [1].

Maichew Particle Board Factory PLC is a private company established under EFFORT with the objectives of producing and supplying high quality three layer particle board products. The factory has annual production capacity of 40,000m3. Its major standard raw material is eucalyptus tree. During processing there is 3360 ton per year emission of dust particles which is captured and collected using an exhaust system connected to a cyclone. But this recovered dust does not used as a fuel for boiler or dryer and also not yet convert to any valuable product. It is simply disposed to the compound and makes a load of materials within last three years operational period. It has been shown that this waste can be converted into valuable product which can bring an economic benefit and reduce the cost incurred on disposition of the dust particles.

Because of this reason, the objective of project studied an integrated briquetting plant which converts the above mentioned waste dust particles generates in Maichew Particle Board Factory to a valuable product so called briquetting.

2. Literature Review

2.1. Agro Residue and Saw Dust as a Fuel Sources

The shortage of fuel in many parts of the developing countries such as Ethiopia caused by an increasing shortage of traditional fuel (firewood, charcoal) creates a need for alternative sources of domestic fuel. Ethiopian has an agricultural based economy where all kind of tropical crops are in cultivation and residues such as coffee husk, cotton stalks, wheat straw, bagasse's, shells and nuts are major potential fuels used in many part of the country. Sawdust, a milling residue is also available in some quantity. Utilization of agricultural and forestry residues is often difficult due to their uneven characteristics, low density, inconvenient shape, high moisture content, low calorific value per unit volume, etc. are some of the hampering factors. This drawback can be overcome by means of densification, i.e. compaction of the residues into products of high density and regular shape. And further processing like carbonization (pyrolysis) is needed to transform the various types of organic waste into an acceptable form of domestic fuel [5].

Densification has aroused a great deal of interest worldwide in recent years as a technique of beneficiation of residues for utilization as energy source. Not only as ultimate solution for the problems of transportation, storage, and handling, but also the direct burning of loose biomass in conventional grates is associated with very low thermal efficiency and widespread air pollution.

2.2. Biomass Briquetting (Densification)

Biomass densification is processing of biomass by-products such as agricultural and forest residues, sawdust, slabs, chips, etc. into uniform sized particles which can be compressed into wood-based fuel products. It is important to mention that biomass densification is simply a physical transformation that does not change the chemical composition of biomass. Thus, the calorific value of biomass is not affected by densification.

Nevertheless, since non-densified products exhibit lower bulk density than pellets and briquettes, fluffy materials (e.g. chips, sawdust, etc.) have lower energy density than densified products. The process offers the following advantages:

- The net calorific value per unit volume is increased,
- Easier handling,
- Lower transportation cost
- The fuel produced is uniform in size and quality,
- Disposal of residue is facilitated, and
- Environmental friendly fuels [7]

Figure 1. Sample of Briquette.

The process of transforming loosen biomass residue in to briquetting requires large amount energy consumption and sustainable supply of raw material. Raw materials for briquetting include waste from wood industries, loose biomass and other combustible waste products. On the basis of compaction, the briquetting technologies can be divided into:

- High pressure compaction,
- Medium pressure compaction with a heating device, and low pressure compaction with a binder.

In all these compaction techniques, solid particles are the starting materials. The individual particles are still identifiable to some extent in the final product. Briquetting and extrusion both represent compaction i.e., the pressing together of particles in a confined volume. If fine materials which deform under high pressure are pressed, no binders are required. The strength of such compacts is caused by van der Waals' forces, valence forces, or interlocking [10].

Natural components of the material may be activated by the prevailing high pressure forces to become binders. Some of the materials need binders even under high pressure conditions. The figure below shows some of the binding mechanisms.

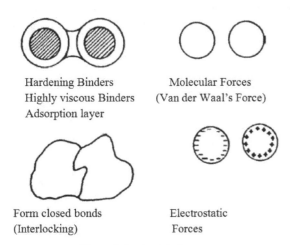

Hardening Binders   Molecular Forces
Highly viscous Binders   (Van der Waal's Force)
Adsorption layer

Form closed bonds   Electrostatic
(Interlocking)   Forces

Figure 2. Binding Mechanisms.

2.3. Binding Mechanisms of Densification

In order to understand the suitability of biomass for briquetting, it is essential to know the physical and chemical properties of biomass which also influence its behavior as a fuel. Physical properties of interest include moisture content, bulk density, void volume and thermal properties. Chemical characteristics of importance include the proximate and ultimate analysis, and higher heating value. The physical properties are most important in any description of the binding mechanisms of biomass densification. Densification of biomass under high pressure brings about mechanical interlocking and increased adhesion between the particles, forming intermolecular bonds in the contact area. In the case of biomass the binding mechanisms under high pressure can be divided into adhesion and cohesion forces, attractive forces between solid particles, and interlocking bonds [6].

High viscous bonding media, such as tar and other molecular weight organic liquids can form bonds very similar to solid bridges. Adhesion forces at the solid-fluid interface and cohesion forces within the solid are used fully for binding. Lignin of biomass/wood can also be assumed to help in binding in this way. Finely divided solids easily attract free atoms or molecules from the surrounding atmosphere. The thin adsorption layers thus formed are not freely movable. However, they can contact or penetrate each other. The softening lignin at high temperature and pressure condition form the adsorption layer with the solid portion. The application of external force such as pressure may increase the contact area causing the molecular forces to transmit high enough which increases the strength of the bond between the adhering partners. Another important binding mechanism is van der Waals' forces. They are prominent at extremely short distances between the adhesion partners. This type of

adhesion possibility is much higher for powders. Fibers or bulky particles can interlock or fold about each other as a result forming interlocking or form-closed bonds. To obtain this type of bond, compression and shear forces must always act on the system. The strength of the resulting agglomerate depends only on the type of interaction and the material characteristics [9].

2.4. Selection of Biomass Residue for Briquetting

In addition to the availability of sustainable supply of raw material for the commercial briquetting plant there are many factors to be considered before a biomass residue qualifies for use as feedstock for briquetting. Some of the characteristics are the following.

2.4.1. Low Moisture Content

Moisture content should be as low as possible, generally in the range of 10-15%. High moisture content will pose problems in grinding and excessive energy is required for drying.

2.4.2. Ash Content and Composition

Biomass residues normally have much lower ash content (except for rice husk with 20% ash) but their ashes have a higher percentage of alkaline minerals, especially potash. These constituents have a tendency to devolatilize during combustion and condense on tubes, especially those of super heaters. The ash content of some types of biomass are given in the below table.

Table 2. Ash Content of Selected Raw Materials.

Biomass	Ash content (%)
Saw dust	1.3
Coffee husk	4.3
Cotton shells	4.6
Rice husk	22.4

The ash content of different types of biomass is an indicator of slagging behavior of the biomass. Generally the greater the ash content, the greater the slagging behavior. But this does not mean that biomass with lower ash content will not show any slagging behavior. The temperature of operation, the mineral compositions of ash and their percentage combined determine the slagging behavior. If conditions are favorable, then the degree of slagging will be greater. Minerals like SiO_2 Na_2O and K_2O are more troublesome.

2.4.3. Flow Characteristics

The material should be granular and uniform so that it can flow easily in bunkers and storage silos. The properties of the solids that are important to densification are:
- Flow ability and cohesiveness
- Particle size (too fine a particle means higher cohesion, causing poor flow)
- Surface forces (important to agglomeration for strength) [8].

2.5. Biomass Briquetting Process

Densification of biomass is a multi-operation manufacturing process. In general, the raw material has to fulfill some specific properties such as: size, uniformity, and moisture content before being densified. For this reason, the raw material needs to be pre-processed to achieve such conditions. This pre-processing stage may include size reduction operations (milling, grinding, chipping, etc.) and drying operations based the physical characteristics of the raw material. Once the raw material is prepared, it is compressed under high temperature and pressure conditions [8, 9].

2.5.1. Size Reduction

Size reduction refers to the reduction of any biomass by mechanical means into higher valued and relatively uniform bulk material to be used for subsequent processing. It is desired to reduce the given biomass particle size to the required final particle size with optimum energy consumption. Except saw dust other agricultural or forest residue should be crushed to 6-8 mm size with 10-20% fines to achieve optimum briquetting results. While many types of crushing and grinding equipment's are available in the market, for biomass materials, hammer mills are considered the most suitable. These are available in various sizes from a few kg/hr. to 10-15 TPH. Maintenance is rather routine and heavy in these machines and it is advisable not to operate these machines for more than 20 hours per day. Some hammer mills are symmetrical so the direction of the rotor can be reversed. In this case, more running time is possible without maintenance [7].

2.5.2. Drying

Drying refers to the reduction of the amount of moist of water in biomass. The main objective is to achieve uniform moisture content in the material at the end of the drying process. Drying is necessary in the production of briquettes due to technological reasons linked to the manufacturing process as well as storing and transportation reasons. Drying is normally not required for materials like coffee husk, groundnut shells and rice husk. If feed is wet and drying becomes essential, integrated drying cum disintegration should be carried out by using the hot flue gases from the thermic fluid preheating furnace. However, drying is essential for sawdust, wet coir pith, bagasse and bagasse pith and some other agro-residues like mustard stalk if their moisture content is greater than 10%.

The types of drier employed for biomass materials are paddle indirect drier, flash, direct type, pneumatic or flash, and direct or indirect type rotary driers. Direct driers are those in which hot air or flue gases are intimately mixed with material and indirect ones are when heat is transferred to materials through a metallic surface and material is not mixed with the hot streams. Indirect driers are normally inefficient and require a large heat transfer area making the equipment bulky and expensive. Rotary driers are highly reliable but tend to be an order of magnitude more expensive

than a flash drier, especially at a capacity less than 3-4 TPH.

All biomass materials are amenable to drying by flash driers with or without disintegration. Even though biomass materials are heat sensitive these can be satisfactorily dried at relatively high temperature because of short drying time. Most of the moisture is removed either in a disintegrator or at the entry point of the feed into the gas stream. Entry temperature of gases up to 300 - 400 ^{0}C can be conveniently employed even though the decomposition temperature of most biomass materials is between 250 - 350 ^{0}C.

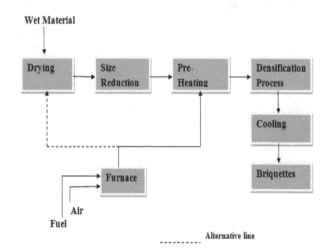

Figure 3. *Flow Diagram of Briquetting Process.*

2.6. Feed Pre-Heating

Material preparation plays a key role in the successful briquetting of biomass. Under the technical aspects, the major bottle neck for disseminating of the briquetting plant in commercial level is due to frequent wearing of the forming screw due to high friction and high power consumption during production. Some of the studies conducted earlier have revealed that the addition of heat benefits by relaxing the inherent fibers in the biomass and apparently softening its structure resulting in release of some bonding or gluing agent on to the surface.

In feed pre heating process the material is heated up to 100 ^{0}C and there is loss of moisture from 10% in feed to 8% in the product. Here furnace is used for preheating. In general feed pre heating results:

- Reduced pressure required for briquetting, resulting in reduction in power consumption
- Reduced frictional forces leading to a reduction of wear to contact parts, particularly the rotating screw
- Reduced resistance to flow leading to an enhanced rate of production.

There are different equipment's which provide direct and indirect heating system. But for biomass pre heating direct heating system is not suitable due to the combustible nature of the material. From the indirect heating systems the most appropriate for developing countries briquetting practice is to deploy high temperature thermic fluid system. The hot oil is heated in separate solid fired furnace and then circulated around the conveyer to heat the biomass [8, 9, and 10].

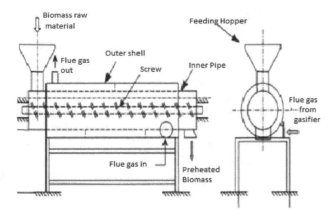

Figure 4. Saw Dust Pre-Heater.

The feed preheating system consists:

2.6.1. A. Thermic Fluid System

This system comprises of circulating pump, oil storage tank, furnace, piping, fittings and instruments. These heat transfer media are mineral oils which are stable up to 350 0C and are widely used for heat transfer applications. This oil is circulated between the heater/furnace and the process equipment and is a very convenient heating system requiring very little maintenance. For applications involving the heating of biomass, oil at a temperature of 200 0C should be adequate. Normally, thermic fluid heating systems employ furnaces fired with oil and gas, but solid fuel fired systems that utilize reject briquettes and other agro-residues and allow the operating costs to be drastically reduced are most suitable for biomass briquetting.

2.6.2. B. Furnace

A furnace is required to heat the heat transfer oil to provide the necessary heat for preheating the biomass. In addition based on wetness of the incoming raw material, it can provide heat for flue gases in case drying of biomass is required. The design of the furnace should incorporate use of solid fuels for heating the heat transfer oils.

2.6.3. Biomass Densification

Biomass densification represents a set of technologies for the conversion of biomass into a fuel.

The technology is also known as briquetting and it improves the handling characteristics of the materials for transport, storing etc. Briquetting is one of several agglomeration techniques which are broadly characterized as densification technologies. Agglomeration of residues is done with the purpose of making them denser for their use in energy production. High and medium pressure compaction normally does not use any additional binder. Normally, the briquetting process bases either on screw press or piston press technology. Other briquetting technologies are less applicable in developing countries because of high investment costs and large throughputs, e.g. roller-presses to produce pellets or briquettes [2].

i.   Screw press

In a screw press or screw extruder, the rotating screw takes the material from the feed port, through the barrel, and compacts it against a die which assists the build-up of a pressure gradient along the screw. Thus, the extruder features three distinct zones: feed, transport, and extrusion zones. The important forces that influence the compaction of the feed material play their role mostly in the compression zone near to the extrusion die.

Briquettes normally have a cylindrical shape although square or hexagonal shapes are possible. Outside diameter vary from 40 to 70 mm, with typical inside hole of 15 to 25 mm.

The frictional forces between feed materials and barrel/screw, the internal friction in the material and external heating device (of the extrusion zone) cause an increase in temperature (up to 300 0C), which softens the feed material. Lignin from the biomass is set free and acts as gliding and binding agent. The speed of densification, the energy consumption of the press and the quality of the briquettes produced depend on:

- Flow ability and cohesion of the feed material
- Particle size and distribution
- Surface forces
- Adhesiveness [2, 3].

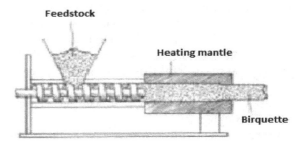

Figure 5. Screw Extruder.

The merits and demerits of this technology are:
- The output is continuous and the briquette is uniform in size.
- The outer surface of the briquette is partially carbonized facilitating easy ignition and combustion. This also protects the briquettes from ambient moisture.
- A concentric hole in the briquette helps in combustion because of sufficient circulation of air.
- The machine runs very smoothly without any shock load.
- The machine is light compared to the piston press because of the absence of reciprocating parts and flywheel.
- The machine parts and the oil used in the machine are free from dust or raw material contamination. The power requirement of the machine is high compared to that of piston press [2, 4].

ii.   Piston Press

The material is fed into a compression chamber where a moving piston forces the material through a slightly taper die. As the material is forced through the die, frictional energy is dissipated, causing the temperature of material to rise beyond the level needed to activate the lignin as a binder. Heat is mainly generated at the outer surface of the briquette. Heat mainly generated as the outer surface of the briquette. Briquettes are building up of layers formed by each stroke of the piston and the briquettes can be broken in to the original by hand. Combustion and mechanically quality is therefore not very good.

The working pressure of hydraulic piston press is lower. Hence the quality and density of the produced briquette is lower. In addition the technology is rather complicated and high initial investment and some technical infrastructures are required.

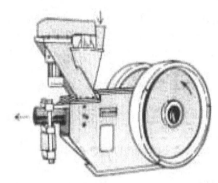

Figure 6. *Mechanical Piston Briquetting Press.*

The merits and demerits of this technology are:
- There is less relative motion between the ram and the biomass hence, the wear of the ram is considerably reduced.
- It is the most cost-effective technology currently offered by the Indian market.
- Some operational experience has now been gained using different types of biomass.
- The moisture content of the raw material should be less than 12% for the best results.
- The quality of the briquettes go down with an increase in production for the same power
- Carbonization of the outer layer is not possible. Briquettes are somewhat brittle.

iii.   Hand Mould Briquetting

Hand moulds are the simplest devices to form small quantities of briquettes. This is a low pressure compaction used to produce households cooking fuels. It is produced from any waste biomass and binding agents. There are various binding agents in use: organic binders (Molasses, Coal tar, Bitumen, and Starch) and inorganic binders (Clay, Cement, Lime, and Sulfite liquor).

2.6.4. Selection of Appropriate Biomass Briquetting Technology

As it stated in the above the principal raw material input for the briquetting plant is the process byproduct of the Maichew Particle Board factory. Maichew particle board factory uses eucalyptus tree as a raw material to produce a three layer particle board. In the material preparation section there is emission of saw dust particle in the screen where the surface material and core material are got separated. This is captured in an exhaust system connected to a cyclone. The under size particles emitted in the preparation section is estimated to be about 1800 ton per year. On the other hand after pressing, the boards are cooled prior to stacking. The 75 % produced particleboard panels are sanded and trimmed to final dimensions. During sanding there is also high emission of particulate matter which is controlled in the same manner as particles from the screen. This is estimated to be about 1550 ton per year.

These saw dust particles are simply captured and collected to control atmospheric emissions, but currently there is no any technology which converts the recovered dust as a by-product fuel for a boiler or dryer or any other valuable product. Rather it is continuously damped in the compound and now days it become a potential hazard of firing for a boiler or dryer or any other valuable product. Rather it is continuously damped in the compound and now days it become a potential hazard of firing.

There are different high compaction technologies or binder less technology, consists of the piston press and the screw press, which convert the biomass residue to high quality briquetting products. At present, screw press and piston press technologies are becoming more important commercially. In a screw extruder press, the biomass is extruded continuously by a screw through a heated taper die. In a piston press the wear of the contact parts e.g., the ram and die is less compared to the wear of the screw and die in a screw extruder press. The power consumption in the former is less than that of the latter. But in terms of briquette quality and production procedure screw press is definitely superior to the piston press technology. The central hole incorporated into the briquettes produced by a screw extruder helps to achieve uniform and efficient combustion and, also, these briquettes can be carbonized. Table: 2 show a comparison between a screw extruder and a piston press [5, 6].

Table 3. *Merits and demerits of piston press and screw press.*

	Piston press	Screw extruder
Optimum moisture content of raw material	10-15%	8-9%
Wear of contact parts	low in case of ram and die	high in case of screw
Output from the machine	in strokes	continuous
Power consumption	50 kWh/ton	60 kWh/ton
Density of briquette	1-1.2 gm/cm³	1-1.4 gm/cm³
Maintenance	High	Low
Combustion performance of briquettes	Not so good	Very good
Carbonization to charcoal	Not possible	Make good charcoals
Suitability in gasifies	Not suitable	Suitable
Homogeneity of briquettes	Nonhomogeneous	Homogeneous

As it has been seen from the above comparison the mechanical or hydraulic piston press technology is rather complicated and high initial investment and some technical infrastructures are required. In addition the quality and density of the briquettes is lower. So, the appropriate technology for this package is a screw press briquetting machine with die heater stove.

3. Preliminary Process Design

3.1. Design Basis

- Annual saw dust generation 3360 ton per year
- Operating days per year = 300
- Operating hours per day = 16
- Moisture content of the saw dust is 8%. Therefore, it

does not need drying
- 1 % Loss of raw material during feed processing
- 11 gm/kg of briquette material is converted into volatile fumes i.e., 1.1% loss of material having 3.5% moisture and
- In preheating and briquetting sub system there is 2% and 2.5 % loss of moisture content.

3.2. Material and Energy Balance

From the above explained, it is decided the operating system and the technology to convert a dried saw dust to briquetting. The ideal representation of the plant to make material balance is:

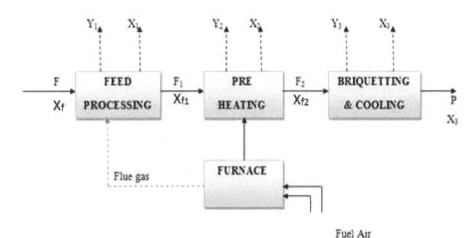

Figure 7. *Schematic Diagram of Ideal Briquetting Plant.*

Overall material balance is given by the following equation:

Output briquettes= Feed raw material – material loss in pre and briquetting process – Moisture loss in

Pre heater & briquetting process

$$P = F - \sum X - \sum Y \ (Kg/hr)$$

Where, F = feed rate with Xf moisture content
X 1, 2, 3 = moisture loss from components
Y1, 2, 3 = material loss from components
P = net production of briquettes
(Xf), (Xf1), (Xf2), (Xf3) = moisture content of different streams on wet basis

Feed Processing Subsystem (1)

Basis: F= 700 kg/hr of dry feed containing X_f= 8 % fraction of moisture content

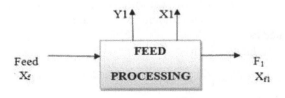

Because of the moisture content of the feed is less than 10%, it doesn't need drying rather there may be loss of material during sieving.

Assume in the feed processing there is 1% loss of material.

F_1= 0.99 (F- X1) assume X_1= 0, because no drying is required

F_1= 693 Kg/hr

Y_1= 7 Kg/hr

Pre-heating Subsystem

In this section the biomass is heated to about $100\ ^0$C. About 2% loss of moisture i.e. from 8% to 6% is expected [5, 7]. Input is:

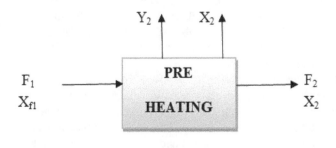

F_1= 693 Kg/hr at 8 % moisture content and there is no dry material loss $Y_2 = 0$.

Loss of moisture content

$$X_2 = F_1 * Xf_1 - (F_1 - X_2) * Xf_2$$

$$X_2 = 693 * 0.08 - (693 - X_2) * 0.06$$

$$X_2 = 13.07 \text{ Kg/hr}$$

$F_2 = F_1 - X_2$
$F_2 = 693\,Kg/hr - 13.07Kg/hr$
$F_2 = 693Kg/hr - 13.07Kg/hr$
$F_2 = 679.3\,Kg/hr$

Briquetting and Cooling Subsystem

During processing of biomass in this subsystem, the moisture is further reduced from 6% to 3. 5% and 11 gm/kg of briquette material is converted into volatile fumes i.e., 1.1% loss of material having 3.5% moisture.

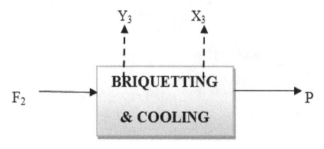

Input F_2 = 679.3 Kg/hr and there is no dry material loss Y_3 = 0

Loss of moisture content X_4

$$X_4 = F_2 * X f_2 - (F_2 - X_3) * X f_3$$

$$X_4 = 679.3 * 0.06\,Kg/hr - (679.3 - X3) * 0.035\,Kg/hr$$

$$X_4 = 8.386\,Kg/hr$$

Loss of material due to emission of volatile fume during briquetting is 11 gm/kg per briquette material.

Product $P = F_2 - X_3 - P * 0.011$

$$P = 679.3 - 8.386 - 0.011 * P$$

$$P = 663.61 \frac{Kg}{hr}$$

3.3. Energy Balance

As it has stated in the above the incoming waste raw material is dried enough for briquetting process. Thus, it doesn't need to incorporate a drier in its process. The furnace is only used for pre heating sub system only. The material is heated up to 80 0C and there is loss of about 2% moisture content during pre-heating.

Basic energy requirements for wood and water heating may be divided into several areas. In general, energy use in drying consists of latent heat of evaporation, sensible heat, heat lost by exhausting of air and heat lost by convection and radiation from the dryer's walls, and other heat losses such as leaks [6].

The latent heat of vaporization can be also calculated from the Clausius Clapeyron Equation in differential form which is stated as (adapted from Flowers and Mendoza 1970 as cited in Humphrey & Bolton, 1989):

$$H_v = X_2 * (2.5^{11} \times 10^6 - 2.48 \times 10^3). \text{T}$$

H_V = latent heat of water (J/kg H_2O)
T = evaporation temperature (^{0}C)
X_2 = evaporated water from pre heater (Kg/hr)

Mathematically, the sensible heating for drying of biomass may be expressed as

$$\Delta H_s = F_1 * C_{BM} * \Delta T$$

ΔHS = sensible heat (KJ/Kg), CBM = specific heat of biomass (KJ/Kg.K), ΔT=temperature change from ambient to evaporation temperature (^{0}F), and F_1= feed flow rate to the pre heater (Kg/hr)

Assumptions:

- Only 50% heat going for preheating
- The remaining 50% accounted for by losses in flue gases (40%) and losses due to radiation in oil circulating system (10%),
- 90% combustion efficiency of furnace which takes into consideration the radiation losses from furnace and uncombusted fuel,
- Ambient temperature 24 0C

Heat required in pre-heater

$$Q = H_v + \Delta H_s$$

$$Q = X_2 * (2.5^{11} x\, 10^6) - (2.48\, x\, 10^3) \cdot T + F_1 * C_{BM} * \Delta T$$

Where biomass with 10% moisture content specific heat is 1.5466 KJ/kg*K (biomass briquetting technology and practices, P.D. Grover & S.K. Mishra April 1996)

$$Q = 13.07 * (2.511\, x\, 10^6 - 2.48\, x\, 10^3)$$

$$100 + 693 * 1.5466 * (353 - 297)$$

$$Q = 89597.86 \text{ KJ/hr}$$

$$Q = 104686.49/0.5\,KJ/kg$$

$$Q = 179195.72\,KJ/kg$$

If it assumes 90 % furnace combustion efficiency, the amount of energy required in furnace for pre heating is:

$$Q = 179195.72/0.9\,KJ/kg$$

$$Q = 199106.35\,KJ/hr$$

If it assumes to use a eucalyptus saw dust briquetting briquette or any biomass with calorific value of 2500 Kcal/kg, the amount of briquetting required to pre heating sub system is calculated as follow:

Total heat duty per year

$$Q = 199106.35 KJ/hr * 16hr/day * 300 days/year$$

$$Q = 955710\,MJ/year$$

Calorific value of briquettes = 10450 KJ/kg
Amount of briquettes required per year
= 955710 MJ/year/10450 KJ/Kg
= 91455 Kg/year

4. Design and Equipment Sizing

The improved briquetting system developed recently consists of the following: a briquetting machine (screw press), a biomass pre-heater, a biomass-fired die-heating stove and a smoke removal system.

4.1. Preliminary Pre Heater Design

Pre-heating biomass before extrusion reduces briquetting energy consumption and also extends the life of the briquetting screw.

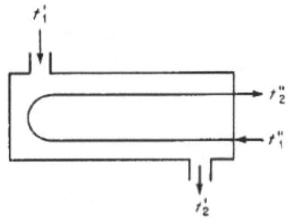

Figure 8. Schematic Diagram of Heat Exchanger.

Therefore, a pre-heater was incorporated in the briquetting package.

The biomass pre-heater essentially consists of two concentric pipes. Biomass is passed through the inner pipe under the action of a screw rotated by a variable speed motor. The raw material is pre-heated while being conveyed through the inner pipe by circulating hot termic oil at 1700c and temperature drop within the heat exchanger is 100C [2, 9].

The rate of heat transfer from hot oil to biomass is affected by individual transfer coefficients on the oil and biomass sides; and to some extent on the thermal resistance of shell thickness, but the controlling transfer resistance is on the biomass side. This is due to the low packing density and the refractory nature of biomass materials [4]. The following data are operational characteristics:

- Total heat duty = 89597.86 KJ/hr
- Feed inlet temperature = 24 0C
- Feed out let temperature = 80 0C
- Hot oil inlet temperature = 170 0C

- Hot oil out let temperature = 160 0C
- Overall heat transfer coefficient = 0.03486 KW/m²*K

$Q = UA\Delta T_{lm}$ Where Q = Rate of heat exchange (KJ/hr)

U = Overall heat transfer coefficient (KJ/m2.K)
A = Heat exchange area (m²)
ΔTlm = mean temperature difference

$$A = Q/U * \Delta T_{lm}$$

$$\text{LMTD} = \Delta T_{lm} = \frac{(t_1' - t_1'') - (t_2' - t_2'')}{\ln\left(\frac{t_1' - t_1''}{t_2' - t_2''}\right)}$$

$= \Delta T_{lm} = (180 - 24) - (160 - 80)/ln(180 - 24)/(160 - 80)$

$$= \Delta T_{lm} = 110^0C\ (383\ K)$$

A = 89597.86 KJ/hr/ 0.03486 KW/m2*K * 383 K
A = 1.86 m2

4.2. Screw Press Briquetting Machine Design

The briquetting machine used in this package is a Thailand manufacture, heated die biomass screw press machine. The main components of the machine shown below is a screw which feeds the saw dust material from the feeding hopper which is driven by the main motor with 30/45 KW capacity, compact it and press it in to a die of square, cylindrical, hexagonal or octagonal cross-section depending on the type die it uses.

An electrical coil is fixed on the outer surface of the die, to heat it to about 300 C. This temperature is required to soften the lignin in the biomass, which acts as a binder. Generally the machine has the following specifications.

General Specifications:
- Capacity: 600 - 700 Kg/h (Varies according to type of raw material & input power)
- Main Motor: 30/45 KW (Voltage & Speed to meet exact customer requirements)
- Agitator Motor: 0.25/1.0 KW (Depending on the raw material & machine capacity)
- Mould Heaters: 6.0 KW (Thermocouple & Temperature controller, included)
- Transmission: Through V Belts & Pulley arrangement
- Speed: 450 - 750 R.P.M (Depending on the Raw Material & Capacity of Machine)
- Profile of Moulds: Cylindrical / Hexagonal / Square or Octagonal
- Size of Briquette: 55mm/65mm (Measured across two parallel sides) [3, 7].

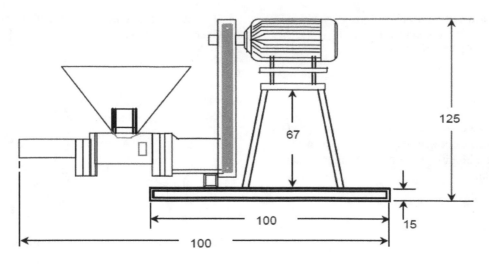

All dimensions are in cm

Figure 9. *Schematic Diagram of the Heated-Die Screw-Press Briquetting Machine.*

5. Economic Analysis of Briquetting

5.1. Estimation of Total Investment

Table 4. *Cost of Major Equipment.*

Equipment	Power rating (horsepower, hp)	Cost (birr)
Screw press briquetting machine	57	144800
Feed pre-heater	3	45000
Main screw conveyer	3	45500
Cooling conveyer	3	71000
Furnace		55000
Tractor and trailer		200000
Total	66	561300

Purchased equipment = F* Cost of major equipment
Where F = 1.3, is a correction factor based on wise guess
Purchased equipment cost = 729,690 birr

Table 5. *Estimation of Fixed Capital Investment (FCI).*

Component	Assumed % of FCI	Cost (birr)
Purchased equipment	32	729,690
Purchased equipment installation	5	114,014
Instrumentation (installed)	3	68,408
Piping (installed)	3	68,408
Electric (installed)	2	45,605
Building (including service)	21	478,859

Component	Assumed % of FCI	Cost (birr)
Yard improvement	2	45,605
Service facilities (installed)	10	228,028
Engineering and supervision	4	91,211
Construction expense	8	182,422
Contract fee	5	114,014
contingency	5	114,014
Total (fixed capital investment)		2,280,281

Total capital investment = fixed capital investment + working capital

Assume working capital is 10% of total capital investment

$$TCI = FCI + 0.1\,TCI$$

$$TCI = (2280281 + 228028)\ \text{Ethiopian Birr}$$

$$TCI = 2,508,309\ Ethiopian\ Birr$$

The total capital investment which will be offered by the Machine Particle Board Company is about 2,508,309 Ethiopian Birr.

5.2. Estimation of Total Production Cost

Total product cost = manufacturing cost + general expenses

I.   Manufacturing cost = direct production costs + fixed charges + plant overhead costs

Table 6. *Estimation of Total Production Cost.*

Components		Qty/ number	Cost (birr)
	1. Raw materials (it is assumed that cost the dust particle is zero)	3360 ton dust/year	0
	2. Operating, supervisory and clerical labor		
	Production head	1	36000
	Shift leader (technician)	2	57600
	Machine operator	3	21600
A. Direct	Mechanical &electrical maintenance	3	43200
production costs	Accountant	1	13000
	Store keeper	1	9000
	Daily laborers	8	57600
	Sub Total	19 (man power)	238000
	3. Utilities		
	Electric power	80 hp (52 Kw)	214848

Components		Qty/ number	Cost (birr)
	Fuel (briquettes)	91455 Kg/year	137182.5
	Sub Total		352030
	4. Maintenance and Repairs and operating Supplies (10% of fixed capital)		228028
B. Fixed Charges	1. Depreciation (10% of fixed-capital investment for machinery and equipment		228028
	2. Local taxes (1 - 4 % of fixed capital investment)		80000
C. Plant-overhead costs	(SO-70% of cost for operating labor, supervision, and maintenance, or 5-15% of total product cost);		90000
	Total		1216086

$Manufacturing\ Cost = Direct\ Production + Fixed\ Charges + Plant\ Overhead\ Costs$

$Manufacturing\ Cost = (238000 + 352030\ + 228028) + (228028 + 80000) + 90000$

$= 1216086\ Birr$

The total manufacturing cost for the plant is 1216086 Ethiopian Birr.

II.   General expenses = administrative costs + distribution and selling costs

Component	Cost (birr)
Administrative costs	70000
Distribution and selling costs	80000
Total	150000

General expenses = Administrative costs + distribution and selling costs

General expenses = 70000 + 80000

General expenses = 150000 Birr

The general expenses cost of for the plant is 150000 Ethiopian Birr.

III.   Total product cost = manufacturing cost + general expenses

$Total\ product\ cost\ =\ 1216086\ +\ 150000$

$=\ 1,381,086\ birr$

The total cost for operating the plant and selling the product is 1,381,086 Ethiopian Birr.

5.3. Profitability Analysis

Profit before tax = revenue – total production cost
*   If it assumes sell price of briquetting is 1.5 birr/kg and a capacity utilization of 75%.

$Total\ production\ per\ year$

$=\ 660\ Kg/hr * 16hr/day * 300day/year * 0.75$

$=\ 2376000\ Kg/year$

$Revenue\ =\ 2376000Kg/year * 1.5\ birr/kg$

$Revenue\ =\ 3564000\ birr/year$

$Profit\ before\ tax\ =\ 3564000\ -\ 1381086$

$Profit\ before\ tax\ =\ 2,182,913\ birr$

$Profit\ after\ tax\ =\ (1 - tR)\ PBT\text{-}Depreciation$

$Where: tR\ (tax\ rate)\ =\ 35\%$

$Profit\ after\ tax\ =\ (1 - 0.35) * 2182913 - 228028$

$Profit\ after\ tax\ =\ 1190865.25\ birr/year$

$Payback\ period\ =\ FCI/Net\ Profit$

$=\ 2280281\ birr/1190865.45\ birr/year$

$Payback\ period\ =\ 1.9\ year$

The above profitability analysis indicates that the total investment can return with almost two years with one single screw press machine.

6. Conclusion

Tough briquetting practice can offer important and substantial environmental and socio economic benefit in developing countries by remedied deforestation and fire wood shortage; it is not developed yet in a desired rate. In Ethiopia briquetting plant set up in early 1980s have failed to practice in commercial wise, mostly because of poor management skill, high power consumption and lack of technical constraints involved and the lack of knowledge to adapt the technology to suit local conditions. By alleviating such problems, the developing countries like Ethiopia can turn their huge agro and forest residue to generate energy with environmental friendly.

* As it observed in the above; from the different available briquetting technology screw press extruder with a pre heater and heated die coils has been chosen with respect to optimum energy consumption, low wear age and quality of briquettes.
* As a matter of fact the saw dust materials from different lines of Maichew Particle Board Factory are in ultimate optimum conditions for briquetting process. It doesn't incur additional cost for pretreatment like chopping and/or drying. They are at the desired range of size and moisture content (6 – 8 %).
* The integrate briquetting plant can turn the environmental unfriendly saw dust materials which was simply disposed to the environment to a valuable high quality briquettes with a calorific value 2800 – 4200 Kcal/Kg.
* Based on the results of this study, manufacturing of briquettes from saw dust residue is feasible with

economic benefit besides to the socio environmental benefits. It is an investment which can recover its capital with almost two years.

Acknowledgment

First of all we would like to thank God for his blessing and guidance throughout our life. We would like to thank to Asgele Gebrekidan from Maichew Particle Board Factory for his continuous offers of different current data from the company. We wonder if this project would be completed without his help. We would also to thank Sileshi Abebe from water and energy minister for his kindness in offers us different materials.

References

[1]   Robert H.Perry; and Don.Green. (2007). Perry's Chemical Engineers' Handbook. 8th Edition, McGraw Hill Professional

[2]   MC Cabe W.L. and J.C.Smith. (1956). Unit Operations in Chemical Engineering, pp943.

[3]   Max Peters and Klaus Timmerhaus. (1991). Plant Design and Economics for Chemical Engineers. 4th Edition, Mcgraw Hill Inc.

[4]   Eriksson, S. and M. Prio., (1990).The briquetting of agricultural wastes for fuel, FA0 Environment and Energy Paper 11, FA0 of the UN, Rome.

[5]   Aqa, S. and Bhattacharya, S.C. (1992). Densification of preheated sawdust for energy conservation, Energy.

[6]   P.D.Grover and S.K.Mishra. (1996). Biomass Briquetting: Technology and Practices, RWEDP.

[7]   S.C. Bhattacharya and S. Kumar. (2005).Technology Packages: Screw-press briquetting machines and Briquette-fired stoves.

[8]   http://www.ronakbriquettingmachine.com/briquetting-plant.html,

[9]   Pietsch, W. (1991). Size Enlargement by Agglomeration, John Wiley & Sons Ltd., England.

[10] Reed, T.B., Trefek, G, and Diaz, L. (1980). Biomass densification energy requirements in thermal conversion solid wastes and biomass, American Chemical Society, Washington D.C.

20

A Experimental Study of Air Heating Plane Solar Panels as Used in Drying Papayes

Abene Abderrahmane

University of Valenciennes, ISTV Science Faculty, Valenciennes, France

Email address:

a.abene@yahoo.fr

Abstract: One of the major problems concerning the use of solar panels for heating is the low level of thermal interchange with air in the dynamic vein of the solar panel. This weakness in such systems does not enable an optimum performance or high level of thermal efficiency to be obtained from their use. There is, however, a very noticeable improvement to thermal transfer when baffles are placed in rows in the ducts. To conduct the experiments, solar energy was simulated, the aim being to improve the ratio between temperature and thermal efficiency of an air heating plane solar panel and to make use of the system to reduce the drying time of papayes.

Keywords: Solar Energy, Simulation, An Air Heating Plane Solar Panel, Effects of Baffles Placed in the Vein, The Ratio Between Temperature and Thermal Efficiency of the Solar Panel, Drying, Papayes

1. Introduction

To improve the performance of solar panels, a wise choice of their components enables thermal losses between the absorber and the environment to be limited. Recent research studies have focussed more particularly on the circulation of the coolant fluid as a means of optimising performances and in which several methods have been proposed to deal with this objective. Zugary and Vulliere [1] sought to limit the losses near the fore part of the solar panel; two other researchers [2, 3] centred their work more on the absorber {the active part of the solar panel} while yet other papers [2-8] have shown that placing baffles in the dynamic air vein of the solar panel enables a turbulent air flow to be created which in turn boosts the interchange of thermal convection between the air and the absorber.

The baffles have to be placed very carefully. They can be fixed either onto the insulating material or under the absorber or indeed in both positions. In all three cases, results are improved because of the reduction of the hydraulic diameter (D_h) when compared with the performances of solar panels without baffles. In the air flow vein, the Reynolds number is calculated by starting from the maximum speed (V_m) corresponding to the minimum air flow section of the duct (S_{min}) and is expressed by :

$$Re = \frac{V_m.D_h}{v} = \frac{Q_v.D_h}{v.S_{min}} \text{ with } D_h = \frac{2d\ell}{d+\ell}$$

$$\text{By } d << \ell \Rightarrow d+\ell \approx \ell \text{ so } D_h = 2d$$

$$\text{Posing } b'=1-\frac{S_{min}}{d\ell} \quad \text{we are } \quad Re = \frac{2Q_v}{v.\ell.(1-b')} \tag{1}$$

D_h : the hydraulic diameter of the duct

b' : the blockage coefficient in the air vein

The coefficient of thermal convection interchange (h_{ccf}) between the absorber and the coolant fluid is dependent on the Reynolds number; it is an increasing Re function, otherwise h does not increase. It can therefore be deduced that when b' increases, both Re, and consequently h_{ccf}, increase. The minimum air flow section of the duct (S_{min}) is dependent on the shape of the baffles, their dimensions and their layout in relation one to another. The following three positional fixings of the baffles have been studied:

- fixed onto the insulating material: according to the shape chosen, the flow becomes turbulent and the fragmentation of swirls takes place very close to the absorber towards which the air is orientated and thermal transfer is improved

- fixed under the absorber: in addition to the aeraulic effect, these baffles also act as blades which means that the thermal transfer capacity emanating from the surface of the absorber increases and which in turn contributes to the improvement in the efficiency of the solar panel. As regards some of those shapes already studied [2], the results obtained have shown that with weak air flow, efficiency increases whereas with air flow stronger than a given value, efficiency tends to decrease. Where the air flow is very strong, swirls are much more evident near the insulation material. This positioning of the baffles results in a less efficient performance. Indeed, at the tips of the blades, the air temperature is lower than it is at the abosrber.

- To improve the performance of this second positioning, some intermediate baffles can be fixed onto the insulating material which will convey air flow towards the absorber [2]. In this case, charge losses in the solar panel will be higher and its thermal inertia will increase.

Given the above findings and to continue with the experiments, it was decided to fix the baffles onto the insulating material. When choosing the geometrical shape of the baffles to be used, certain criteria have to be satisfied. Indeed, both the layout and the shape of the baffles influence air flow during its trajectory. The baffles ensure that the absorber is well irrigated, reduce the zones of inertia, create turbulence and lengthen the course of the air by increasing the time it remains in the solar panel. A meticulous and systematic study was then undertaken of several different methods of arrangement of the air flow veins in the solar panels.

The first part of this paper deals with a comparison of the results obtained, initially using solar panels without baffles (SC) and then with baffles. Of the latter, two types were selected, namely Delta-shaped Curved Longitudinally (DCL1) and Ogival-shaped Curved Longitudinally (OCL1) baffles. The second part concerns the results obtained when firstly using solar panels without baffles and then with DCL1 baffles for drying papayes. In addition, to conduct experiments that would highlight the effects of baffles even further, papayes were dried by using a solar panel provided with Transversal-Longitudinal ones (TL) of the same type as those already studied [4]. A comparison of the results obtained shows that a solar panel provided with baffles is far more efficient than one without them.

2. Experimental Device

The solar panel with a single passage of air (Fig. 2) consists of:

- a transparent, alveolar, 1 cm thick polycarbonate cover. The coefficients of transmission (τ_c) and emissivity (ε_c) are respectively 83% and 90% for different wavelengths
- an absorber made of a 0.4 mm thick aluminium sheet painted in mat black on the insulation side. The thermal conductivity (α_a) and absorption (K_a) coefficients are respectively 95% and 205W/m.K. The distance (d) on each side of the absorber is 0.025m.
- a 5 cm thick polystyrene plate of insulating material which can resist temperatures higher than 90°C. Its

coefficient of thermal conductivity (K_{is}) is equal to 0.04 W/m.K.

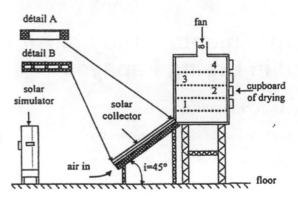

Figure 1. *Experimental device.*

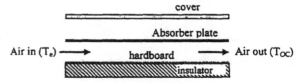

Figure 2. *Solar air flat plate collector without obstacles.*

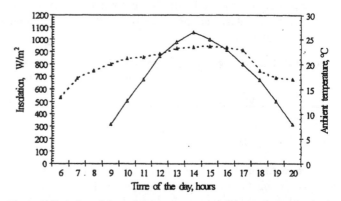

Figure 3. *Variation of the ambient temperature and the insolation during the characteristic day of July, at Valenciennes* —△— *global hourly insolation* (I_{GS}); ··▲·· *AMBIENT temperature* (T_a).

Too, the shape of the inlet (details A) and outlet (details B) of air of the solar panel have to be carefully arranged so as to avoid heating any dead zones. The baffles are fixed on a hardboard sheet just above the polystyrene plate. The experiments took place at Valenciennes in North-East France, the co-ordinates of which are : latitude : $\varphi = 50.3°$; altitude : Z = 60 m; longitude : L = 3.5°, and on a day in July which was considered typical of mean solar time flux (Fig. 3) and which corresponds with the average for the years 1998, 1999 and 2000 [9].

3. Results and Analysis

3.1. Improvements to the Ratio Between Temperature and Thermal Efficiency

A means of extracting the maximum of heat stored in the absorber is to place baffles in the mobile vein of air; they can be fixed either on the underside of the absorber or onto the insulating material, or indeed on both places. The objective is thereby to raise the temperature at the solar panel outlet (T_{SC}), *i.e.* increasing thermal efficiency, and to reduce charge losses to a minimum [2, 3 and 4, 6, 7 and 8]. Results have been obtained using the solar panel without baffles and subsequently provided with baffles in two stages, firstly with the DCL1 type and then with the OCL1 (Photographs 1 and 2).

Photograph 1. *Layout with DCL1 type baffles.*

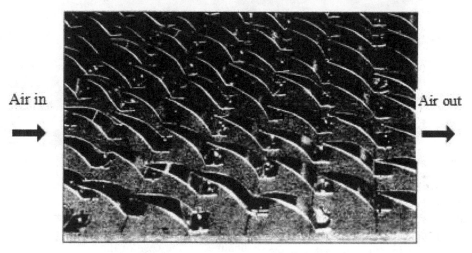

Photograph 2. *Layout with OCL1 type baffles.*

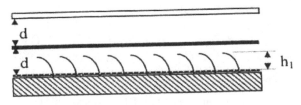

Fig. 4. *Solar panel equipped with baffles.*

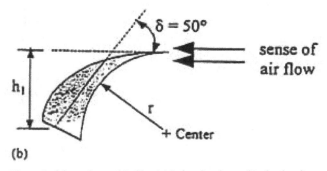

(b)

Figure 5. *delta and ogival baffles (a) before bending ; (b) after bending.*

The baffles selected for use from the range of shapes already experimented with [10] are formed by bending the otherwise straight delta and ogival wings (Fig. 5a and 5b) [10-12, 27, 29], and fixing them onto the insulating material (Fig. 4). The apex angle β of these baffles is 45° (Preferential angle) [13].

The index (1) referring to the DCL1 and OCL1 types of

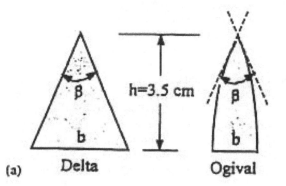

(a) **Delta** **Ogival**

baffles indicates that the air flow takes place near their tips (Fig. 5b).

Experiments carried out in a wind tunnel [14] have shown that the increase in incidence enables swirling fragmentation to progress continuously on the upper surface of the curved wing. The flow ends in a total disorganisation of the swirling systems at the leak edge of the wing which promotes the creation of a flow of considerable turbulence and, consequently, a better convective thermal interchange, which in turn improves the ratio between temperature and thermal efficiency. Total fragmentation of the swirls takes place at an incidence higher than 65°. The nature of the flow obtained as observed in the wind tunnel has been highlighted

(Photographs 3a and 3b). Other such visualisations of other shapes of wings [15, 26, 28] have confirmed results of these differing observations concerning the progressive fragmentation of swirls.

Table 1. Specifications of DCL1 and OCL1 baffles.

	DCL1	OCL1
	45°	45°
H_1(cm)	1.47	1.47
Et (cm)	2.3	2.0
E (cm)	3.5	3.5
b (cm)	3.0	1.4
r (cm)	4.0	4.0
Ncr	15	22
Nr	44	44

Photograph 3. Visualisation of air flow on the upper surface of the Delta baffle bent longitudinally at an incidence i of 60°, viewed from above.

Prior to a presentation of the results obtained in this first part, an explanation of the mathematical expression used to calculate thermal efficiency is called for. The Letz model [16] has been used as it is one of the most recent and complete formulae, being so because it takes into account not only the relative humidity of the air and the leak-flow of air as it is sucked into the sensor by the ventilator but also the temperatures at the inlet and outlet points of the solar panel. According to prior enthalpy assessment, made by the authors of this paper, of the different modes selected for the experiments, thermal efficiency (η) is determined by :

$$\eta = \frac{\rho.C_p.Q_{V_a}.(T_{sc} - T_e)}{I_{GS}.A_C} \quad (2)$$

$$\eta = \frac{\rho.C_p.Q_{V}.(T_{sc} - T_e)}{I_{GS}} \quad (3)$$

$$\rho = \rho_o \frac{273}{273 + T_{SV}} \frac{P(Z)}{P_0} \quad (4)$$

where $P(Z)/PO = (0.88)^Z$ and for Valenciennes $P(Z)/PO \approx 1$
$\rho_o = 1.293$ Kg/m3 and $C_P = 1005$J/Kg.K.

The captive surface A_C is 1.28m². In our case, thermal efficiency (η) is given for a constant global solar flow (I_{GS}) of 1063.5W/m² corresponding to solar midday. Therefore, for a specific flow of 35m³/h.m², 54% thermal efficiency was obtained in the case of the solar panel provided with DCL1 baffles. By increasing the flow to 70m³/h.m², 80% thermal efficiency was achieved. In the case of the solar panel without baffles, the respective thermal efficiency percentages obtained with each of the two flows were only 35% and 55% (Fig. 6) for temperatures at the solar panel outlet (T_{WB}) of 61.7°C and 53.3°C as opposed to 82.8°C and 66.3°C in the case of a solar panel provided with DCL1 baffles. However, under the same conditions as above, but using OCL1 baffles, slightly lower results were obtained than those with DCL1 ones. This can be explained by the fact that the air flow section differs and so, and in particular, the progressive fragmentation of swirls occurs a little earlier. Indeed, for the same flows, the respective thermal efficiency percentages are 52% and 76% and correspond to temperatures (T_{WB}) of 81.3°C and 64.8°C. With these configurations using DCL1 and OCL1 type baffles, 50% thermal efficiency is obtained respectively with flows of 32m³/h.m² and 33m³/h.m² as opposed to 58m³/h.m² when using an SC solar panel where relative flow reductions are respectively 44.8% and 43.1%. The respective temperatures (T_{WB}) are 84.5°C and 82°C as opposed to 55°C. The latter correspond to respective temperature rises ($\Delta T = T_{WB} - T_e$) of 56.1°C and 54°C as opposed to 27.2°C. In both cases, the amounts of charge losses are acceptable.

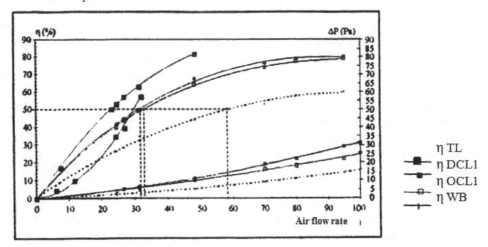

Fig. 6. *Evolution of thermal efficiency (where I_{GS} = 1063.5W/m²) and charge losses in relation to air flow as concerns WB solar panels and those provided with different types of baffles (i.e. DCL1, OCL1 and then TL), as at the Valenciennes site.*

However, the blocking effect of TL baffles enables a very turbulent flow to be created and, consequently, provides a very good level of thermal interchange. It is worth noting that the resultant charge losses are very high because the air flow through the duct is very weak compared with that attained with other types of baffles. A thermal efficiency of 50% obtained with a specific flow of 23m³/h.m² corresponds to a temperature (T_{WB}) of 104°C, *i.e.* an improvement in temperature (T_e) of 75°C at the solar panel inlet. These results are decidedly better than those obtained when using DCL1 or OCL1 type baffles. The lengthening of the distance covered by the air in the solar panel duct results in an even better interchange of heat between the coolant air and the absorber.

3.2. Improving Drying Time

Dating from the early research work of Lewis in 1921 [17] and Sherwood in 1929 [18], techniques of drying have been the subject of many scientific publications and continues to be a priority field of research, especially in respect of countries where traditional methods remain in use and are essential for want of better. As it is readily available at little or no running costs compared with other sources of energy such as electric resistors [19], solar energy is obviously an alternative. At Valenciennes, simulated solar energy was used to carry out experiments applied to drying pre-dried papayes; the simulator was conceived to provide conditions of a typical July day.

For solar energy to be harnessed effectively, certain difficulties have first to be overcome and to be achieved with the help of technically viable and economically profitable systems. The choice of the type of dryer is conditioned by whether or not the product in question can withstand solar radiation; it also has to be made between direct or indirect dryers and depends, too, on the commercial value of the product. Since the performances of absorbers are higher than the thermal conversion capacity of the product, the use of an indirect dryer is the more effective. The system under study is therefore an indirect solar dryer functioning by thermal forced convection.

The construction of a drying installation is very complex and requires the taking into account of a number of parameters and the mastering of many phenomena before it can be devised. For it to work efficiently, it is necessary, first of all, to estimate the quantities of products to be treated and then to carry out a thorough study of the design of the system. What is important, from the thermal point of view and to ensure that the components of the installation are optimised, is to evaluate the various modes of transfer and to assess the energy-giving potential while taking into account the coupling between the warm air generator and the drying unit with a view to its dimensional set-up. In our experiment, the device (Fig. 1), being constructed with only one plane solar panel and a "drying cupboard" holding four trays {*i.e.* a simplified version of a sort of kiln}, has been designed to treat small quantities of products and, consequently, equipping it also with means for storing energy; with an auxiliary heating system and with a device for recycling air is therefore unnecessary.

The quantities of heat (Q_u) recovered by the coolant fluid, as far as the absorber is concerned, depends on the efficiency of the solar panel used. Given that these quantities are proportional to the variations in temperature between the inlet and the outlet of the solar panel, the results presented above show that a solar panel provided with baffles functions more efficiently and so baffles are essential fittings because they reduce drying times.

In our experimental work, the objective is to carry out drying by a simulation of solar energy. However, for a given air flow, we wanted to study the variations in certain parameters of the drying process as at different times during the typical day under consideration. In view of the considerable expenditure that could be involved in setting up a real-life operation, the use of thermal forced convection would seem to be less suitable in applying the findings of our small-scale experiment to a large-scale situation. Nevertheless, it would be profitable to take advantage of natural convection in a solar chimney. Its application is, of course, all the more valid in geographical zones deprived of electrical power. The choice between forced and natural convection depends on several factors, in particular on the quantity of the product to be treated, on the capillary structure of that product and its nutritional value while not neglecting the financial budget available. Drying time is indeed of paramount importance. As regards large-scale {industrial} concerns, an external source of energy is required. Where electrical power is available, even if weak but at an affordable rate, it is logical to make use of it to actuate the ventilators, blowers or other devices necessary to increase the efficiency of the system. Where a system functions with natural convection in a solar chimney, the driving force of gravity is created by differences in the density of air between the exterior {ambient conditions} and the interior of the chimney. The height of the chimney, which influences the efficiency of extracting air, is a factor that has therefore to be adequately investigated. Pasumarthi and Sherif [20] have shown that for a given height and an increasing solar flux, the temperature at one and the same given point in the chimney also increases. Heat interchange improves but the total charge losses of the system, which are proportional to the

height of the chimney and to the differences in air density, increase considerably.

Prior to setting out the findings of our experiments, a brief description of the type of TL baffles used is called for. The height of the large (transversal) baffle is 2.5 cm and that of the small (longitudinal) one is 2 cm (Fig. 7). The surface A_C is 1.28 m².

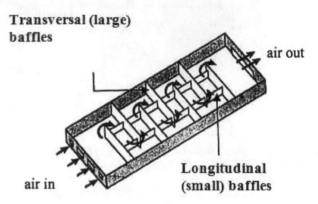

Fig. 7. *Solar panel equipped with TL type baffles [Kge/KgMS : kilograms of water per kilograms of dried mass of the product].*

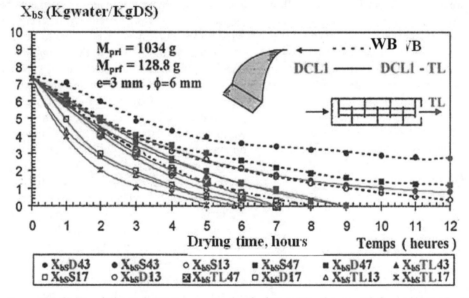

Fig. 8. *Evolution of the water content (X_{bS}) in relation to drying time measured at the first and fourth trays with flows of 31.3m³/h.m² and 70m³/h.m² using WB, DCL1 and TL type solar panels, data recorded on a typical July day at the Valenciennes site.*

To study the influence of the flow of drying air on drying time, it was considered of value to use two flows, one of 31.3m³/h.m², the other of 70m³/h.m². This adjustment is made with the help of a ventilator (solar simulator). The air flow is measured using a "Jules et Richard" anemometer with a 10 cm diameter propeller. With a flow of 31.3m³/h.m², drying times at the first (bottom) tray (Fig. 1) take up to 8 hours as regards the solar panel fitted with DCL1 baffles, 6 hours 35 minutes for that with TL baffles while the longest time taken was with the WB solar panel. There are therefore reductions of drying time of 49% and 59% in comparison with the WB solar panel. The final content of water collected in the WB solar panel is only attained after 14 hours 10 minutes of drying time (Fig. 8). The air coming from the level of the first tray is still heavy

with moisture and, consequently, for this same air flow, drying time at the level of the fourth (top) tray takes longer for all three types of solar panels used. By increasing the flow to 70m³/h.m², drying times decrease in each of the three solar panels. As drying is brought about by force of speed of the flow, this faster flow results in a more rapid evacuation of the moist air.

By increasing the flow from 31.3m³/h.m² to 70m³/h.m², and as regards the solar panel with DCL1 baffles, the drying time at the first tray is reduced by one hour, *i.e.* a relative reduction of 15 % whereas a relative reduction of 13.8% is attained using a solar panel with TL baffles. The drying times at the level of the fourth tray are respectively 10 hours (DCL1 baffles) and 8 hours (TL baffles). Comparing these results with the performance of the solar panel without baffles, and with a flow of 70m³/h.m², the reductions in drying times at the first tray are respectively 27% and 39.5%.

A graph (Fig. 9) plots the evolution of the loss of mass (ΔM) at each hour, for each of the two flows and for each of the types of solar panel used. Fig. 10 shows the evolution in temperature of the product (T_{Pr}) in relation to the passage of time during the drying process. It is to be noted that for every type of solar panel used, drying takes place at temperatures that vary in accordance with the solar time flux particular to the day on which the experiment is conducted. In every case, a constant phase of drying cannot therefore exist (Fig. 11).

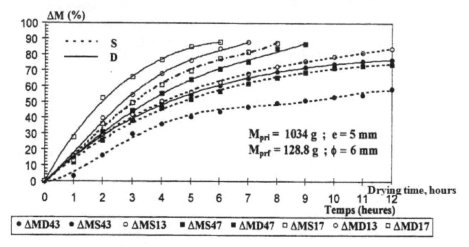

Fig. 9. Evolution of the loss of mass (ΔM) of plums in relation to drying time measured at the first and fourth trays with flows of 31.3m³/h.m² and 70m³/h.m² using DCL1 and TL type solar panels, data recorded on a typical July day at the Valenciennes site.

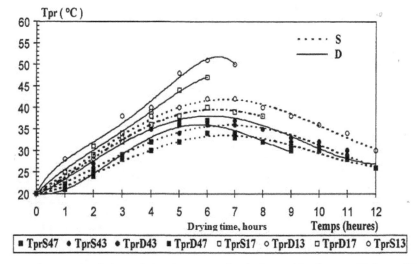

Fig. 10. Evolution of the temperature of papayes in relation to drying time measured at the first and fourth trays with flows of 31.3m³/h.m² and 70m³/h.m² using WB, DCL1 and TL type solar panels.

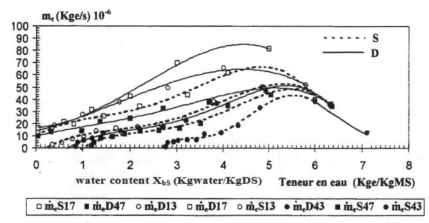

Fig. 11. Evolution of the drying speed (m_e) in relation to water content to the point when the product is completely dried with flows of 31.3m³/h.m² and 70m³/h.m² using WB and DCL1 type solar panels.

Analysis of the findings relative to the WB solar panel (without baffles), reveals that functioning with a low air flow is considerably more efficient because of the reduction in drying time. The mechanical power consumed (P_{mc}) by the ventilator is proportional to charge losses and to the air flow in the dynamic air vein of the panel. This same power is expressed by :

$$P_{mc} = \Delta P.Q_V = \zeta.Q_V^3 \qquad (5)$$

where ζ is the factor of friction, characteristic of artificial rough places (baffles)

As the relationship between the two flows is 2.24, the power is therefore increased by a factor of 11.24, a fact which further highlights preference for using a low flow. In spite of the recommendations made by some research workers not to exceed a drying air temperature of 55°C, higher temperatures were used in our experiments (at around solar midday). However, at temperatures above 70°C, reddening spots (i.e. signs of burning) appeared on the products. Indeed, the quality, colour, savour and nutritional value of the product are all closely subjected to conditioning by the thermal process. Consequently, to create ideal drying conditions at temperatures lower than those recommended for the product in question, some precautions can be taken such as :

• install a temperature indicator at the inlet to the "drying cupboard" and use a higher air flow if necessary to reduce the temperature while bearing in mind that should the increase in the air flow become imperative at midday, it would not be so when the sun is less high and its rays more oblique in relation to the position of the solar panel, i.e. in the early morning and late afternoon.

• install a temperature regulator adjusted to provide a constant drying air temperature of 55°C.

The quantities of heat available for use and reclaimed at the solar panel outlet are much higher when using solar panels equipped with TL baffles than those with DCL1 baffles. Variations in those quantities (Q_u), in global quantities of drying heat (Q_s) and their differences ($\Delta Q = Q_u - Q_s$) are shown in Fig. 12 (for WB solar panels without baffles) and Fig. 13 (solar panels equipped with DCL1 baffles). Worthy of note is the fact that the quantities of heat available for use are increased by a factor of approximately 1.65 as regards the performance of the SC solar panel. The differences in quantities (ΔQ) are of some consequence because they are, in fact, surplus to normal requirements for the drying process and can therefore be stored and made available for use, for example, during the night or on days when sunlight is mediocre [22, 23, 24, 25]. This excess of heating needs can be kept in underground ducts and thus ready for use when needed.

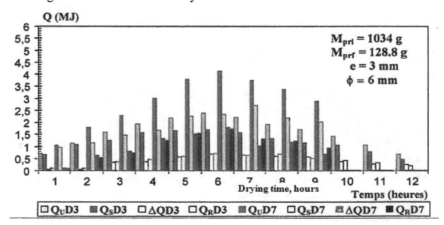

Fig. 12. *Variations in quantities of heat (Q) in relation to drying time with flows of 31.3m³/h.m² and 70m³/h.m² using an WB type solar panel (without baffles).*

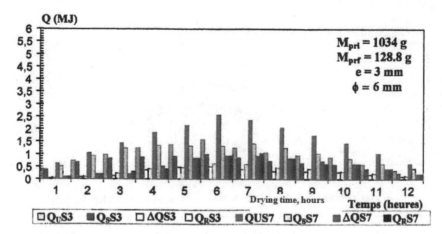

Fig. 13. *Variations in quantities of heat (Q) in relation to drying time with flows of 31.3m³/h.m² and 70m³/h.m² using a solar panel equipped with DCL1 baffles.*

4. Conclusion

It can de deduced from the findings of our experiments using various types of solar panels that the placing of baffles in the air vein is a very significant factor which serves to improve the performance of a given solar panel. Several determinants have, of course, to be taken into consideration to include the shapes and dimensions of the baffles, the number of rows and their layout. The study has shown that a solar panel equipped with baffles not only appreciably improves the ratio between temperature and thermal efficiency but also reduces drying time of the product. Also noteworthy is the fact that a reduction in transversal (Et) and longitudinal (El) spaces contributes considerably to the quality of the results. Moreover, an increase in the angle (Δi) brings about even better results. However, some constraints imposed by the nature of the finished product such as its quality, savour, colour and nutritional value, have to be taken into account in determining what constitutes the ideal temperature of drying air.

Nomenclature

A_C: surface activates plane solar collector [m^2]
b: width of the baffles at the base
b': coefficient of blocking
C_p: heat capacity of the air [J/Kg.K]
d: outdistance between the absorber and the cover or the insulator
El : longitudinal space between lines of baffles
Et: space transverse between two baffles of the same line
h: initial height of the baffles
h_1: swing-over bed of the baffles compared to the insulator
h_{ccf}: coefficient of convectif heat exchange enters the air and the absorber [W/K.m^2]
I_{GS}: total time solar flow of simulation [W/m^2]
K_a, K_{IS}: thermal conductivity of the absorber and the insulator [W/m.K]
L: longitude of the place [degrees]
Z: altitude of place [km]
ℓ: width of the vein [m]
m_e: speed of drying [Kg/s]
N_{Cr}, N_r: a number of baffles per line and a number of lines
Pmc: consumed mechanical power of the ventilator [W]
Po, P(Z): pressure atm on the sea level (10.13 105 Pa) and with altitude Z of place [Pa]
Q_{Va}: volume throughput of the air in the sensor [m^3/h]
Q_V: volume throughput of the air per unit of area [m^3/h.m^2]
Q_u: quantity of useful heat on the outlet side of sensor [MJ]
Q_S: quantity of heat of drying of product [MJ]
r: ray of bending of the baffles
Re: Reynolds number
S_{mini}: minimal bypass section of the air in the vein [m^2]
T_{SV}, T_a: temperature at the exit of the ventilator and the ambient air [°C]

T_e, T_{SC}: temperature of the air at the entry and the outlet side of the sensor [°C]
T_{Pr}: temperature on the level of the surface of the product [°C]
V_m: maximum speed of the air flow in the vein [m/s]
X_{bS}: water content at base dries of the product [Kg/KgMS]
X_{ObS}, X_{fbS} : water content initial and final at base dries [Kg/KgMS]
Greek Letters:
α : angle of inclination of the sensor compared to the ground [degrees]
β : angle of apex (or at the top) of the baffles [degrees]
ε_c : coefficient of emissivity of the cover
φ : latitude of the place [degrees]
ν : viscosity kinematic of the air[m^2/s]
ρ_0, ρ : masse voluminal of the air on the sea level and altitude Z of the place [Kg/m^3]
η, η_t : output of the plane solar collector and thermal efficiency of the system of drying [%]
τ_C : coefficient of transmission of cover
ζ : factor of friction characterizing artificial roughnesses (obstacles)
Δi : angle of bending of the baffles [degrees]
ΔM : loss of mass of the product [%]
ΔP : pressure losses in the vein of sensor [Pa]
ΔQ :quantity heat excédentaire[MJ]
Notations:
DCL1 : Forme of the baffles in the mobile vein of air: Delta and Curved Longitudinally with the flow attacking by point
OCL1 : Forme baffles in the mobile vein of air: Ogival Curved Longitudinally with the flow attacking by point
WB, TL : Without Baffles and Transversal-Longitudinal baffles
ηS : Rendement of sensor WB
ηD : Rendement of the sensor provided with baffles DCL1
ηTL : Rendement of the sensor provided with baffles TL
X_{bS}TL13 : Tenor out of water at dry base in the case with the sensor provided with baffles TL, on the level of the 1st tray, using the flow of 31.3 m^3/h.m^2
ΔMD13:Perte of mass of the product in the case of the sensor provided with baffles DCL1, on the level of the 1st tray, using the flow of 31.3 m^3/h.m^2
T_{pr}S17: Temperature of the product in the case of sensor WB, on the level of the 1st tray, using the flow of 70 m^3/h.m^2.
m_eD47: Speed of drying in the case of the sensor provided with baffles DCL1, on the level of the 4th tray, using the flow of 70 m^3/h.m^2
Q_uS3: Quantity of useful heat, in the case of sensor WB, using the flow of 31.3 m^3/h.m^2
Q_SD7: Quantity of heat of drying, in the case of the sensor provided with baffles DCL1, using the flow of 70 m^3/h.m^2.

References

[1] Zugary M.R., Vullierne J.J., "Amélioration des performances thermiques d'un capteur solaire par l'utilisation d'une structure à lamelles", Entropie,n°176,(1993), pp25-30.

[2] Ouard S., "Optimisation des formes et disposition d'obstacles dans la veine mobile du fluide des capteurs solaires plans à deux couches d'air en vue de la maximisation du couple rendement et température", thèse doctorat, Valenciennes, (1989).

[3] Moummi N., "Prévisions systématiques et optimisation des performances des capteurs solaires plans à air dans divers sites de climats Méditerranéens ou sahariens et avec ou sans altitudes", thèse doctorat, Valenciennes, (1994).

[4] Ben Slama R., "Contribution à l'étude et au développement de pompes et capteurs solaires", thèse doctorat, Valenciennes, (1987).

[5] Choudhury C., Gary H.P., Performance of air heating collectors with packed airflow passage", Solar Energy, V 50,3, (1993), pp 205-221.

[6] Gbaha P., "Etude et optimisation des échanges thermiques et des performances des capteurs solaires plans à deux veines d'air", thèse doct, Valenciennes, (1989).

[7] Hachemi A., "Contribution à l'optimisation des performances thermiques des insolateurs plans à air à lit garni de rangées d'obstacles aux pas serrés, amélioration du rendement par interaction entre rayonnement et convection", thèse doctorat, Valenciennes (1992).

[8] Moummi A., "Etude globale et locale du rôle de la géométrie dans l'optimisation des capteurs solaires plans à air", thèse doctorat, Valenciennes, (1994).

[9] Météo France, Direction Régionale, Villeneuve d'Ascq, Lille, (2002).

[10] Ahmed-Zaid A., "Optimisation des capteurs solaires plans à air. Application au séchage de produits agricoles et de la pêche", thèse doctorat, (1998)

[11] Messaoudi H., Ahmed-Zaid, Le Ray M. "Le rôle de la géométrie dans l'amélioration ou la diminution des échanges thermiques turbulents dans les capteurs solaires à air", 3ème congrès de mécanique, Société Marocaine des Sciences Mécaniques et de l'université Abdelmalek Essaadi, Faculté Science de Tétouan, VI,(22-25 avril 97),pp637-644.

[12] Ahmed-Zaid A., Messaoudi H., Benoyounes R., Le Ray M, "Etude expérimentale de l'effet des chicanes sur les capteurs solaires plans à air, par simulation de l'énergie solaire", 3èmes Journées Maghrébines sur la Mécanique, Institut de Mécanique, Guelma (Algérie), VI, (29-30 avril 1997), pp 212-218.

[13] Le Ray M., Deroyon J.P., Deroyon M.J., Minair C., " Critères angulaires de stabilité d'un tourbillon hélicoïdal ou d'un couple de tourbillons rectilignes, rôle des Angles Privilégiés dans l'optimisation des ailes, voiles, coques des avions et navires", Communication à l'Association Technique Maritime et Aérodynamique, Bulletin de l'ATMA, Paris, (avril 1985), pp 511-529.

[14] Messaoudi H.,"Structures tourbillonnaires induites par des obstacles triangulaires complexes et interactions ailes delta minces-corps de révolution. Application des Angles Privilégiés, en Aérodynamique et Energétique", thèse doct, Valenciennes, (1996).

[15] Abene A., "Etude systématique des positions et de la stabilité des structures tourbillonnaires au-dessus d'ailes ogivales et de cônes", thèse doctorat, Valenciennes, (1990).

[16] Letz T., "Modélisation et dimensionnement économique d'un système de chauffage domestique biénergie", thèse doct, INSA Lyon, (1985).

[17] Lewis W.K., "The rate of drying of solid materials", Journal of. Ind and Eng. Chem, Vol 13, 5, (1921), pp 427-432.

[18] Sherwood T.K., "The drying of solid", V 21, 10, (1929), pp 12-16.

[19] Kondé G., "Résolution des équations de Luikov appliquée au séchage de l'oignon en vu de la réalisation de séchoirs solaires dans les pays sahéliens", thèse doct, Perpignan,(1983).

[20] Pasumarthi N., Sherif S.A., "Experimental and theoritical performance of a demonstration solar chimney model-part II : Experimental and theorical results & economic analysis", Internat Journal Of Energy Research, 22, (1998), pp 443-461;

[21] Desmons J.Y., "Formulation et résolution numérique des problèmes aux limites appliquées, aux générateurs de chaleur tubulaires enterrés", thèse doctorat, Valenciennes, (1984).

[22] Benyounes R., "Prévision numérique et expérimentale de la réponse intrinsèque d'un échangeur bitubulaire enterré en régime continu", thèse doctorat, Valenciennes, (1993).

[23] Portales B., "Etude et expérimentation du chauffage solaire des serres à stockage thermique souterrain", thèse doctorat, Valenciennes, (1984).

[24] N'Dongo M.M., "Etude théorique d'un générateur solaire à air chaud constitué d'une serre agricole et d'un stockage de chaleur en sous-sol à l'aide de conduits. Application au séchage", thèse doctorat, Clermont-Ferrand II. Blaise Pascal, (1989).

[25] Dubois V., " Etude détaillée des structures tourbillonnaires développées à l'extrados de corps élancés simples, complexes et de corps de révolution. ", thèse doctorat, Valenciennes, (2005).

[26] Abene A., Dubois V., Michel Le Ray M., A. Ouagued A., "Study of a solar air flat plate collector: use of obstacles and application for the drying of grape", Journal of food engineering 65, (2004).

[27] Abene A., Dubois V., Michel Le Ray M., Ouagued A., « Etude expérimentale des performances thermiques de diverses configurations de chicanes placées dans la veine d'écoulement », Les Technologies Avancées n°16, Alger, (2003).

[28] Abene A., Dubois V., Michel Le Ray M., Si-Youcef M., Ouagued A., « Etude expérimentale de capteurs solaires à air : le séchage de la figue » Les Technologies Avancées n°17, Alger, (2005).

[29] A. ABENE "Visualisation of Vortex Structures Developed on the Upper Surface of Double-Delta Wings", J Aeronaut Aerospace Eng 2:118, doi: 10.4172/2168-9792.1000118. September 2013

[30] A. ABENE An experimental study of the drying of papayes by solar panels -journal of sustainable development studies , ISSN 2201-4268 volume 7,number 1,2014,84-108 octobre2014

[31] A. ABENE An vortex structures on the upper surface of a pointed gothic wing observed at low ,median and high incidences -journal of vortex science and technology,ISSN 2090-8369 volume2-1, janvier 2015.

21

An innovative similarity transformation for in-depth research of convection heat and mass transfer

De-Yi Shang[1], Bu-Xuan Wang[2], Liang-Cai Zhong[3]

[1]136 Ingersoll Cres., Ottawa, ON, Canada K2T 0C8
[2]Dept. of Thermal Engineering, Tsinghua University, Beijing 100084, China
[3]Department of Ferrous Metallurgy, Northeastern University, Shenyang 110004, China

Email address:

deyishang@yahoo.ca (De-Yi S.)

Abstract: Our innovative similarity transformation for in-depth research of convection heat and mass transfer is presented. For solving convection heat and mass transfer issues, the boundary layer analysis method is used, and meanwhile, the Falkner-Skan transformation is currently popular to treat the core similarity variables for velocity field similarity. But this type of transformation is inconvenient to do this core work, for similarity transformation of velocity field, because it is necessary to first induce flow function and group theory to derive an intermediate function for an indirect similarity transformation of the velocity field. This case also allows a difficult situation on consideration of variable physical properties. With our innovative similarity transformation, the above inconvenient and difficult situations are avoided, and the velocity components can be directly transformed to the related dimensionless ones. Then, the similarity analysis and transformation of the governing partial differential equations can be simplified greatly. Furthermore, our innovative similarity transformation can conveniently treat variable physical properties and their coupled effect on heat and mass transfer for enhancement of the practical value of convection heat and mass transfer, and so is a better alternative transformation method to the traditional Falkner-Skan transformation. It was proved that the above two innovative methods have a wide practical application in industry.

Keywords: Innovative similarity transformation, Convection heat transfer, Boundary layer analysis, Falkner-Skan Transformation, Core similarity variables, Flow function, Group theory

1. Introduction

1.1. Background of Our Innovative Similarity Transformation

Since Prandtl proposed his famous boundary layer theory [1], many complex issues on convective hydrodynamics and heat and mass transfer could be solved more conveniently. Meanwhile, the similar analysis method is usually used to transform the dimensional physical variables into the dimensionless physical parameters, and the governing partial differential equations are transformed equivalently into the governing ordinary ones. Since the transformed dimensionless parameter is the group of the dimensional physical variables, the transformed governing ordinary differential equation will be greatly simplified. This is the reason the boundary layer theory greatly accelerated the development of the research of convection heat and mass transfer.

Actually, convection heat and mass transfer is strongly affected by fluid variable thermo-physical properties and their interactions within the boundary layer. In fact, the core work of similarity analysis of the governing partial differential equations lies in the similarity transformation of the velocity field of the boundary layer. However, currently popular Falkner-Skan transformation [2] is difficult to do this job, especially with consideration of variable physical properties, because it is necessary to first induce flow function and group theory to derive an intermediate function for an indirect transformation of the velocity field. It is a complicated process, and allows even complicated issue for consideration of variable physical properties. Therefore, most of the related traditional theoretical research on convection heat and mass transfer did not consider or did not consider well the coupled effect of variable physical properties. (For more detailed discussion about it, please refer to books [3,4]). It lead to that a series of traditional related researches lack practical application value on convection heat and mass transfer, for a simple example, a lot of reported correlations on convection heat and mass transfer application are empirical or semi-empirical obtained from experiment. It is background for development of the innovative similarity transformation.

1.2. Current Application of the Innovative similarity transformation

Before introduction of the innovative similarity transformation for in-depth research on convection hydrodynamics and heat and mass transfer, here it is necessary to mention its current application areas. So far, this innovative similarity transformation has been applied in investigation of convection hydrodynamics and heat and mass both without and with phase change, including the following broad areas:

- Free/forced convection [3-7];
- Free film boiling convection [4,8];
- Free/forced film condensation convection of pure vapour [3, 4, 9, 10];
- Free/forced film condensation convection of vapour-gas mixture [3, 4,11];
- Accelerating falling film flows of non-Newtonian power law fluids [4, 12-14].

1.3. Expanded Application of the Innovation similarity transformation

This innovative similarity transformation has got its wide application so far, since they were proposed on in-depth research for convection hydrodynamics and heat and mass transfer, for example as below:

Studies [15,16] applied our formulization equations on heat transfer coefficient of free convection [4,5] for temperature distribution assessment during radiofrequency ablation related to surgical medical treatment. It is seen that even for the lead phantom close to the edges, for the maximal discrepancy from their test data, a satisfactory agreement (2%) was achieved. Our formulization equations on heat transfer coefficient were just obtained by using the present innovative similarity transformation.

Our innovative similarity transformation was applied for study on deposition of SiO2 nanoparticles in heat exchanger during combustion of biogas [17]. In this study, our set of dimensionless similarity physical parameters with their formulae including our innovative core similarity variables for similarity transformation of velocity field [3] were applied for similarity transformation of the governing partial differential equations of a laminar forced convection. Based on our innovative similarity transformation, the governing ordinary differential equations were obtained for numerical solution.

Ref. [18] successfully applied our heat transfer equations of free convection of gas with consideration of variable

physical properties [4,5] for analytical and experimental study on multiple fire sources in a kitchen. It follows that our formulization equation of Nusselt number combined with the gaseous temperature parameter has wide application value for evaluation of heat transfer with large temperature differences. Our formulization equations on heat transfer coefficient were just obtained by using the present innovative similarity transformation.

The above examples also reflect that our innovative method has special theoretical and practical value and broad application prospects in more industrial departments.

2. The Innovative Similarity Transformation

Our innovative similarity transformation was originally developed for extensive investigation of coupled effect of temperature- dependent physical properties of gases and liquids on free/forced convection heat transfer [4-7]. The detailed theoretical analysis and mathematical derivation for this transformed of the governing differential equations are reported in our books [3,4]. Due to the space limitation here we have to introduce this transformation briefly through its application. Here, free and forced convection boundary layer issues are taken as examples for introduction of our innovative similarity transformation. First, it is necessary to show its governing partial differential equations.

2.1. For Free Convection Boundary Layer

2.1.1. Governing Partial Differential Equations

Governing partial differential equations of free convection boundary layer with consideration of variable physical properties are

$$\frac{\partial}{\partial x}(\rho w_x) + \frac{\partial}{\partial y}(\rho w_y) = 0 \tag{1}$$

$$\rho(w_x \frac{\partial w_x}{\partial x} + w_y \frac{\partial w_x}{\partial y}) = \frac{\partial}{\partial y}(\mu \frac{\partial w_x}{\partial y}) + g|\rho_\infty - \rho| \tag{2}$$

$$\rho c_p(w_x \frac{\partial t}{\partial x} + w_y \frac{\partial t}{\partial y}) = \frac{\partial}{\partial y}(\lambda \frac{\partial t}{\partial y}) \tag{3}$$

where Eqs.(1) to (3) are mass, momentum, and energy conservation equations respectively. Here, it is seen that the effect of variable thermo-physical properties, such as density, absolute viscosity, and thermal conductivity are considered. However, for most of common gases and liquids, the variation of specific heat with temperature can be omitted.

2.1.2. Similarity Physical Variables (or Parameters)

The detailed physical analysis and mathematical derivation about our innovative similarity analysis method are reported in [4-7] for treatment of above free convection boundary layer. For saving space, here we only show the set of the related dimensionless similarity physical variables (or parameters). However, the core work for the similarity analysis and transformation issues is to treat the similarity problem of velocity field. To this end, we proposed the following innovative core similarity variables $W_x(\eta)$ and $W_y(\eta)$ and their formulae as follows for similarity transformation of velocity field :

$$W_x(\eta) = [2\sqrt{gx} \left| \frac{\rho_\infty}{\rho_w} - 1 \right|^{\frac{1}{2}}]^{-1} w_x \qquad (4)$$

$$W_y(\eta) = [2\sqrt{gx} \left| \frac{\rho_\infty}{\rho_w} - 1 \right|^{\frac{1}{2}} (\frac{1}{4} Gr_{x,\infty})^{-\frac{1}{4}}]^{-1} w_y \qquad (5)$$

where the local Grashof number is defined as

$$Gr_{x,\infty} = \frac{g|\rho_\infty / \rho_w - 1|x^3}{v_\infty^2} \qquad (6)$$

The similarity coordinate variable is defined as

$$\eta = \frac{y}{x}(\frac{1}{4} Gr_{x,\infty})^{1/4} \qquad (7)$$

The similarity temperature is given as

$$\theta = \frac{t - t_\infty}{t_w - t_\infty} \qquad (8)$$

Here, innovative equations (4, 5) for similarity velocity variables are derived in our book [4]. It is emphasized that the similarity velocity variables $W_x(\eta)$ and $W_y(\eta)$ are core similarity variables for similarity transformation of velocity field. Then, it is expected that they will dominate the transformed governing ordinary differential equations.

2.1.3. Governing Similarity Ordinary Differential Equations

The whole similarity transformation process for governing similarity ordinary differential equations of free convection boundary layer is reported in our book [4]. Here, we only give the transformed   governing ordinary differential equations as below to save space:

$$2W_x(\eta) - \eta \frac{dW_x(\eta)}{d\eta} + 4 \frac{dW_y(\eta)}{d\eta} - \frac{1}{\rho}\frac{d\rho}{d\eta}(\eta W_x(\eta) - 4W_y(\eta)) = 0 \quad (9)$$

$$\frac{v_\infty}{v}(W_x(\eta)(2W_x(\eta) - \eta\frac{dW_x(\eta)}{d\eta}) + 4W_y(\eta)\frac{dW_x(\eta)}{d\eta})$$
$$= \frac{d^2W_x(\eta)}{d\eta^2} + \frac{1}{\mu}\frac{d\mu}{d\eta}\frac{dW_x(\eta)}{d\eta} + \frac{v_\infty}{v}\frac{\frac{\rho_\infty}{\rho} - 1}{\frac{\rho_\infty}{\rho_w} - 1} \qquad (10)$$

$$Pr\frac{v_\infty}{v}(-\eta W_x(\eta) + 4W_y(\eta))\frac{d\theta}{d\eta} = \frac{1}{\lambda}\frac{d\lambda}{d\eta}\frac{d\theta}{d\eta} + \frac{d^2\theta}{d\eta^2} \qquad (11)$$

It is found that the dimensionless velocity similarity variables W_x and W_y as the core similarity variables dominate the transformed governing ordinary differential equations.

equations.

2.2. For Forced Convection Boundary Layer

2.2.1. Governing Partial Differential Equations

Our innovative similarity transformation for forced convection boundary layer was proposed in   [3]. The governing partial differential equations are as follows for consideration of variable thermo-physical properties:

$$\frac{\partial}{\partial x}(\rho w_x) + \frac{\partial}{\partial y}(\rho w_y) = 0 \qquad (12)$$

$$\rho(w_x\frac{\partial v_x}{\partial x} + w_y\frac{\partial v_x}{\partial y}) = \frac{\partial}{\partial y}(\mu\frac{\partial v_x}{\partial y}) \qquad (13)$$

$$\rho c_p(w_x\frac{\partial t}{\partial x} + w_y\frac{\partial t}{\partial y}) = \frac{\partial}{\partial y}(\lambda\frac{\partial t}{\partial y}) \qquad (14)$$

where Eqs.(12) to (14) are mass, momentum, and energy conservation equations respectively. Here, it is seen that the effect of variable thermo-physical properties, such as density, absolute viscosity, and thermal conductivity are considered. However, for most of common gases and liquids, the variation of specific heat with temperature can be omitted.

2.2.2. Similarity Physical Variables (or Parameters)

The detailed physical analysis and mathematical derivation of our innovative similarity analysis method are reported in [3] for similarly equivalent transformation of forced boundary layer convection. For saving space, here we only show the set of equations with the related dimensionless similarity physical variables (or parameters). However, the core work for the similarity analysis and transformation issues is to treat the similarity transformation of velocity field. To this end, we proposed the innovative core similarity variables and their formulae as below for the similarity transformation of the velocity field [3]:

$$W_x(\eta) = \frac{w_x}{w_{x,\infty}} \qquad (15)$$

$$W_y(\eta) = (\frac{1}{2}Re_{x,\infty})^{1/2}\frac{w_y}{w_{x,\infty}} \qquad (16)$$

where the local Reynolds number is defined as

$$Re_{x,\infty} = \frac{w_{x,\infty}x}{v_\infty} \qquad (17)$$

The similarity coordinate variable is defined as

$$\eta = \frac{y}{x}(\frac{1}{2}Re_{x,\infty})^{1/2} \qquad (18)$$

The similarity temperature is given as

$$\theta = \frac{t - t_\infty}{t_w - t_\infty} \qquad (19)$$

Here, Innovative equations (15, 16) for similarity transformation of velocity field are originally proposed and derived in our book [3]. It is emphasized that the similarity velocity variables $W_x(\eta)$ and $W_y(\eta)$ are core similarity variables for whole similarity transformation. Obviously, they will dominate the transformed governing ordinary differential equations.

2.2.3. Governing transformed Ordinary Differential Equations

The whole similarity transformation process for governing similarity ordinary differential equations is presented in our book [3]. Here, we only give the transformed governing ordinary differential equations as below:

$$-\eta\frac{dW_x(\eta)}{d\eta}+2\frac{dW_y(\eta)}{d\eta}+\frac{1}{\rho}\frac{d\rho}{d\eta}[-\eta\cdot W_x(\eta)+2W_y(\eta)]=0 \quad (20)$$

$$\frac{v_\infty}{v}(-\eta W_x(\eta)+2W_y(\eta))\frac{dW_x(\eta)}{d\eta}=\frac{d^2W_x(\eta)}{d\eta^2}+\frac{1}{\mu}\frac{d\mu}{d\eta}\frac{dW_x(\eta)}{d\eta} \quad (21)$$

$$\Pr\frac{v_\infty}{v}(-\eta W_x+2W_y)\frac{d\theta}{d\eta}=\frac{1}{\lambda}\frac{d\lambda}{d\eta}\frac{d\theta}{d\eta}+\frac{d^2\theta}{d\eta^2} \quad (22)$$

It is found that the core similarity variables $W_x(\eta)$ and $W_y(\eta)$ for similarity transformation of velocity field dominate the whole transformed governing ordinary differential equations.

3. Extensive Application of the Innovative Similarity Transformation

Since the innovative similarity transformation was developed for hydrodynamics and heat transfer investigation of free and forced convection, its application has been extended to our study on convection hydrodynamics and heat and mass transfer with film boiling/condensation convection, such as free film boiling convection [4,8], free/forced film condensation convection of pure vapour [3,4,9,10], free/forced film condensation convection of vapour-gas mixture [3, 4,11]. It is indicated that in all of these studies, the complicated physical factors including the coupled effect of variable physical properties were well considered, then, the heat and mass transfer results demonstrate their practical value. To save space, here we only introduce briefly the application of our innovative similarity transformation in extensive study on free film condensation convection of vapour-gas mixture as below:

3.1. Governing Partial Differential Equations for Free Film Condensation Convection of Vapour-Gas Mixture

Free film condensation convection of vapour-gas mixture is attributed to multiple phase-film convection. It consists of three parts: condensate liquid film flow, vapour-gas film flow, and the interface with multiple physical balance conditions. The detailed derivation of the related similarity equations of free convection condensation of vapour-gas mixture are

reported in our book [4]. For saving space, the interfacial boundary condition equations are omitted, and only the governing partial differential equations of condensate liquid film flow and vapour-gas mixture film flow are described as below with consideration of couples effect of variable thermo-physical properties:

For condensate liquid film flow

$$\frac{\partial}{\partial x}(\rho_l w_{xl})+\frac{\partial}{\partial y}(\rho_l w_{yl})=0 \quad (23)$$

$$\rho_l(w_{xl}\frac{\partial w_{xl}}{\partial x}+w_{yl}\frac{\partial w_{xl}}{\partial y})=\frac{\partial}{\partial y}(\mu_l\frac{\partial w_{xl}}{\partial y})+g(\rho_l-\rho_{m,\infty}) \quad (24)$$

$$\rho_l[w_x\frac{\partial(c_{p_l}t)}{\partial x}+w_y\frac{\partial(c_{p_l}t)}{\partial y}]=\frac{\partial}{\partial y}(\lambda_l\frac{\partial t}{\partial y}) \quad (25)$$

where Eqs.(23)-(25) are continuity, momentum and energy conservation equations respectively.

For vapor-gas mixture film

$$\frac{\partial}{\partial x}(\rho_m w_{xm})+\frac{\partial}{\partial y}(\rho_m w_{ym})=0 \quad (26)$$

$$\rho_m(w_{xm}\frac{\partial w_{xm}}{\partial x}+w_{ym}\frac{\partial w_{xm}}{\partial y})=\frac{\partial}{\partial y}(\mu_m\frac{\partial w_{xm}}{\partial y})+g(\rho_m-\rho_{m,\infty}) \quad (27)$$

$$\begin{aligned}\rho_m c_{p_m}(w_{xm}\frac{\partial t_m}{\partial x}+w_{yv}\frac{\partial t_m}{\partial y})\\=\frac{\partial}{\partial y}(\lambda_m\frac{\partial t_m}{\partial y})+\frac{\partial}{\partial y}[\rho_m D_v(c_{pv}-c_{pg})\frac{\partial C_{mv}}{\partial y}t_m]\end{aligned} \quad (28)$$

$$\frac{\partial(w_{xm}\rho_m C_{mv})}{\partial x}+\frac{\partial(w_{ym}\rho_m C_{mv})}{\partial y}=\frac{\partial}{\partial y}(D_v\rho_m\frac{\partial C_{mv}}{\partial y}) \quad (29)$$

where Eqs.(26)-(28) are continuity, momentum and energy conservation equations respectively, while Eq.(29) is species conservation equation.

3.2. Dimensionless Similarity Parameters and Variables

Based on our innovative similarity trsmdformation, a set of equations for similarity analysis and transformation are described as below:

For condensate liquid film flow

For convection issue, the core work for the similarity analysis and transformation is always the similarity of velocity field. Then, the core similarity variables and their formulae are as follows for the similarity analysis and transformation of the velocity field [4]:

$$W_{xl}(\eta_l)=(2\sqrt{gx}(\frac{\rho_{l,w}-\rho_{m,\infty}}{\rho_{l,s}})^{1/2})^{-1}w_{xl} \quad (30)$$

$$W_{yl}(\eta_l)=(2\sqrt{gx}(\frac{\rho_{l,w}-\rho_{m,\infty}}{\rho_{l,s}})^{1/2}(\frac{1}{4}Gr_{xl,s})^{-1/4})^{-1}w_{yl} \quad (31)$$

In addition, the *local Grashof number* $Gr_{xl,s}$ is set up as

$$Gr_{xl,s} = \frac{g(\rho_{l,w} - \rho_{m,\infty})x^3}{v_{l,s}^2 \rho_{l,s}} \quad (32)$$

The Dimensionless coordinate variables η_l is set up as follows:

$$\eta_l = (\frac{1}{4}Gr_{xl,s})^{1/4} \frac{y}{x} \quad (33)$$

Dimensionless temperature is assume as

$$\theta_l = \frac{t_l - t_{s,int}}{t_w - t_{s,int}} \quad (34)$$

For vapour-gas mixture film flow

In view of that the core work for the similarity analysis and transformation is always to treat the similarity problem of velocity field for convection issue, based on our innovative similarity transformation, we proposed the following innovative dimensionless similarity variables and their formulae for the similarity treatment of the velocity field of vapour-gas mixture film flow [4]:

$$W_{xm}(\eta_l) = (2\sqrt{gx}(\rho_{m,s}/\rho_{m,\infty} - 1)^{1/2})^{-1} w_{xm} \quad (35)$$

$$W_{ym}(\eta_l) = (2\sqrt{gx}(\rho_{m,s}/\rho_{m,\infty} - 1)^{1/2}(\frac{1}{4}Gr_{xm,\infty})^{-1/4})^{-1} w_{ym} \quad (36)$$

Additionally, the local Grashof number $Gr_{xm,\infty}$ of vapour-gas mixture film is set up as

$$Gr_{xm,\infty} = \frac{g(\rho_{m,s}/\rho_{m,\infty} - 1)x^3}{v_{m,\infty}^2} \quad (37)$$

The dimensionless coordinate variables η_m of vapour-gas mixture film flow is set up as

$$\eta_m = (\frac{1}{4}Gr_{xm,\infty})^{1/4} \frac{y}{x} \quad (38)$$

The dimensionless temperature of vapour-gas mixture flow film is defined as

$$\theta_m = \frac{t_m - t_\infty}{t_{s,int} - t_\infty} \quad (39)$$

The vapor relative mass fraction is defined as

$$\Gamma_{mv}(\eta_m) = \frac{C_{mv} - C_{mv,\infty}}{C_{mv,s} - C_{mv,\infty}} \quad (40)$$

3.3. Governing Ordinary Differential Equations

By means of Eqs.(30) to (34), the governing partial differential equations (23) to (25) are equivalently transformed to the following ordinary partial equations:

3.3.1. For Liquid Film Flow

$$2W_{xl}(\eta_l) - \eta_l \frac{dW_{xl}(\eta_l)}{d\eta_l} + 4\frac{dW_{yl}(\eta_l)}{d\eta_l}$$
$$- \frac{1}{\rho_l}\frac{d\rho_l}{d\eta_l}(\eta_l W_{xl}(\eta_l) - 4W_{yl}(\eta_l)) = 0 \quad (41)$$

$$\frac{v_{l,s}}{v_l}[W_{xl}(\eta_l)(2W_{xl}(\eta_l) - \eta_l \frac{dW_{xl}(\eta_l)}{d\eta_l}) + 4W_{yl}(\eta_l)\frac{dW_{xl}(\eta_l)}{d\eta_l}]$$
$$= \frac{d^2 W_{xl}(\eta_l)}{d\eta_l^2} + \frac{1}{\mu_l}\frac{d\mu_l}{d\eta_l}\frac{dW_{xl}(\eta_l)}{d\eta_l} + \frac{\mu_{l,s}}{\mu_l}\frac{(\rho_l - \rho_{m,\infty})}{\rho_{l,w} - \rho_{m,\infty}} \quad (42)$$

$$Pr_l \frac{v_{l,s}}{v_l}(-\eta_l W_{xl}(\eta_l) + 4W_{yl}(\eta_l))\frac{d\theta_l}{d\eta_l} = \frac{d^2\theta_l}{d\eta_l^2} + \frac{1}{\lambda_l}\frac{d\lambda_l}{d\eta_l}\frac{d\theta_l}{d\eta_l} \quad (43)$$

where Eqs.(41)-(43) are continuity, momentum and energy conservation equations respectively (see our book [4] for the detailed mathematical derivation). Here, it is seen that the core similarity variables, dimensionless similarity velocity variables dominate the governing ordinary differential equations of condensate liquid film.

3.3.2. For Vapour-Gas Mixture Film Flow

By means of Eqs.(35) to (40), the governing partial differential equations (26) to (29) are equivalently transformed to the following ordinary partial equations respectively:

$$2W_{xm}(\eta_m) - \eta_m \frac{dW_{xm}(\eta_m)}{d\eta_m} + 4\frac{dW_{ym}(\eta_m)}{d\eta_m}$$
$$- \frac{1}{\rho_m}\frac{d\rho_m}{d\eta_m}(\eta_m W_{xm}(\eta_m) - 4W_{ym}(\eta_m)) = 0 \quad (44)$$

$$\frac{v_{m,\infty}}{v_m}[W_{xm}(\eta_m)(2W_{xm}(\eta_m) - \eta_m \frac{dW_{xm}(\eta_m)}{d\eta_m}) + 4W_{ym}(\eta_m)\frac{dW_{xm}(\eta_m)}{d\eta_m}]$$
$$= \frac{d^{2(\eta)} W_{xm}(\eta_m)}{d\eta_m^2} + \frac{1}{\mu_m}\frac{d\mu_m}{d\eta_m}\frac{dW_{xm}(\eta_m)}{d\eta_m} + \frac{\mu_{m,\infty}}{\mu_m}\cdot\frac{\rho_m - \rho_{m,\infty}}{\rho_{m,s} - \rho_{m,\infty}} \quad (45)$$

$$\frac{v_{m,\infty}}{v_m}[-\eta_m W_{xm}(\eta_m) + 4W_{ym}(\eta_m)]\frac{d\theta_m}{d\eta_m}$$
$$= \frac{1}{Pr_m}(\frac{d^2\theta_m}{d\eta_m^2} + \frac{1}{\lambda_m}\frac{d\lambda_m}{d\eta_m}\cdot\frac{d\theta_m}{d\eta_m})$$
$$- \frac{1}{Sc_{m,\infty}}\frac{v_{m,\infty}}{v_m}(C_{mv,s} - C_{mv,\infty})\frac{c_{pv} - c_{pg}}{c_{p_m}} \quad (46)$$
$$[\frac{d\theta_m}{d\eta_m}\frac{d\Gamma_{mv}(\eta_m)}{d\eta_m} + (\theta_m + \frac{t_\infty}{t_{s,int} - t_\infty})\frac{d^2\Gamma_{mv}(\eta_m)}{d\eta_v^2}$$
$$+ (\theta_m + \frac{t_\infty}{t_{s,int} - t_\infty})\frac{1}{\rho_m}\frac{d\rho_m}{d\eta_m}\frac{d\Gamma_{mv}(\eta_m)}{d\eta_m}]$$

$$[-\eta_m W_{xm}(\eta_m) + 4W_{ym}(\eta_m)]\frac{d\Gamma_{m,v}(\eta_m)}{d\eta_m}$$
$$= \frac{1}{Sc_{m,\infty}}[\frac{d^2\Gamma_{m,v}(\eta_m)}{d\eta_v^2} + \frac{1}{\rho_m}\frac{d\rho_m}{d\eta_m}\frac{d\Gamma_{m,v}(\eta_m)}{d\eta_m}] \quad (47)$$

where Eqs.(44)-(46) are continuity, momentum, and energy conservation equations, while Eq.(47) is the species conservation equation with mass diffusion (see our book [4] for the detailed mathematical derivation). Here, it is seen that core similarity variables for similarity transformation of velocity field dominate the governing ordinary differential equations of vapour-gas mixture film flow.

4. Conclusions

For solving the governing partial differential equations, the similarity transformation of the equations is usually done. For such transformation, some dimensionless similarity variables (or parameters) are formed, and each similarity variable (or parameter) as the group of the dimensional physical variables to reduce the number of the independent variables of the physical phenomena. These similarity variables (or parameters) will simplify the governing partial differential equations for getting the solutions more easily.

Our innovative similarity transformation, was developed for in-depth research on convection hydrodynamics and heat and mass transfer, covering convection heat transfer, as well as heat and mass transfer of film boiling/condensation convection. Obviously, it covers very wide area of convection heat and mass transfer

Currently popular Fakner-Skan transformation on convection heat and mass transfer has some inconvenience for similarity transformation of the core variable on velocity field. By using this transformation, it is necessary to first induce flow function and group theory to derive an intermediate function for an indirect similarity transformation of the velocity field. It is a complicated process and allows difficult consideration of variable physical properties, and then confines the practical value of research results on convection heat and mass transfer.

Compared the Falkner-Skan transformation, a big advantage of our innovative similarity transformation lies in resolving the above core work. With our similarity transformation, the velocity variables can be directly transformed to the corresponding dimensionless similarity velocity variables. Then, all above inconvenient actions caused by using Falkner-Skan transformation can be avoided. In addition, our innovative similarity transformation can transform the variable physical properties into the physical property factors, organically combined with the transformed governing ordinary differential equations, and then simplify greatly the treatment of variable physical properties. It demonstrates that our innovative similarity transformation is a better alternative over the traditional Falkner-Skan transformation for in-depth research on convection heat and mass transfer. It was found that this innovative similarity transformation has wide application in engineering.

Nomenclature

C_{mg} — gas mass fraction in vapour-gas mixture
C_{mv} — vapor mass fraction in vapour-gas mixture
$C_{mv,\infty}$ — bulk vapor mass fraction

c_p — specific heat at constant pressure, J/(kg k)
D_v — vapor mass diffusion coefficient in gas, m²/s
g — gravity acceleration, m/s²
$Gr_{x,\infty}$ — local Grashof number
Pr — randtl number
$Re_{x,\infty}$ — local Reynolds number
$Sc_{m,\infty}$ — defined local Schmidt number, defined as $Sc_{m,\infty} = \frac{v_{m,\infty}}{D_v}$
t — temperature, °C
T — absolute temperature, K
$w_x, w_y,$ — velocity components in x and y direction respectively, m/s
$W_x(\eta)$, $W_y(\eta)$ — dimensional velocity component in x, y, z direction respectively, respectively
$w_{x,\infty}$ — velocity component beyong the boundary layer

Greek symbols
ρ — mass density, kg/m³
λ — thermal conductivity, W/(m K)
μ — absolute viscosity, kg/(m s)
v — Kinetic Viscosity, m²/s
η — dimensionless coordinate variable for boundary layer and film flow
θ — dimensionless temperature
Γ_{mv} — vapor relative mass fraction
$(\frac{d\theta}{d\eta})_{\eta=0}$ — dimensionless temperature gradient on the plate
$\frac{1}{\rho}\frac{d\rho}{dx}$ — density factor
$\frac{1}{\lambda}\frac{d\lambda}{d\eta}$ — thermal conductivity factor
$\frac{1}{\mu}\frac{d\mu}{d\eta}$ — viscosity factor

Subscripts
f — film
g — gas
l — liquid
m — vapour-gas mixture
v — vapour
w — at wall
x,y — coordinate
∞ — far from the wall surface

References

[1] L. Prandtl, ÜbDie Flussigkeitsbewegung bei Sehr Kleiner Reibung, Proc. 3d Intern. Math. Koug. Heidelberg, 1904.

[2] V.M. Falkner, S.W. Skan, Some approximate solutions of the boundary layer equations, Phil. 264 Mag. 12, pp.865, 1931.

[3] D.Y. Shang, "Theory of heat transfer with forced convection film flows", Series (book) of Heat and Mass Transfer, Springer-Verlag, Berlin, Heidelberg, New York, 1st ed., 346 p. 2011.

[4] D.Y. Shang, " Free Convection Film Flows and Heat Transfer - Models of Laminar Free Convection with Phase Change for Heat and Mass Transfer", Series (book) of Heat and Mass Transfer, Springer-Verlag, Berlin, Heidelberg, NewYork, 2nd ed., 535 p, 2013.

[5] D. Y. Shang and B. X. Wang, Effect of variable thermophysical properties on laminar free convection of gas, Int. J. Heat Mass Transfer, Vol.33, No.7, pp.1387-1395, 1990.

[6] D. Y. Shang and B. X. Wang, Effect of variable thermophysical properties on laminar free convection of polyatomic gas, Int. J. Heat Mass Transfer, Vol. 34, No.3, pp. 749-755, 1991.

[7] D.Y. Shang, B.X. Wang, Y. Wang and Y. Quan, Study on liquid laminar free convection with consideration of variable thermophysical properties, Int. J. Heat Mass transfer, Vol.36, No.14, pp.3411-3419, 1993.

[8] D. Y. Shang, B. X. Wang and L. C. Zhong, A study on laminar film boiling of liquid along an isothermal vertical plate in a pool with consideration of variable thermophysical properties, Int. J. Heat Mass Transfer, Vol.37, No.5, PP. 819-828, 1994.

[9] D.Y. Shang and T. Adamek, Study on laminar film condensation of saturated steam on a vertical flat plate for consideration of various physical factors including variable thermophysical properties, Warme-und Stoffübertragung 30,pp. 89-100, 1994.

[10] D.Y. Shang and B.X. Wang, An extended study on steady-state laminar film condensation of a superheated vapor on an isothermal vertical plate, Int. J. Heat Mass Transfer, Vol.40, No.4, pp.931-941, 1997.

[11] D.Y. Shang anf L.C. Zhong, Extensive study on laminar free film condensation from vapor-gas mixture, Int. J. Heat and mass Transfer, Vol.51, pp. 4300-4314, 2008.

[12] H. Andersson and D.Y. Shang, An extended study of hydrodynamics of gravity-driven film flow of power-law fluids Fluid Dynamics Research.pp.345- 357,1998.

[13] D.Y. Shang and H. I. Andersson , Heat transfer in gravity-driven film flow of power-law fluids, Int. J. Heat Mass Transfer 42, No.11, pp. 2085-2099, 1999.7.

[14] D.Y. Shang and J. Gu, Analyses of pseudo-similarity and boundary layer thickness for non-Newtonian falling film flow, Heat Mass Transfer, Vol. 41,No.1, pp. 44–50, 2004.

[15] G. Taton, T. Rok1,2, E. Rokita， Temperature distribution assessment during radiofrequency ablation ， IFMBE Proceedings 22, pp. 2672–2676, 2008

[16] G. Taton, T. Rok, E. Rokita, Estimation of temperature distribution with the use of a thermo- camera, Pol J Med Phys Eng.,14(1), pp. 47-61, 2008.

[17] AA Turkin, M Dutka, D Vainchtein, S Gersen , etc., Deposition of SiO2 nanoparticles in heat exchanger during combustion of biogas, Applied Energy, Vol. 113, pp. 1141–1148, 2014.

[18] Y.Gao, Q.K.Liu, W.K.Chow, M.Wu, Analytical and experimental study on multiple fire sources in a kitchen, Fire Safety Journal 63, pp. 101–112, 2014.

22

Numerical analysis of the thermal behaviour of a photovoltaic panel coupled with phase change material

G. Santiago-Rosas[1, 2], M. Flores-Domínguez[2], J. P. Nadeau[2]

[1]Center for Research and Advanced Studies, National Polytechnic Institute, Mexico City, Mexico
[2]Trefle Laboratory, Arts and Trade Paris-Tech (ENSAM), Talance, France

Email address:

mfdguez@gmail.com (M. Flores-Domínguez)

Abstract: Elevated operating temperatures reduce the solar to electrical conversion efficiency of photovoltaic devices. In this paper, a numerical study is conducted to understand the thermal behavior of a photovoltaic panel with phase change materials. The analysis of the PV/PCM system was realized in three phases: temperature rise, start of melting of PCM and molten PCM so, three mathematical models for the resolution of thermal model of PV/PCM system during the solar insolation were used. First, an analytical model was used to study the behavior of PCM in state solid, after, an explicit and an implicit model were used to study globally the transfers. Different simplifications are introduced in the heat balances to realize a simple numerical approach. The Matlab software was used to solve different proposed models in this analysis. The results obtained are considered as a first approach to the study of a PV/PCM system.

Keywords: Photovoltaic Panel, Phase Change Material, Numerical Analysis

1. Introduction

The production of clean energy has allowed the development of the photovoltaic industry. The photovoltaic devices operate at elevated temperature and it reduces their efficiency. Similarly, the solar panels are simultaneously a heat source that penalizes the production of electric power due to the fact that, during operation of the photovoltaic cell, only about 15 % of the solar radiation that incident on them is transformed into electricity and the remainder is transformed into heat. So that, the performance of the photovoltaic modules is closely related to its temperature [1].

The most important research works in solar cells focus on its performance, as production decreases as its temperature increases. The loss coefficient is estimated at approximately 0.45 % per degree Celsius of increase, with reference to 25°C [2].

In order to use solar cells at low operating temperature, researchers have focused on cooling and recovery of the dissipated heat. However, despite all efforts realized for optimizing the performance of solar panels, the actual temperature reaching is still not enough, so that the relentless pursuit of solutions for cooling continues nowadays.

New materials and systems have been studied in cooling of panels, such is the case of the phase change materials (PCMs).

The PCMs absorb a large amount of energy as latent heat at a constant phase transition temperature and are thus used for passive heat storage. The thermal energy transfer occurs when the material changes from solid to liquid, or liquid to solid phase. The PCMs are of great interest for their use in various applications as the thermal energy storage and the thermal control systems. The energy storage capacity of the PCMs is dependent on its properties, methods of heat transfer and the system configuration [3].

The PCM cooling is a natural method used since antiquity in the cooling of buildings with thermal energy storage [4]. When there is a temperature gradient within a system exists a transfer of energy that allows release or transfer of energy in the zone of state change. The advantage in this fact lies in that is possible to store large amounts of thermal energy in relatively small masses and with small variations of temperature.

To incorporate the PCM into the photovoltaic (PV) panel to form a PV/PCM system can allow that the efficiency of the conversion of solar power of the panel to be closer at characterization conditions of the standard cell. In the study conducted by Huang [5] was validated numerically that use of PCM in photovoltaic panels moderates the increase of temperature of thereof. The model gives a detailed explanation of the thermal performance of the phase transition from the solid to the liquid state of the material.

In this work, a numeric analysis to understand the thermal behavior of a photovoltaic panel with material change of phase is proposed. The thermal analysis of PV/PCM system to know the behavior of the temperature of the panel was realized. The system analysis was realized in three phases: 1) temperature rise, 2) start of melting of PCM and, 3) molten PCM; so that, three mathematical models for the resolution of thermal model of PV/PCM system were used. First, an analytical model was used to study the behavior of PCM solid state, after, an explicit and an implicit model were used to study the transfers jointly. The Matlab software was used to solve different proposed models in this analysis.

2. Modeling of System

A schematic of the system configuration discussed in this paper is shown in Fig. 1. The system is comprised of a photovoltaic panel and of a material change of phase. Air circulates freely over the PV panel, the PCM is in direct contact with the PV panel and, the back wall of the container of PCM is considered adiabatic. The solar insolation received by the PV panel during the day is a sinusoidal function, with a maximum value of 750 Wm^{-2} [6].

The physical phenomena taken into account for this analysis were:
- Sun/PV interaction: Solar radiation.
- Air/PV interaction: Natural convection.
- PV: Source of heat due to the operation of the photovoltaic cell (Conduction).
- Liquid PCM: Conduction (natural convection negligible).
- PV/Solid PCM Interface: Start of melting.
- Liquid PCM/Solid PCM Interface: phase change.
- PCM/Air Interaction: Heat transfer does not exist, adiabatic wall.

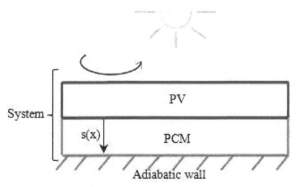

- External environments: sun and air
- s(x) is Melted thickness of the PCM

Figure 1. *Schematic representation of PV/PCM system*

The system analysis was carried out in three phases:
Phase 1: Rise of temperature.
At the beginning, the PV panel and the solid PCM are considered that are at ambient temperature. The PV panel temperature increases over time due to the absorption of the

solar energy it receives directly on its front surface. This energy is converted into heat and is transferred by conduction to the solid PCM whose temperature also increases. In this phase, the system was analyzed through an analytical model, where the hypothesis of a uniform temperature in the solid PCM was made. Furthermore, the hypothesis of $T_{PCMs}=T_{PV}$ was raised.

Phase 2: Start of melting
Once the solid PCM reached the melting temperature, its temperature depends on the molten fraction of the same one. This temperature remains almost constant during the period required for that phase-change process is performed. Thus, to analyze the behavior of the temperatures in this phase, the hypothesis of a constant temperature in the PV/PCM interface was posed, and thus, an explicit method was used. For a more accurate approximation, an implicit method that allows to follow the actual behavior of the temperature of the material was used. To simplify calculations, the hypothesis of uniformity of the temperature of the liquid PCM was made.

Phase 3: Melted PCM
Finally, when there is no more PCM in solid state, its temperature begins to increase, and the temperature of the PV depends on the conductive transfer with the PCM in liquid state.

For the resolution of thermal model of PV/PCM system different hypothesis were raised. The models are valid if the following hypothesis are true.

1. The PCM is in direct contact with the PV.
2. The temperature at the liquid/solid interface is constant in the explicit model and is uniform in the implicit model.
3. The latent heat is constant.
4. The container is adiabatic.
5. The ambient temperature is $21 \pm 1°C$.
6. There is no convection in the liquid PCM (the liquid PCM is viscous $\rightarrow$ conductive transfer).

3. Analysis of the System

3.1. Analytical Model

In this section, the development of the temperature of the PCM solid state is analyzed. Considering that the temperature is almost uniform in the material, the only variables were the time and the temperature of the PCM and PV. For the analysis of the system, the resistance to heat transfer of the specimen was characterized, that is, the thermal resistance of the surface and the internal conduction resistance of the specimen was compared [7]. The dimensionless Biot number (Bi), which compares the convective flow (boundary condition) with the conductive flux that matter can emit or absorb, is used. If Bi<0.1, then the hypothesis of temperature uniformity in the specimen is verified.

For the proposed system, the value of Bi=03 is used, so that it was necessary to take into account the temperature gradient existing in the first hours of solar insolation (when the PCM is even solid).

The analysis to estimate, roughly, the value of the temperature gradient was carried out in this section. An analytical model that used the method of the Laplace transform was used and, to represent the temperature change with respect to time was possible.

Through the Stehfest method combined with the thermal quadrupole method, the simple heat equation is a linear partial differential equation of parabolic type, where the particular solutions can be overlapped to obtain the general solution of the system [8].

If the case of unidirectional heat transfer through a wall of thickness e is considered, a thermal quadrupole method for the analysis of this particular case can be used, as shown in Fig. 2.

In general, it is observed that there is a linear relationship

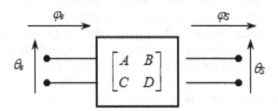

Figure 2. *Quadrupole of a passive wall*

between the input values and the output values, given by:

$$\begin{bmatrix} \theta_e \\ \varphi_e \end{bmatrix} = \begin{bmatrix} A & B \\ C & D \end{bmatrix}\begin{bmatrix} \theta_e \\ \varphi_e \end{bmatrix} \tag{1}$$

where θ_e and θ_s are the Laplace transform of temperature T at $x=0$ and $x=e$ respectively; φ_e and φ_s are the Laplace transform of flow Q en $x=0$ and $x=e$ respectively; and the matrix is the reverse transfer matrix of the quadrupole associated to wall. For the particular case of study, is given by the following:

Balance of solid PCM

The configuration of the system is shown in Fig. 3. The flow $\phi(t)$ directly affects the PV, which is designated with the number 1. The panel/material interface is denoted by the number 2 and, the adiabatic wall is the number 3.

From the linear relation shown in (1), the terms of the matrix are defined as:

$$A = D = \cosh(\alpha\, e) \quad , \quad B = \frac{1}{\lambda A}\frac{\sinh(\alpha\, e)}{a}$$
$$C = \lambda A \sinh(\alpha\, e)$$

with

$$\alpha = \sqrt{\frac{p}{a}} \quad \text{and} \quad p = \frac{j\ln(2)}{t}$$

where λ is heat conductivity, p is Laplace variable, and a is thermal diffusivity.

Thus, (2a) and (2b) are, respectively, for the wall 1 and 2, these are:

$$\begin{bmatrix} \theta_1 \\ \varphi_1 \end{bmatrix} = \begin{bmatrix} A_{PV} & B_{PV} \\ C_{PV} & D_{PV} \end{bmatrix}\begin{bmatrix} \theta_2 \\ \varphi_2 \end{bmatrix} \tag{2a}$$

$$\begin{bmatrix} \theta_2 \\ \varphi_2 \end{bmatrix} = \begin{bmatrix} A_{PCM} & B_{PCM} \\ C_{PCM} & D_{PCM} \end{bmatrix}\begin{bmatrix} \theta_3 \\ \varphi_3 \end{bmatrix} \tag{2b}$$

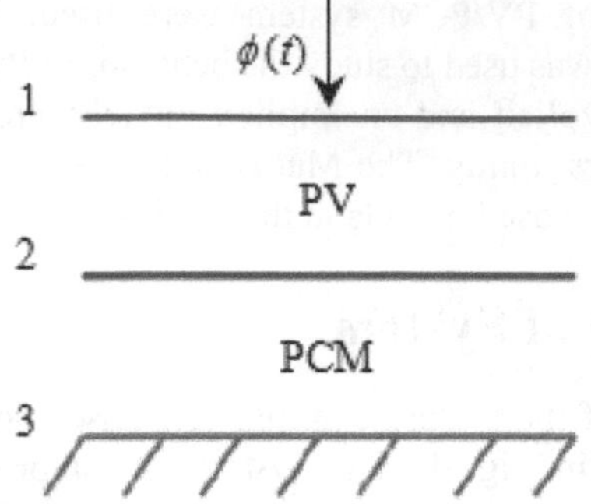

$\phi(t)$

1 ─────────────────────

PV

2 ─────────────────────

PCM

3 ///////////////

Figure 3. *The PV/ PCM system interfaces*

where $\varphi_3 = 0$ since the system is adiabatic on the side of the PCM, the equation can be rewritten as follows:

$$\begin{bmatrix} \theta_1 \\ \varphi_1 \end{bmatrix} = \begin{bmatrix} A_{PV} & B_{PV} \\ C_{PV} & D_{PV} \end{bmatrix}\begin{bmatrix} A_{PCM} & B_{PCM} \\ C_{PCM} & D_{PCM} \end{bmatrix}\begin{bmatrix} \theta_3 \\ 0 \end{bmatrix} \tag{3}$$

Solving for θ_3, yield:

$$\theta_3 = \frac{\varphi_1}{\left(C_{PV}A_{PCM} + A_{PV}C_{PCM}\right)} \tag{4}$$

and because the flow that is received by the panel is of sinusoidal form, θ_3 is:

$$\theta_3 = \frac{\varphi_1}{\left(C_{PV}A_{PCM} + A_{PV}C_{PCM}\right)}\frac{\omega}{p^2 + \omega^2} \tag{5}$$

Balance of PV

To calculate the temperature of the PV/PCM interface, the flow in the wall 2 was used, which is represented by (2b), that is:

$$\theta_2 = A_{PCM}\theta_3 \tag{6}$$

3.2. Explicit Model

In this section, the behavior of the temperature of the PV panel and of the PCM with respect to the time was analyzed, so the explicit model was divided into three stages:

3.2.1. When $T_{PCM} = T_{PV}$

Initially, the temperature of the PV panel (T_{PV}) is equal to the ambient temperature (T_{amb}) and, the temperature of the solid PCM (T_{PCM}) is equal to the temperature of the panel.

Energy balance in the PV

(It was considered that there is no variation in the temperature of the solid PCM). When the temperature of the PV panel is less than the melting temperature ($T_{PV} < T_M$), the difference between the source of energy of the system and the loss in the air/PV interface is the variation of the thermal mass, this is:

$$(1-\eta) AI - Ah(T_{PV} - T_{amb}) =$$
$$(M_{PV} Cp_{PV} + M_{PCMs} Cp_{PCMs}) \frac{dT_{PV}}{dt} \qquad (7)$$

where A is the surface area, I is the incident energy absorbed by the PV/PCM system, h is the coefficient of natural convection, M is the mass, and Cp is the specific heat capacity.

Balance of heat energy in the PCM

$$T_{PCM} = T_{PV}$$

$$s(x) = 0$$

3.2.2. When $T_{PV} = T_M$

In this stage, the PV panel reaches the melting temperature of the PCM. The PCM begins to melt.

Energy balance in the PV panel

The difference between the source of energy, the loss in the air/PV interface and the transfer at the PV/PCM interface is the variation of the thermal mass, that is:

$$(1-\eta) AI - Ah(T_{PV} - T_{amb}) - A\frac{\lambda_L}{s(x)}(T_{PV} - T_M) =$$
$$(M_{PV} Cp_{PV}) \frac{dT_{PV}}{dt} \qquad (8)$$

Balance of heat energy in the PCM

(It was considered that there is no sensible heat in the liquid PCM). When $T_{PCM} = T_M$, the difference between the source of energy and the loss in the loss in the air/PV interface is transfer at the PV/ liquid PCM interface, this is:

$$(1-\eta) AI - Ah(T_{PV} - T_{amb}) = \frac{s(x)}{dt} A\rho_L \Delta h_M \qquad (9)$$

where ρ is the reference density.

3.2.3. When $T_{PCM} > T_F$

Finally, when the volume of the solid PCM has melted its temperature begins to rise and, as result, the temperature of the panel also increases due to conductive transfer that exist between the liquid PCM and the PV panel, is given as:

$$T_{PCM_L} = (T_{PV} + T_M)/2 \qquad (10)$$

3.3. Implicit Model

To increase the accuracy in the analysis, the values of next or new step of time for all nodes were calculated, simultaneously.

The variation of the enthalpy H allows to express the amount of heat involved during the transformation to constant pressure of a thermodynamic system [9]. Furthermore, the enthalpy depends on the temperature and this is a function of time.

$$\frac{\partial H_{PCM}}{\partial t} = \frac{\partial H_{PCM}}{\partial T_{PCM}} \cdot \frac{\partial T_{PCM}}{\partial t} \qquad (11)$$

Balance of heat energy in the PCM

The variation of the thermal mass is equal to the transfer at the PV/PCM interface, that is:

$$V_{PCM} \frac{\partial H_{PCM}}{\partial t} = Ah(T_{PV} - T_{PCM}) \qquad (12)$$

where V is the volume. In this case, the enthalpy is a function of the fraction of molten PCM, is given as:

$$H_{PCM} = (1-f) \cdot \rho_S \cdot Cp_S \cdot T_M +$$
$$f \cdot \rho_L \cdot Cp_L \cdot T_M + f \cdot \rho_L \cdot \Delta h_M \qquad (13)$$

when f is molten fraction of material and may be expressed as:

$$f = \frac{s(x)}{V_{PCM}} \qquad (14)$$

The solid-to-liquid phase transformation takes place at the melting temperature, if the temperature of the PCM is greater that its melting temperature, the PCM is completely liquid, such that f is given as:

$$f = \begin{cases} f = 0 & T_{PCM} < T_M \\ f = 1 & T_{PCM} < T_M \end{cases} \qquad (15)$$

Due to the fact that it is not possible to express the enthalpy H with respect to the temperature T as a linear function during the phase of latent heat, the equivalent heat capacity C_{eff} in which the melting temperature T_M is in the interface between the solid and liquid fraction was used [10]. The relation between enthalpy and temperature is shown in Fig. 4. when

$$C_{eff} = \begin{cases} C_{eff} = \rho_S Cp_S \quad ; & T_{PCM} < T_{min} \\ C_{eff} = \rho_L Cp_L \quad ; & T_{PCM} > T_{max} \\ C_{eff} = \frac{\Delta h_M}{T_{max} - T_{min}} ; & T_{PCM} < T_M \text{ and } T_{PCM} > T_M \end{cases} \qquad (16)$$

In summary, the energy balance of PCM is as follows:

$$V_{PCM} C_{eff} \frac{\partial T_{PCM}}{\partial t} = A\frac{\lambda_L}{s(x)}(T_{PV} - T_{PCM}) \qquad (17)$$

Energy balance in the PV

The variation of the thermal mass is equal to the sum of the loss in the air/PV interface, the transfer at the PV/PCM interface and the source of energy, that is:

$$V_{PV}\rho_{PV}Cp_{PV}\frac{\partial T_{PV}}{\partial t} = Ah\left(T_{amb} - T_{PV}\right) +$$

$$A\frac{\lambda_L}{s(x)}\left(T_{PCM} - T_{PV}\right) + (1-\eta)AI \qquad (18)$$

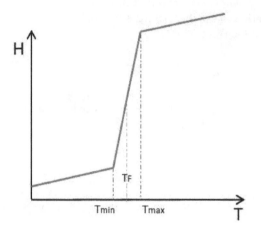

Figure 4. *Enthalpy–temperature curve. The relation between enthalpy and temperature by phase change.*

4. Results

The thermal evaluation of the PV/PCM system during the solar insolation was conducted. The temperature of the panel and the material was calculated using the propose models. The numerical simulation of the models was carried out in the Matlab software.

In the analyses of the energy balance, it can be noted that in the explicit and implicit models the variable s(x) is used, which is the thickness of the phase change material that is melted with regard to the time. In the models, the commercial paraffin Rubitherm RT42, with melting temperature of 42°C, was used as PCM to filling the back of solar panel.

4.1. Analytical Model

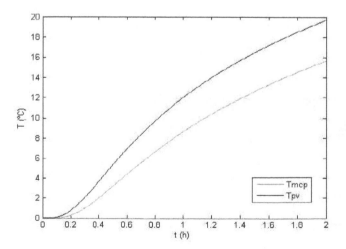

Figure 5. *T-t curve of solid PCM and PV panel by analytical model.*

The temperature of the PV/PCM interface was obtained of (6). The T-t curve is presented in Fig 5, in this is shown the

difference of the temperature between the PCM in solid state and the PV, with regard to the time, during first hours of solar insolation. In this graph is observed that, when the temperature of PCM is of 14°C, the temperature of PV is of 18°C, that is, there is a temperature gradient of 4°C only for two hours of solar insolation, due to the fact that to the beginning of the day, the values of flow are small.

Based on the foregoing, if the hypothesis that the PCM and the PV are at the same temperature is done, the error is of 4°C. The following models were considered this hypothesis.

4.2. Explicit Model

In the Fig. 6, the behavior of the temperature of the PCM as well as the temperature of the PV, which reached 111°C after 12 hours of solar insolation, is observed.

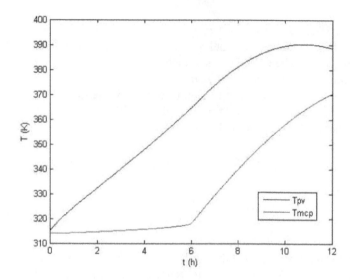

Figure 6. *T-t curve of PCM and PV by the explicit model.*

4.3. Implicit Model

To solve this problem, the implicit Euler method was used. At each time-step, an iterative procedure was performed and the convergence was assured by a criterion on the enthalpy of the PCM.

For analyze the behavior of the temperature of the PV, the simulation was carried out with two different PCMs. In principle, for the numerical simulation, the model was tested using a PCM with a melting temperature of 42 °C. Furthermore, the commercial paraffin with a melting temperature of 27°C was used, because it is already known that for best performance, the maximum temperature of the photovoltaic panel must be 25 °C, then this is a condition for the use of the material RT27 as a different PCM in the analysis of the behavior of panel.

In the Fig. 7 and Fig 8, the behavior of the temperature of the PCM (RT42 and RT27 respectively) and the temperature of the PV is shown. In both figures, the temperature behavior of the material can be observed. In the Fig. 7 is observed that the temperature of the solar panel reached 107°C after 12 hours of exposure; this is 5°C more than the with material RT27 (see Fig 8).

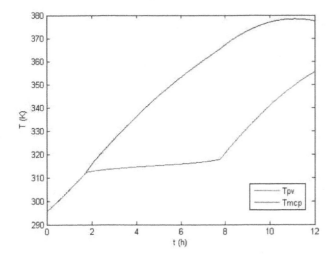

Figure 7. T-t curve of PCM RT42 °C and PV by the implicit model.

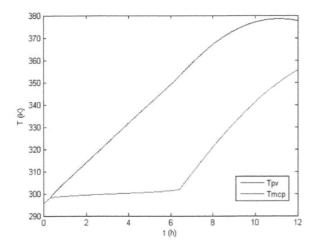

Figure 8. T-t curve of PCM RT27 °C and PV by the implicit model.

The results of the explicit and implicit mathematical models can be compared since they express the same phenomenon and use the same PCM. The Fig. 6, 7 and 8 show that the initial temperature of the panel is equal to the ambient temperature and with the time, it increases to the temperature of melting of the PCM. Depending on the quantity of material, the temperature remains almost constant during a time due to its ability to store energy, that is, to the latent heat. Later, when the PCM is completely liquid, its temperature continues to increase and in the same way the temperature of the PV, this is due to the phenomenon of conduction that exists between the liquid fraction of the material and the panel.

5. Conclusions

A thermal analysis of the cooling of a PV panel with PCM was carried out. In this study, three different models: analytical, explicit and implicit, were used to understand the behavior of the temperature of the system during the solar insolation. The following conclusions can be drawn:

- The temperature behavior, of both the PCM and the PV panel in the explicit and implicit models, follows the same trend. The temperature of the PV continued

increasing even during the melting of the PCM.
- The conduction that exists between the material and the panel prevents the real cooling, that is, it exists a low efficiency of the panel.
- The size of the surface of exchange should be greater to increase the heat transfer and to avoid the increase of temperature of the panel.
- The complete melting of the material must be avoided. The PCM in solid state must always be in contact with the panel to absorb the heat generated and to keep the panel to a low temperature.
- The results obtained from the numerical analysis presented in this paper, should be considered as a first approach to the study of a PV / PCM system due to numerous simplifications introduced in the heat balances, as well as in the numerical method defined.

Acknowledgements

G. Santiago-Rosas wishes to express their gratitude to the Mexican National Council of Science and Technology (CONACYT) for its financial support throughout of Master of Science Program.

M. Flores-Domínguez would like to thank for the financial support received from Council of Science and Technology of Mexico State (COMECYT), Mexico during the realization of this work.

Nomenclature

C_{eff}	Equivalent thermal capacity
Cp	Specific heat capacity
f	Molten fraction
h	Coefficient of natural convection
H	Enthalpy
I	Solar insolation
M	Mass
$s(x)$	Thickness of material
S	Surface
t	Time
T	Temperature
V	Volume
Δh	Latent heat
Greek letters	
ρ	Density
λ	Heat conductivity
η	Performance of PV
Subscripts	
L	Liquid phase
S	Solid phase
M	Molten phase

References

[1] H. G. Teo, H., P. S. Lee, M. N. A Hawlader, "An active cooling system for photovoltaic modules," Applied Energy, vol. 90, no 1, pp. 309-315, 2012.

[2]   W. He, Y. Zhang, J. Li, "Comparative experiment study on photovoltaic and thermal solar system under natural circulation of water," Applied Thermal Engineering, 31(16), pp. 3369-3376, Nov. 2011.

[3]   M. J. Huang, P. C. Earnes, B. Norton, "Natural convection in an internally finned phase change material heat sink for the thermal management of photovoltaics," Solar Energy and Solar Cells, vol. 95, no 7, pp. 1598-1603, 2011.

[4]   B. Vidal Jiménez, "Modelización del cambio de fase sólido-líquido. Aplicación a sistemas de acumulación de energía térmica," s.l.: Centro Tecnológico de Transferencia de Calor, Depto.de Máquinas y motores térmicos, Universidad Politécnica de Cataluña, 2007.

[5]   M. J. Huang, P. C. Earnes, B. Norton, "Thermal regulations of building-integrated photovoltaics using phase change materials," International Journal of Heat and Mass Transfer, vol. 47, no 12, pp. 2415-2733, 2004.

[6]   J. P. Braun, B. Faraggi, A. Labouoret, "Les cellules solaires", ETSF (Editions Techniques et Scientifiques Françaises) – Paris, 1996.

[7]   J. P. Nadeau, "Energetique Industrielle". Talence: Art et Métiers-ParisTech, 2009.

[8]   Y. Jannot, "Transformation de Laplace inverse", En: Transferts thermiques, p. 113, 2002.

[9]   H. Mehling, L. F. Cabeza, Heat and cold storage with PCM Handbook, Publisher Springer, Germany, 2008.

[10] F. Rouault, "Analyse expérimentale et modélisation d'un échangeur-stockeur contenant des matériaux à changement de phase," Congrès SFT (Société Française de Thermique), Talence, France, 2012.

23

Key risk index identification of UHV project construction based on improved rough set model

Sen Guo[*], Huiru Zhao

School of Economics and Management, North China Electric Power University, Changping District, Beijing 102206, China

Email address:

guosen324@163.com (Sen Guo), guosen@ncepu.edu.cn (Sen Guo)

Abstract: Compared with the conventional power grid project, the UHV project construction faces more challenges and risks. Identifying the key risk indicators of UHV project construction can improve the level of project risk management and reduce risk-related loss. Taken the data missing of risk indicators into consideration, an improved rough set model for key risk indicators identification of UHV project construction is employed with the introduction of information content. Firstly, the information content of conditional attributes set and information significance of each conditional attribute are calculated; Secondly, the reduced core attribute matrix is formed; Then, the discernibility matrix is built; Finally, the final core attribute set is determined. After building the risk index system, the key risk indicators identification of a certain UHV project construction is performed. The calculation result shows that "land requisition and logging policy risk", "project security management risk" and "land requisition, removing and crop compensation risk" are the key risk indicators.

Keywords: UHV Project Construction, Key Risk Indicators, Index Identification, Improved Rough Set Model

1. Introduction

Currently, the long-distance and high-capacity transmission capacity of 500kV power grid in China is limited. The electricity energy produced by the Northwest coal base and Southwestern hydropower base cannot be delivered to the load-intensive eastern regions of China in a cost-effective manner. In this case, the eastern region of China has to construct coal base to meet the electricity demand, which exacerbates the tensions of coal supply and transportation. Therefore, build the UHV transmission systems with the characteristics of large-capacity, long-distance and low-loss transmission capacity will be conducive to the intensive development of China's large coal, large hydro and nuclear power base, which can also optimize the layout of energy production and consumption and promote the coordinated development of regional economy [1]. The UHV construction holds the characteristics of tight time, heavy task, difficult construction and complicated environment, which face many issues, such as difficult stakeholder coordination, tight local government relations, and so on [2]. Compared with the conventional construction projects, the UHV construction project faces more challenges and risks. Therefore, to identify the key risk indicators of

UHV project construction can provide some references for the project risk management, improve risk management level, and reduce the risk loss during the UHV construction process.

Currently, there are some researchers that employ the Multiple Attribute Decision Making (MADM) to evaluate the risk of some engineering construction projects. Ref. [3] used the AHP and Monte Carlo simulation method to evaluate the water conservancy construction risk; Ref. [4] applied expert investigation and AHP method to evaluate the subway construction risk; Ref. [5] identified the risk of oil pipeline construction by using catastrophe progression evaluation method. With respect to the power grid construction, most research papers mainly focus on power grid project management. Ref. [6] built the smart grid project management model using System Dynamics; Ref. [7] studied the application of project file management network system on the power grid construction projects; Ref. [8-9] introduced the typical design overview of American transmission line, the typical design overview of France and South Korea transformer substations, and the inspiration for the design and construction of China's power grid project. With respect to the power grid construction risk, the current research literatures are relatively few. Ref. [2] employed the AHP and FCE method to evaluate the risk of UHV power transmission construction project; Ref. [10] studied the power transmission and distribution project construction risk by employing the Fault Tree and Set Pair Analysis theory; Ref. [11] analyzed the risk factors in various stages of power transmission and distribution projects from the perspective of power grid enterprise, and then proposed a checklist of risk factors.

With respect to UHV project, to the best of our knowledge, the current research mainly focuses on the technical aspect other than management aspect, such as the power calculation method of UHV power transmission lines [12], the wind load transferring mechanism in the UHV transmission tower-line system [13], and so on. Therefore, it is very necessary to explore the research related to the risk management of UHV construction project.

In this paper, the key risk indicators of UHV project construction were identified based on Improved Rough Set Model (IRS). Taken a 1000kV UHV construction project as the example, the risk index system was firstly built, and then the IRS method was employed to identify the key index. The risk identification result can provide some references for the UHV construction management, which can also improve the risk management level and reduce the risk loss.

2. Risk Index System of UHV Project Construction

When building the risk index system of UHV project construction, the basic principles of scientific, comprehensiveness, hierarchy and operability should be adopted. Meanwhile, the characteristics of UHV project construction should also be considered. This paper combined the Checklist method and Delphi method to find out the important risk index, and then built the risk index system of UHV project construction. The built risk index system of UHV project construction is shown in Fig. 1, which includes 5 first-level risk criteria and 12 second-level risk indicators.

3 The Basic Theory of Improved Rough Set Model

3.1. Rough Set Theory

The rough set theory (RS), proposed by Polish computer scientist Zdzisław I. Pawlak in 1982, is a kind of mathematic analysis methods which can describe the incompleteness, inaccuracy and uncertainty. The rough set theory can effectively analyze the random information with the inconsistent and inaccurate characteristics, and can also reason the data to explore the implicit potential law of data without any priori information [14, 15]. Owing to the above merit, the rough set theory has been applied in the field of electricity, such as the fault diagnosis rule extraction of

power grid [16], fault diagnosis of transformer substation [17], multi-objective power grid planning optimization model [18], and so on.

Let the nonempty set U be a finite set, called domain field, and R be an equivalence relation on U. For any subset $X \in U$, the pair $T = (U, R)$ is called an approximation space.

Define two subsets:

$$\underline{R}X = \left\{ x \in U \middle| [x]_R \subseteq X \right\}; \overline{R}X = \left\{ x \in U \middle| [x]_R \bigcap X \neq \phi \right\} \quad (1)$$

are called the R-lower and R-upper approximation of X, respectively.

$$BN_R(X) = \overline{R}X - \underline{R}X \quad (2)$$

is called R-boundary region of X. If $BN_R(X) = 0$, this set is a crisp set, and if $BN_R(X) > 0$, this set is a rough set.

$POS_R(X) = \underline{R}X$ denotes the R-positive region of X, just as shown in blackened cells in Fig. 2. $NEG_R(X) = U - \overline{R}X$ denotes the R-negative region of X, just as shown in the white cells in Fig. 2. The relationship between different sets in RS is shown in Fig. 2.

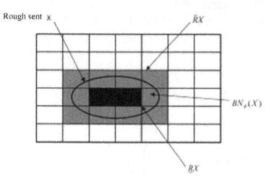

Figure 2. *Schematic diagram of rough set theory.*

4-tuple $S = (U, C \bigcup D, V, f)$ denotes a knowledge representation system. Of which, U is the universe; C and D are the conditional and decision attribute sets for U; is the value range of attribute a; $f : U \times (C \bigcup D) \to V$ is an information function, which assigns information value to each object property according to certain rules, namely $\forall x \in U$, if $a \in C \bigcup D$, then $f(x, a) \in V_a$.

3.2. Improved Rough Set Mode

When there is missing information at the risk information system decision table of UHV construction, the conventional rough set theory cannot be used to explore the potential information of data. When performing the questionnaire, the respondents sometimes maybe miss to give the evaluation scores on some risk indicators, which will lead to the missing information at the information system decision table. Bai C and Sarkis J proposed an improved rough set (IRS) model with the introduction of 'information capacity' concept to overcome this issue [19]. This paper employed the IRS model to identify the key indicators of UHV project construction.

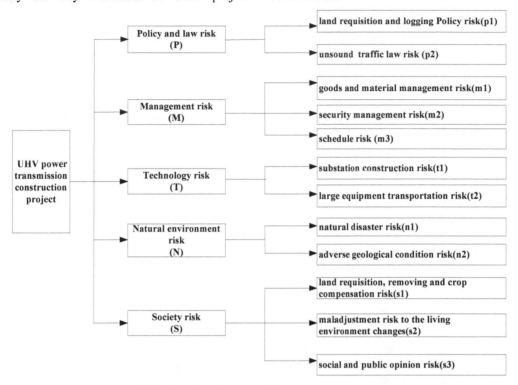

Figure 1. *Risk index system of UHV project construction*

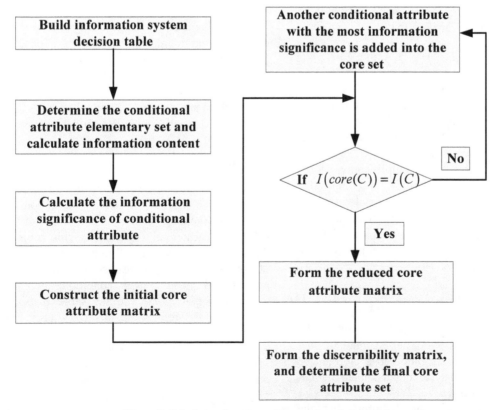

Figure 3. *Calculation flow chart of improved rough set model.*

The concept steps of employing *IRS* model to identify the key indicators of UHV project construction are as follows.

Step 1: Build the information system decision table

According to the goal of research object and the relative

evaluation index system, the information system decision table is built, which includes the missing information, denoted by '*'.

Step 2: Determine the conditional attribute elementary set and calculate information content

The set consisted of objects with the same conditional attribute value is called conditional attribute elementary set.

$$SIM(C) = \left\{ (i,j) \in U \times U \mid \forall c \in C, c(i) = c(j), \text{or } c(i) = *, \text{or } c(j) = * \right\} \quad (3)$$

where $c(i)$ and $c(j)$ denote the value of conditional attribute c for object i and j; C is the conditional attribute set; * denotes the value missing of conditional attribute.

At the information system decision table, the information capacity of conditional attribute can be calculated by Equation (4).

$$I(C) = 1 - \frac{1}{|U|^2} \sum_{i=1}^{|U|} |X_i| \quad (4)$$

where $I(C)$ is the information capacity of conditional attribute; $|U|$ is the total number of all the objects; $|X_i|$ is the number of objects with similar attributes levels across all the conditional attributes for object i.

Step 3: Calculate the information significance of conditional attribute

The information significance of conditional attribute will help to reduce the unimportant or irrelevant conditional attribute and save the conditional attribute that has close relationship with the decision attribute [20]. The information significance $Sig_{C\setminus\{c\}}(c)$ of conditional attribute can be calculated by Equation (5)

$$Sig_{C\setminus\{c\}}(c) = I(C) - I(C\setminus\{c\}) \quad (5)$$

Equation (5) identifies the difference between the information content of the full set and the information content of the reduced set that does not include attribute c.

Step 4: Construct the initial core attribute matrix and then form the reduced core attribute matrix

Rough set theory requires some simplification of the data to be able to arrive at overall rules. The knowledge reduction seeks to remove superfluous knowledge from information systems while preserving the consistency of classifications, which is also called attribution reduction.

The initial core attribute matrix of object can be constructed according to Equation (6).

$$core(C) = \left\{ c \in C \mid Sig_{C\setminus\{c\}}(c) > 0 \right\} \quad (6)$$

Then, compare the information capacity between initial core attribute set and original conditional attribute set, and judge whether this initial core attribute set is the reduced core attribute set of information system decision table or not. The judgment rules are as follows.

(1) If $I(core(C)) = I(C)$, we can determine that the initial core attribute set is the reduced core attribute set, and then the reduced core attribute matrix can be formed;

(2) If $I(core(C)) = I(C)$, another conditional attribute with the most information significance is added into the core set (as shown in Equation (7)), and then calculate the

The conditional attribute elementary set with the missing information can be defined as:

formation content of the new conditional attribute set. If it equals to $I(C)$, the reduced core attribute matrix can be formed; otherwise, loop this step until forming the reduced core attribute matrix.

$$\text{if } c_0 \in C \setminus (core(C)), \text{let } Sig_{core(C)}(c_0) = \max_{c \in C \setminus core(c)} \left\{ Sig_{core(C)}(c) \right\},$$
$$\text{and } core(C) \bigcup \{c_0\} \rightarrow core(C) \quad (7)$$

Step 5: Form the discernibility matrix, and determine the final core attribute set

Discernibility matrix is an important tool in RS to express the reduction algorithm. The discernibility matrix with the missing information can be defined as:

$$(c_{ij}) = \begin{cases} \{a \in A \setminus a(x_i) \neq a(x_j) \text{ and } a(x_i) \neq * \text{ and } a(x_j) \neq *\}, D(x_i) \neq D(x_j) \\ \phi, \qquad\qquad\qquad\qquad\qquad\qquad D(x_i) = D(x_j) \end{cases} \quad (8)$$

where c_{ij} is the ith-row j-th column element of discernibility matrix; A is the conditional attribute set; $a(x)$ is the value of attribute a for object x; D is the decision attribute set; $D(x)$ is the value of attribute D for object x.

After obtaining the discernibility matrix, the matrix with only one conditional attribute element is determined as the final core attribute matrix; and then, the conditional attribute sets that not include this core attribute is converted to the final core attribute set by using the conjunctive normal form and disjunctive normal form.

The calculation flow of IRS model is as shown in Fig. 3.

4. Case Study

There is a 1000kV UHV power transmission construction project which starts at the 1000kV substation located in the north of Zhejiang province and ends at the 1000kV substation located in Fuzhou cit. This project will build three new UHV transformer substations and two 603km-length AC transmission lines. Due to the complicated geographical condition, high construction technical requirement, and the large resistance of land acquisition and relocation, this UHV project faces many challenges and risks.

In this paper, the object set is consisted of 15 random selected questionnaire respondents, namely $U = \{I_i, i = 1, 2, \cdots, 15\}$. The conditional attribute is $C = \{c_i, i = 1, 2, \cdots, 12\}$, which reflects the risk index attribute of UHV construction project consisted of 12 second-level risk indicators. The decision attribute is $D = \{d\}$, which reflects the overall risk situation of this UHV project construction.

The concrete calculation details of the key index

identification of 1000kV UHV project construction by employing the IRS method are as follows.

Step 1: Build the information system decision table

Based on the built risk evaluation index system, the risk questionnaire can be formed, just as list in Table 1. Of which, both the likelihood of risk occurrence and the degree of risk influence adopt three-level scale standards, namely '1' represents the small likelihood of risk occurrence and low degree of risk influence; '2' represents the occasional likelihood of risk occurrence and medium degree of risk influence; '3' represents the frequent likelihood of risk occurrence and high degree of risk influence. The questionnaires were dispatched to the experts, academics and project staff. The random selected 15 questionnaire results were used to the calculation sample. Then, the risk information system decision table of UHV project construction was formed, which is listed in Table 2.

Table 1. *Risk questionnaire of UHV project construction*

Risk Index		Likelihood of risk occurrence			Degree of risk influence		
		frequent	occasional	small	high	medium	low
P	p1						
	p2						
	m1						
M	m2						
	m3						
T	t1						
	t2						
N	n1						
	n2						
	s1						
S	s2						
	s3						

Table 2. *Risk information system decision table of UHV project construction*

Respondents	Conditional attribute (C)												Decision attribute (D)
	p1	p2	m1	m2	m3	t1	t2	n1	n2	s1	s2	s3	d
I1	2	1	*	1	1	1	2	1	1	2	1	2	1
I2	2	1	1	2	2	1	1	1	1	1	2	1	1
I3	2	1	1	2	1	*	2	1	1	2	1	2	2
I4	2	1	1	1	1	1	2	1	1	2	1	2	1
I5	3	3	3	1	3	3	3	3	1	3	2	1	3
I6	3	2	2	2	3	2	3	*	2	3	3	2	2
I7	1	2	1	2	1	2	1	2	2	2	3	2	2
I8	1	1	1	1	2	2	2	2	2	2	1	2	1
I9	3	1	*	2	1	2	2	1	1	2	1	2	1
I10	3	1	1	1	1	1	1	2	2	1	2	2	2
I11	2	2	2	2	2	1	2	2	2	3	3	2	2
I12	2	1	2	2	1	2	3	*	2	1	3	2	2
I13	1	2	2	2	1	2	1	2	2	1	2	2	2
I14	1	2	2	2	1	2	1	2	2	2	*	2	2
I15	2	2	2	*	2	2	2	2	2	3	3	2	2

Note: 1. The result of D is calculated based on the questionnaire results in term of 'the degree of risk influence';
2. * denotes the missing data

Step 2: Determine the conditional attribute elementary set and calculate information content

According to Equation (3), the conditional attribute elementary set of UHV project construction can be obtained, which is listed in Table 3.

Table 3. *Elementary sets based on similar conditional attribute values*

Element set category	Element set object	Number of element set$^{|X|}$
1	I1、I4	2
2	I2	1
3	I3	1
4	I5	1
5	I6	1
6	I7	1
7	I8	1
8	I9	1
9	I10	1
10	I11	1
11	I12	1
12	I13	1
13	I14	1
14	I15	1

Then, the information capacity of conditional attribute can be calculated according to Equation (4).

$$I(C)=1-\left(\frac{1}{15^2}(2+1+1+2+1+1+1+1+1+1+1+1+1+1+1)\right)=0.9244$$

Step 3: Calculate the information significance of conditional attribute

According to Equation (4), the information capacity of

induced attribute set $C \setminus \{p_1\}$ without conditional attribute p1 can be calculated.

$$I(C \setminus \{p_1\}) = 1 - \frac{1}{15^2}(2+1+2+2+1+1+1+1+2+1+1+1+1+1+1) = 0.9156$$

Then, the information significance of conditional attribute p1 can be calculated according to Equation (5).

$$Sig_{C \setminus \{p_1\}}(p_1) = I(C) - I(C \setminus \{p_1\}) = 0.9244 - 0.9156 = 0.0088$$

Similarly, the information significance of other conditional attributes can also be calculated, the results of which is listed in Table 4.

Table 4. *Information significant of each conditional attribute*

Conditional attribute	p1	p2	m1	m2	m3	t1	t2	n1	n2	s1	s2	s3
Information significance	0.0088	0	0.0088	0.0177	0	0.0088	0	0	0	0.0088	0	0

Step 4: Construct the initial core attribute matrix and then form the reduced core attribute matrix.

Based on the Step 3 and Equation (6), the initial core attribute set can be obtained, namely $core(C) = \{P_1, M_1, M_2, T_1, S_1\}$. According to Equation (4), we can get $I(core(C)) = 0.9244 = I(C)$. Therefore,

conditional attributes P_1, M_1, M_2, T_1, S_1 are the reduced core attribute of information decision table, and the induced core attribute set is $core(C) = \{P_1, M_1, M_2, T_1, S_1\}$. The induced core attribute matric is shown in Table 5.

Table 5. *Reduced core attribute matrix*

Respondents	Conditional attribute(C)					Decision attribute (D)
	p1	m1	m2	t1	s1	d
I1	2	*	1	1	2	1
I2	2	1	2	1	1	1
I3	2	1	2	*	2	2
I4	2	1	1	1	2	1
I5	3	3	1	3	3	3
I6	3	2	2	2	3	2
I7	1	1	2	2	2	2
I8	1	1	1	2	2	1
I9	3	*	2	2	2	1
I10	3	1	1	1	1	2
I11	2	2	2	1	3	2
I12	2	2	2	2	1	2
I13	1	2	2	2	1	2
I14	1	2	2	2	2	2
I15	2	2	*	2	3	2

Step 5: Form the discernibility matrix, and determine the final core attribute set

According to Equation (8), the discernibility matrix of UHV project construction can be calculated, just as shown in Table 6.

Table 6. *Discernibility matrix*

	I1	I2	I3	I4	I5	I6	I7	I8	I9	I10	I11	I12	I13	I14	I15
I1	0	0	m2	0	p1 t1 s1	p1 m2 t1 s1	p1 m2 t1	0	0	p1 s1	m2 s2	m2 t1 s1	p1 m2 t1 s1	p1 m2 t1	t1 s1
I2		0	s1	0	p1 m1 m2 t1 s1	p1 m1 t1 s1	p1 t1 s1	0	0	p1 m2	m1 s1	m1 t1	p1 m1 t1	p1 m1 t1	m1 t1 s1
I3			0	m2	p1 m1 m2 s1	0	0	p1 m2	p1	0	0	0	0	0	0
I4				0	p1 m1 t1 s1	p1 m1 m2 t1 s1	p1 m2 t1	0	0	p1 m2 t1	m1 m2 s1	m1 m2 t1 s1	p1 m1 m2 t1 s1	p1 m1 m2 t1	m1 t1 s1

I5	0	m1 m2 t1	p1 m1 m2 t1 s1	p1 m1 t1 s1	m2 t1 s1	m1 t1 s1	p1 m1 m2 t1	p1 m1 m2 t1 s1	p1 m1 m2 t1 s1	p1 m1 m2 t1 s1	p1 m1 t1
I6		0	0	p1 m1 m2 s1	s1	0	0	0	0	0	0
I7			0	m2	p1	0	0	0	0	0	0
I8				0	0	p1 m1 t1 s1	p1 m1 m2 t1 s1	p1 m1 m2 t1 s1	p1 m1 m2 s1	m1 m2	p1 m1 s1
I9					m2 t1 s1	0	p1 s1	p1 s1	p1 s1	p1	p1 s1
I10						0	0	0	0	0	0
I11							0	0	0	0	0
I12								0	0	0	0
I13									0	0	0
I14										0	0
I15											0

From the discernibility matrix, the conditional attributes p1, m2, s1 are the final core attributes. The conditional attribute sets that do include the final core attributes are {M1, T1}. The conditional attribute combination free of final core attribute is displayed in the form of conjunctive normal form, namely $P = \{M_1 \vee T_1\}$, and then is transformed by disjunctive normal form to be $P = \{M_1\} \vee \{T_1\}$. Therefore, $\{P_1, M_1, M_2, S_1\}$ and $\{P_1, M_2, T_1, S_1\}$ can be considered as the reduced attribute set of risk identification of UHV project construction.

From the calculation result, we can see that 'land requisition and logging policy risk', 'security management risk' and 'land requisition, removing and crop compensation risk' are the key risk indicators of this UHV project construction, which should be mainly focused on when managers perform the risk management and control on this project.

5 Conclusions

In this paper, the key risk indicators are identified by employing the IRS model with the introduction of 'information capacity' concept, which can identify the key risk index when including the missing information in the index system. Taken a certain 1000kV UHV project as the example, the key risk indicators were identified. The following conclusions can be safely drawn:

(1) 'land requisition and logging policy risk', 'security management risk' and 'land requisition, removing and crop compensation risk' are the key risk indicators of this UHV project construction, which should be mainly focused on when managers perform the risk management and control on this project;

(2) The risk index sets {land requisition and logging policy risk, goods and material management risk, security management risk, land requisition, removing and crop compensation risk} and { land requisition and logging policy risk, security management risk, substation construction risk, land requisition, removing and crop compensation risk} can be used as the reduced index system of UHV project construction, which simplifies the risk index system and risk evaluation calculation under the premise of preserving vital information.

Acknowledgements

The authors thank the editor and reviewers for their comments and suggestions.

References

[1] Liu Zhenya. China's electricity and energy M] .Beijing: China Electric Power Press, 2012

[2] Zhao H, Guo S. Risk Evaluation on UHV Power Transmission Construction Project Based on AHP and FCE Method [J]. Mathematical Problems in Engineering, 2014, 2014.

[3] Wei Minghua, Lu Shibao, Zheng Zhihong. Risk analysis on construction of agricultural water conservancy projects [J]. Transactions of the CSAE, 2011, 27(Supp.1): 233−237.

[4] Huang Hong-wei, Zhu Lin, Xie Xiong-yao. Risk assessment on engineering feasibility of key events in Shanghai metro line No. 11 [J]. Chinese Journal of Geotechnical Engineering, 2007, 29(7): 1103-1107.

[5] Xu Wei-ping, Wang Xiu-ying. Environmental risk assessment of oil field pipeline construction project based on mutation series method [J]. Chinese Journal of Systems Science, 2012, 20(1): 89-93.

[6] Zhou Li-sha, Li Chen, Yu Shun-kun. System dynamics simulation research on project management model for china's smart grid [J]. East China Electric Power, 2012, 40(1): 31-34.

[7] Zhang Qiang, Wang Ning. Application of network management system for project files in power grid construction [J]. Power System Technology, 2007, 31(S2): 83-86.

[8] Guo Ri-cai, Xu Zi-zhi, Qi Li-zhong, et al. General situation of typical transmission line design in usa and its enlightenment to design and construction of power grids in china [J]. Power System Technology, 2007, 31(12): 33-41.

[9] Guo Ri-cai, Yuan Zhao-xiang, Li Bao-jin. A survey on typical design of substations in france and korea and relevant suggestion to power network engineering in china [J]. Power System Technology, 2006, 30(6): 73-76.

[10] An lei, Wang Mian-bin, Tan Zhong-fu. Risk assessment model based on set pair-fault tree method for power transmission project [J]. East China Electric Power, 2011, 39(1): 12-18.

[11] Wu Yunna, Liu Yarui, Wang Weibing. Risk factor analysis on power grid transmission and transformation project [J]. China Rural Water and Hydropower, 2009, (2): 133-139.

[12] Wang Liping, Wang Xiaoru. Differential protection based on calculated power for uhv transmission lines [J]. Proceedings of the CSEE, 2013, 33(19): 174-182.

[13] Xie Qiang, Li Jiguo, Yan Chengyong, et al. Wind tunnel test on wind load transferring mechanism in the 1000 kV UHV transmission tower-line system [J]. Proceedings of the CSEE, 2013, 33(1): 109-116.

[14] GAO Shuang, DONG Lei, GAO Yang, et al. Mid-long term wind speed prediction based on rough set theory [J]. Proceedings of the CSEE, 2012, 30(1): 32-37.

[15] Wang Gang, Luo Huihui, Huang Min. A rough-set based detection method for detuning components in a triple-tuned DC filter[J]. Automation of Electric Power Systems,, 2011, 35(20): 81-87.

[16] Zhang Zhiyi, Yuan Rongxiang, Yang Tongzhong, et al. Rule Extraction for Power System Fault Diagnosis Based on the Combination of Rough Sets and Niche Genetic Algorithm [J]. Transactions of China Electrotechnical Society, 2009, 24(1): 158-163.

[17] Su Hong-sheng, Li Qun-zhan. Substation fault diagnosis method based on rough set theory and neural network model [J]. Power System Technology, 2005, 29(16): 66-70.

[18] Liu Si-ge, Cheng Hao-zhong, Cui Wen-jia. Optimal model of multi-objective electric power network planning based on rough set theory [J]. Proceedings of the CSEE, 2007, 27(7): 65-69.

[19] Bai C, Sarkis J. Green supplier development: analytical evaluation using rough set theory [J]. Journal of Cleaner Production, 2010, 18(12): 1200-1210.

[20] Liang J., Shi Z., Li D.. Information entropy, rough entropy and knowledge granulation in incomplete information systems [J]. International Journal of General Systems, 2006, 35 (6): 641–654.

Permissions

List of Contributors

Dessie Tarekegn Bantelay
School of Mechanical & Industrial Engineering, Bahir Dar Institute of Technology, Bahir Dar, 26, Ethiopia

Nigus Gabbiye
School of Food & Chemical Engineering, Bahir Dar Institute of Technology, Bahir Dar, 26, Ethiopia

Ravindra B. Sholapurkar
Department of Chemical Engineering, DR. Babasaheb Ambedkar Technological University, Lonere, Tal Mangaon, Dist. Raigad, Maharashtra, India

Yogesh S. Mahajan
DR. Babasaheb Ambedkar Technological University, Lonere, Tal Mangaon, Dist. Raigad, Maharashtra, India

Huseyin Murat Cekirge
Department of Mechanical Engineering, Prince Mohamed Bin Fahd University, Al Khobar, KSA

Huseyin Murat Cekirge
Department of Mechanical Engineering, Prince Mohamed Bin Fahd University, Al Khobar, KSA

Veronica Kavila Ngunzi
Department of Engineering and Innovative Technology, Kisii University, Kisii, Kenya

Hualong Cai
State Key Laboratory of Water Resource and Hydropower Engineering Science, Wuhan University, Wuhan, China
Yalong River Hydropower Development Co., Ltd., Chengdu, China

Yuanshou Liu and Yunfeng Liu
State Key Laboratory of Water Resource and Hydropower Engineering Science, Wuhan University, Wuhan, China
Yalong River Hydropower Development Co., Ltd., Chengdu, China

Mahendra Kumar Trivedi, Rama Mohan Tallapragada, Alice Branton, Dahryn Trivedi and
Gopal Nayak
Trivedi Global Inc., Henderson, USA

Rakesh Kumar Mishra and Snehasis Jana
Trivedi Science Research Laboratory Pvt. Ltd., Bhopal, Madhya Pradesh, India

Ahmed Farouk AbdelGawad
Mech. Eng. Dept., Umm Al-Qura Univ., Makkah, Saudi Arabia

Hasibur Rahman Sardar
Department of Electronics & Communication Engineering, P.A College of Engineering, Karnataka, India

Abdul Razak Kaladgi
Department of Mechanical Engineering, P.A College of Engineering, Karnataka, India

Shah Shahood Alam and Ahtisham Ahmad Nizami
Pollution and Combustion Engineering Lab. Department of Mechanical Engineering, Aligarh Muslim University, Aligarh-202002, U.P. India

Tariq Aziz
Department of Applied Mathematics, Aligarh Muslim University, Aligarh-202002, U.P. India

Ahmed F. Abdel Gawad, Muhammad N. Radhwi, Asim M. Wafiah and Ghassan J. Softah
Mech. Eng. Dept., College of Eng. & Islamic Archit., Umm Al-Qura Univ., Makkah, Saudi Arabia

Nader A. Nader, Bandar Bulshlaibi, Mohammad Jamil, Mohammad Suwaiyah and Mohammad Uzair
Department of Mechanical Engineering, Prince Mohammad Bin Fahd University, Al Khobar, Kingdome of Saudi Arabia

Huseyin M. Cekirge
Department of Mechanical Engineering, Prince Mohammad Bin Fahd University, Al Khobar, KSA

Omar K. M. Ouda
Department of Civil Engineering, Prince Mohammad Bin Fahd University, Al Khobar, KSA

Ammar Elhassan
Department of Information Technology, Prince Mohammad Bin Fahd University, Al Khobar, KSA

Ehsan Fadhil Abbas Al-Showany
Refrigeration & Air Conditioning Engineering Department, Kirkuk Technical College, Northern Technical University, Kirkuk, Iraq

Kherchouche Younes, Savah Houari
Faculty of Electrical Engineering, University Djillali Liabes , ICEPS (Intelligent Control Electrical Power System) Laboratory, Sidi Bel Abbes, Algeria

Takashi Oda
System Design of Tokyo Metropolitan University, Tokyo, Japan
Nissin Sangyo Co., Ltd., Tokyo, Japan

Kimihiro Yamanaka
System Design of Tokyo Metropolitan University, Tokyo, Japan

Mitsuyuki Kawakami
Human Sciences of Kanagawa University, Kanagawa, Japan

Yanlin Qu
Business College of Honghe University, Mengzi County, Yunnan Province, 661100, China

Yilan Su
College of Life Science and Technology, Honghe University, Mengzi County, Yunnan Province, 661100, China

Temesgen Gebrekidan, Yonas Zeslase
Department of Chemical Engineering, Adigrat University, Adigrat, Ethiopia

Abene Abderrahmane
University of Valenciennes, ISTV Science Faculty, Valenciennes, France

De-Yi Shang
136 Ingersoll Cres., Ottawa, ON, Canada K2T 0C8

Bu-Xuan Wang
Dept. of Thermal Engineering, Tsinghua University, Beijing 100084, China

Liang-Cai Zhong
Department of Ferrous Metallurgy, Northeastern University, Shenyang 110004, China

G. Santiago-Rosas
Center for Research and Advanced Studies, National Polytechnic Institute, Mexico City, Mexico
Trefle Laboratory, Arts and Trade Paris-Tech (ENSAM), Talance, France

M. Flores-Domínguez and J. P. Nadeau
Center for Research and Advanced Studies, National Polytechnic Institute, Mexico City, Mexico
Trefle Laboratory, Arts and Trade Paris-Tech (ENSAM), Talance, France

Sen Guo and Huiru Zhao
School of Economics and Management, North China Electric Power University, Changping District, Beijing 102206, China

Index